Peter Anderson
19, Ashbrook Drive
Belfast BT4 2FG
Ph. (0232) 652202

Peter Anderson
lab. 236
School of Chemistry
David Keir Building
Stranmillis Road
The Queens University
of Belfast

Molecular Electronics

Molecular Electronics

Edited by
Geoffrey J. Ashwell
Cranfield Institute of Technology, U.K.

RESEARCH STUDIES PRESS LTD.
Taunton, Somerset, England

JOHN WILEY & SONS INC.
New York · Chichester · Toronto · Brisbane · Singapore

RESEARCH STUDIES PRESS LTD.
24 Belvedere Road, Taunton, Somerset, England TA1 1HD

Marketing and Distribution:

Australia and New Zealand:
JACARANDA WILEY LTD.
GPO Box 859, Brisbane, Queensland 4001, Australia

Canada:
JOHN WILEY & SONS CANADA LIMITED
22 Worcester Road, Rexdale, Ontario, Canada

Europe, Africa, Middle East and Japan:
JOHN WILEY & SONS LIMITED
Baffins Lane, Chichester, West Sussex, England

North and South America:
JOHN WILEY & SONS INC.
605 Third Avenue, New York, NY 10158, USA

South East Asia:
JOHN WILEY & SONS (SEA) PTE LTD
37 Jalan Pemimpin #05-04
Block B Union Industrial Building, Singapore 2057

Library of Congress Cataloging in Publication Data

Molecular electronics / edited by Geoffrey J. Ashwell.
    p. cm.
  Includes bibliographical references and index.
  ISBN 0-86380-125-0 (Research Studies Press). — ISBN 0-471-93386-4
(Wiley)
  1. Molecular electronics. I. Ashwell, Geoffrey J., 1947–
TK7874.8.M65 1992
620.1'1297—dc20   91-40616
                                                          CIP
British Library Cataloguing in Publication Data

A catalogue record for this book is available
from the British Library

  ISBN 0 86380 125 0 (Research Studies Press Ltd.)
  ISBN 0 471 933864 (John Wiley & Sons Inc.)

Printed in Great Britain by SRP Ltd., Exeter

Dedicated to my parents

Kit and Joe

Preface

The interdisciplinary area of Molecular Electronics, defined here in its broadest sense as a subject concerning active materials with potential photonic and electronic applications, bridges the gaps between physics, chemistry and materials science. Its conception is difficult to pinpoint as so many areas of research led to its growth. However, at the molecular level, Aviram and Ratner paved the way in 1974 with their proposal of molecular rectification and by the early 1980's Forest Carter advocated the idea of molecular switches and molecular gates. These areas of molecular electronics are still at an early stage and may remain as no more than curiosities. However, along the way there has been considerable progress in the design and synthesis of materials with applicable properties: the moderately high T_c carbon-based superconductors; polymeric conductors; photochromics; electrochromics; piezo, pyro and ferroelectric materials; liquid crystals; materials for nonlinear optics. The success of Molecular Electronics is dependent upon their exploitation.

Molecular Electronics covers a wide interdisciplinary area and many working in the area have contributed significantly without understanding the wider aspects outside their own specialisation. The aim of this book is to provide an overview of the areas with sufficient background information to bridge these gaps. The topics covered are the materials, in particular photochromics, organic conductors and liquid crystals (Chapter 1), conducting polymers (Chapter 2), Langmuir-Blodgett films (Chapter 3), nonlinear optics (Chapter 4), piezoelectricity, pyroelectricity and ferroelectricity (Chapter 5) and holography (Chapter 6). To the contributors of these chapters, Dr Simon Allen (ICI), Dr Ken Firth (GEC), Dr Ian R. Peterson (Mainz), Professor Michael F. Rubner (MIT), Dr Ian Sage (Merck), Professor Juliusz Sworakowski (Wroclaw) and Dr Clive Trundle (GEC), I wish to express my sincere gratitude for their considerable effort. In addition, Harry Block is acknowledged for helpful discussions, Sylvia Skevington and Peter Thompson for generating the diagrams in Chapter 1 and Tracey Buchan for compiling the data for Figure 10 of the chapter.

Special thanks are also due to my wife Sue and children Emma-Jane and Robert for their patience and understanding.

G.J.Ashwell

October 1991

Contributors

SIMON ALLEN
ICI Wilton Materials Research Centre, PO Box 90, Wilton, Middlesbrough,
Cleveland TS6 8JE, UK

GEOFFREY J. ASHWELL
Centre for Molecular Electronics, Cranfield Institute of Technology,
Cranfield MK43 0AL, UK

KEN FIRTH
Marconi Research Centre, West Hanningfield Road, Great Baddow,
Chelmsford, Essex CM2 8HN, UK

IAN R. PETERSON
Institut für Physikalische Chemie, Johannes Gutenberg Universität,
D6500 Mainz, Germany

MICHAEL F. RUBNER
Department of Materials Science and Engineering, Massachusetts Institute of
Technology, Cambridge, Massachusetts 02139, USA

IAN SAGE
Organic Development Department, Merck Ltd., Broom Road, Poole,
Dorset BH12 4NN, UK

JULIUSZ SWORAKOWSKI
Institute of Organic and Physical Chemistry, Technical University of Wroclaw,
Wybrzeże Wyspiańskiego 27, 50-370 Wroclaw, Poland

CLIVE TRUNDLE
GEC Marconi Materials Technology Ltd., Caswell, Towcester,
Northants NN12 8EQ, UK

Contents

CHAPTER 1
Molecular Electronic Materials

G. J. Ashwell, I. Sage *and* C. Trundle

The field of *Molecular Electronics* has grown dramatically in the past ten years and its definition has broadened from electronics at the molecular level [1] to include molecular materials with potential electronic and photonic applications. Examples include conducting polymers, organic superconductors, molecular ferromagnets, electrochromics and photochromics, nonlinear materials for rectification and second harmonic generation, piezoelectric and pyroelectric materials and, of course, liquid crystals. In many of these areas the advances have been serendipitous but sustained future progress is dependent upon the modelling and synthesis of advanced materials with specific properties; an evolution of techniques for organising the molecules into supramolecular assemblies and, at the molecular level, the emergence of methods of addressing individual molecules such as the advancement of scanning tunnelling microscopy.

In this first Chapter the molecular requirements are reviewed and materials, old and new, with properties which may be utilised to store or process information are outlined, with particular emphasis on photochromics and liquid crystals. A brief review of organic conductors is also included, the topic being covered in greater depth for conjugated polymeric systems in Chapter 2. In addition, some recent developments in the field of optically nonlinear Langmuir-Blodgett film forming materials are discussed. Emphasis is placed on the fabrication of non-centrosymmetric assemblies of interlocking molecules for enhanced second harmonic generation, whereas the physics and general material requirements are discussed at length in Chapter 4.

1. PHOTOCHROMISM

Materials which undergo reversible changes in their optical transmission characteristics due to a photochemical stimulation are classified as photochromic (for reviews, see [2-4]). Microcrystals of silver halide in glass undergo photochromic colouration when exposed to ultraviolet light and revert to colourless *via* a thermal process when light is excluded. This photochromic system has been successfully applied to sunglasses; it outperforms its organic competitors with $>10^6$ switching cycles. However, it is the organic materials which are of particular interest. They may be switched at one wavelength and interrogated at another and, thus, they offer potentially exciting applications as optical data storage media. Many revert to the unswitched form in the dark, the rate being determined by temperature, but compounds with thermally stable switched forms are also known. These may be used for long term data storage and examples are found among the azobenzenes [5], dianthracenes [6–8], succinic anhydrides [9], spiropyrans [10] and a series of donor-π-acceptor zwitterions [11].

Other characteristics of photochromic compounds which determine their potential use are as follows: (*i*) the quantum yield of the photo reaction; (*ii*) contrast between the switched and unswitched forms which results from changes in conjugation brought about by molecular reorganisation; (*iii*) the absorption coefficient and, for many optical applications, its variation (and that of the refractive index) with wavelength; (*iv*) the extent of non-reversible reactions during irradiation and/or heating and their limiting effect on the number of switching cycles.

For an ideal photochromic system the reactions occurring can be described by the general formula:

$$A \xrightleftharpoons[h\nu']{h\nu} B$$

where A represents the unswitched form (usually but not necessarily colourless) and B the switched form. This ideal system has been the goal of many organic chemists but, for the majority of materials, either A or B (*or both*) yields fatigue products, and of the many types of photochromic compounds so far reported only the succinic anhydrides (fulgides) [9] closely represent the fatigue free system upon repetitive switching.

There are numerous examples of photochromic compounds but the majority may be classified into six main classes dependent upon the molecular changes that occur upon irradiation. These categories, some having several hundred representative compounds, are listed below:

  (a) cis-trans isomerisation;

  (b) geometrical isomerisation;

  (c) dimerisation;

  (d) hydrogen shift;

  (e) bond cleavage;

  (f) charge transfer.

Several articles cover the subject area in much greater depth than is appropriate in this section; in particular, "*Photochromism*" edited by Dürr and Bouas-Laurent [2], "*Perspectives in Photochromism*" [3] and "*Materials for Optical Data Storage*" [4], provide excellent general reviews.

1.1 CIS-TRANS ISOMERISATION

Isomerisation contributes greatly to the thermal stability of the switched form in other categories of photochromic compounds, notably the spiropyrans which initially undergo heterolytic bond cleavage. Most materials which undergo photo-induced *cis-trans* isomerisation, for example, stilbene [12] and thioindigo [13] derivatives, are of little practical interest. Changes to the absorption characteristics are usually relatively minor and the materials thermally revert to the unswitched form. However, examples have been reported where isomerisation serves an important function: *e.g.*, in natural systems such as retinal in the eye and in azobenzene derivatives where the colour changes may be used for optical data storage.

Liu, Hashimoto and Fujishima [5] have reported a stable recording medium based upon the two-step photoelectrochemical reduction of 4-octyl-4'-(5-carboxypentamethyleneoxy)azobenzene deposited as a Langmuir-Blodgett film. The photochromic properties of the azobenzenes are well established [14]; isomerisation occurs when they are irradiated with a UV source with reversal occurring either thermally or at visible wavelengths. Liu et al. [5] have demonstrated that the less stable *cis* form (but not the *trans* form) may be electrochemically reduced to hydrazobenzene; this is stable in an inert atmosphere but may be electrochemically oxidised to *trans*-azobenzene (Fig. 1), the photoelectrochemical process being stable for several hundred write/read cycles.

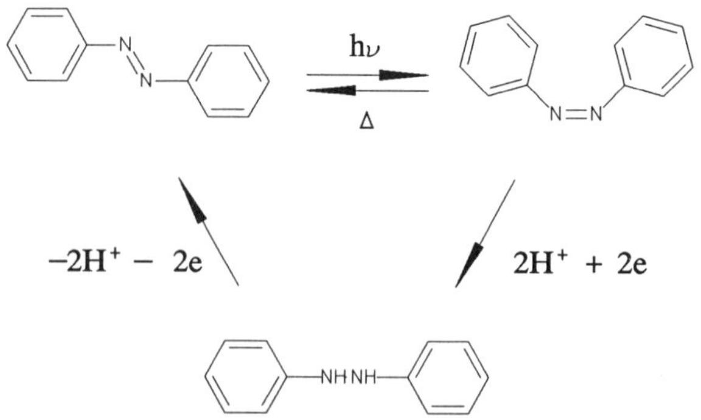

Figure 1. Photochemical switching of azobenzene.

Although at an early stage this work reveals the potential of the azobenzenes as erasable optical data storage media, the limit being *ca.* 10^8 bits/cm^2

(proportional to $1/\lambda^2$), but as a practical device a one-step writing process would be preferable and the number of write/read cycles falls below that of magnetic tape (10^5 to 10^6). However, Liu *et al.* [5] have suggested that by uniformly irradiating the film, to convert it to the *cis* form, storage densities as high as 10^{12} bits/cm^2 may be feasible by using an electrochemical scanning tunnelling microscopy probe to write on the *cis*-azobenzene film.

1.2 GEOMETRICAL ISOMERISATION

Isomerisation, resulting in the formation of cyclic structures, is responsible for the photochromic behaviour of many materials, the simplest example to exhibit this type of rearrangement being the *all cis*-1,3,5-hexatriene which forms 1,3−cyclohexadiene on exposure to ultra violet radiation. This type of pericyclic reaction forms the basis of the photochromic switching in many other classes of compounds, the most important being the succinic anhydrides (fulgides), reported by Stobbe as long ago as 1911 [15] and more recently by Heller [9]. Irradiation at UV wavelengths causes the fulgides to colour; the mechanism was originally assigned to *cis-trans* isomerisation but was later attributed by Santiago *et al.* [16] to photocyclisation, the mechanism being equivalent to that of 1,3,5-hexatriene. Fatigue processes in fulgides involve hydrogen shifts (see Fig. 2) but these may be eliminated by substituting the offending hydrogen by, for example, an alkyl group. In such substituted compounds steric hindrance, encountered in the thermal bleaching process, results in the coloured form being thermally stable at temperatures to about 150°C.

The colour of the switched form may be varied by substitution at either the *3* or *5-position* of the phenyl ring. Electron donating groups (*e.g.* alkoxy) cause the band to be bathochromically shifted [17] whereas electron withdrawing groups (*e.g.* acetyl) result in a hypsochromic shift [18]. The acetyl group also reduces thermal fatigue in compounds, unsubstituted at the α-position, possibly by hydrogen bonding.

Significant advances in fulgide chemistry have been made by Heller [9] resulting in fatigue resistant furyl, thienyl and pyrryl analogues (see Fig. 3) which show a high conversion efficiency with excellent thermal stability of the coloured form. Substitution of a methyl of the photoactive isopropylidene group affects the rate of bleaching at visible wavelengths, the order being hydrogen < < methyl < adamantyl. In addition, in these compounds the nature of the heteroatom significantly affects the colour of the switched form; the peak wavelength increases from 490 nm (*furyl*) to 520 nm (*thienyl*) to 630 nm (*pyrryl*). The band may be further shifted by substitution. For example, congeners based upon the γ–butyrolactone structure are similarly photochromic but the absence of one of the carbonyl groups (*note there are two in the fulgides*) causes the wavelength to be hypsochromically shifted compared with the parent fulgide (see Fig. 3).

Heller's fatigue resistant fulgide and lactone derivatives show a wide range of colours and when deposited as thin films, for example, by solvent assisted indiffusion or spin coating, they have potential applications in security printing.

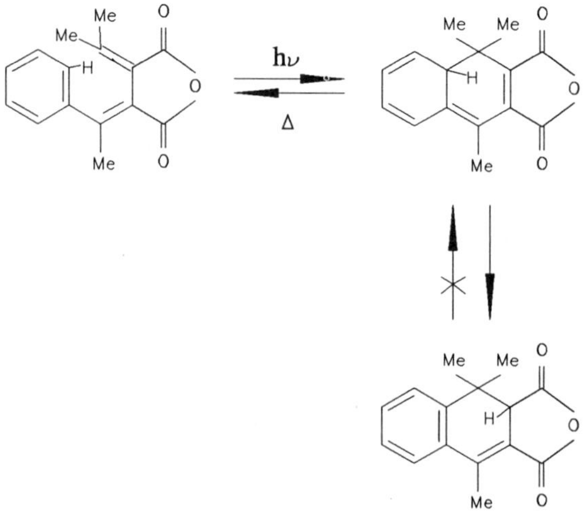

Figure 2. Photochromic switching and fatigue reaction of a fulgide.

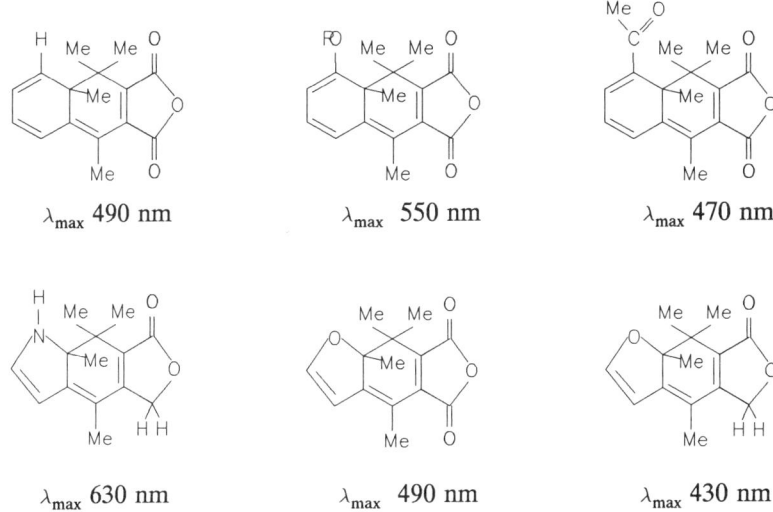

Figure 3. Effect of substitution on the colour of the switched fulgides.

1.3 DIMERISATION

Photodimerisation of conjugated systems, such as annulated aromatic structures, reduces the conjugation length causing a hypsochromic shift of the π-π^* band. As these monomeric compounds generally do not absorb in the visible region this form of photochromism does not result in the normal colouration expected in photochromism. However, as a result of the thermal stability of both the dimer and monomer and the difference in refractive index between the two forms, such dimerisation has excited interest for optical information storage which will be discussed later.

Thomlinson et al. [19] first directed attention to the importance of this form of photochromism, reporting the photoreaction as occurring between an excited and a ground state molecule, the generated excimer subsequently losing energy to yield a dimer.

$$M + h\nu \rightarrow M^*$$
$$M + M^* \rightarrow (M\text{-}M^*) \rightarrow Dimer$$

8

Photodimerisation readily occurs in single crystals whereas in solution the reaction is not efficient; the molecules are not held in proximity and the excited monomer (M*) must collide with a second molecule (M) before the energy is lost by some other pathway. This may be overcome by constraining the molecules in a rigid solution such as a low temperature glass or polymer film and, for the molecules to be sufficiently close, the monomers may be photodimerised prior to inclusion. Thus, subsequent photoreactions occur as if the monomer pair were held in a crystal lattice since the matrix restricts macroscopic motion. This technique still results in a dilution of the overall effect since the bulk of the matrix is inactive.

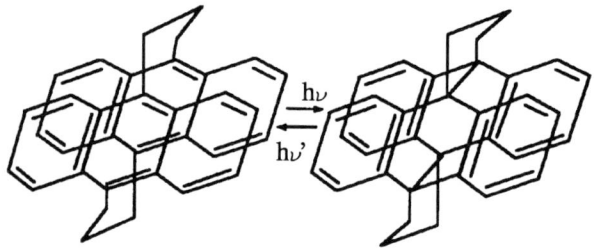

Figure 4. A photodimerisable bridged anthracene.

Wasserman and Keehn [6] and Castellan *et al.* [7] and more recently Usui *et al.* [8] have reported an alternative technique for overcoming the problems of achieving dimerisation in solid solution. In this case the two 'monomer' units are held in proximity, not by the rigidity of the matrix but by connecting chains (Fig. 4). These bridges do not play any part in the photochemistry of the molecule but, for example, can be used to latch the molecule onto polymer chains or to adjust the solubility of the molecule in a given solution. This locking together of monomers is of significant importance since it renders the molecules photochromic in isolation which, allied to the absence of radicals in the reaction, reduces fatigue.

1.4 HYDROGEN SHIFT

The majority of photochromic hydrogen shift compounds are *ortho* substituted aromatic molecules which form orthoquinonoid structures when irradiated at UV wavelengths, important classes of materials being the salicylidineamino-pyridines [20] and salicylidine anilines [21,22] (see Fig. 5).   They are photochromic as crystals but in solution reversion is rapid and they are only visibly photochromic at low temperatures, typically below *ca.* $-75\,^{\circ}$C. Recently, Kawamura et al. [22] have shown that the rate constant for thermal bleaching of Langmuir-Blodgett films is similar to that of the crystals (k $\approx$ 10^{-3} s^{-1}), a factor of 10^6 smaller than for solution. However, the instability of the coloured species reduces this system to one of little practical use.

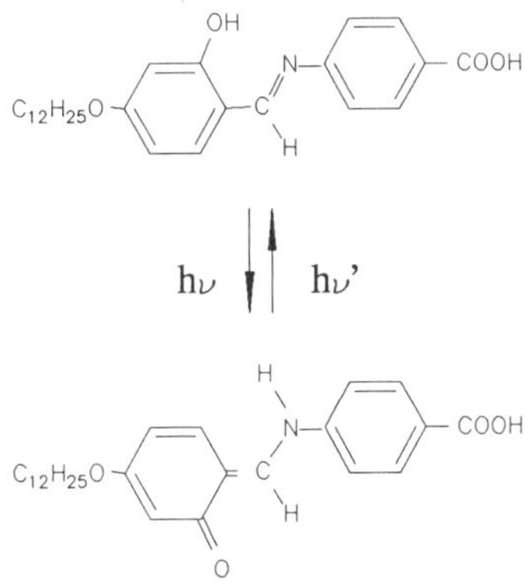

Figure 5. Photochromic switching of a salicylidine aniline.

1.5 BOND CLEAVAGE

Many photochromic processes involve bond cleavage, in some instances with the formation of radicals, examples being the octaaryl-1,1'-bipyrrolyl and hexaaryl-1,1'-biimidazolyl derivatives [23].   The coloured forms are readily produced in solution by UV irradiation but, in the dark, the colour fades. The materials are also photochromic in the crystalline state where the rate of thermal fading is much reduced and, in the absence of oxygen and water, they can be cycled between the *bis* and radical forms without loss of photochromic behaviour. However, the difficulty of eliminating such impurities, particularly considering the rate at which water and gases diffuse through polymer films, reduces any potential use of these compounds.

Alternative photochromic processes involve σ-bond fission where the product is a coloured zwitterion; this is best demonstrated by the spiropyran family which contains several different parent ring forms [24], the materials and properties being recently reviewed in [2,3]. The colouring reaction of these compounds results from the breaking of the σ-bond between the heteroatom and the site of the spiro linkage (Fig. 6). The electrons are retained by the heteroatom resulting in a charge-separated merocyanine-type species. This initial zwitterionic form is short-lived, rearranging in less than 20 ps, *via trans-to-cis* isomerisation, to a planar form which is stabilised by reduced steric interaction.

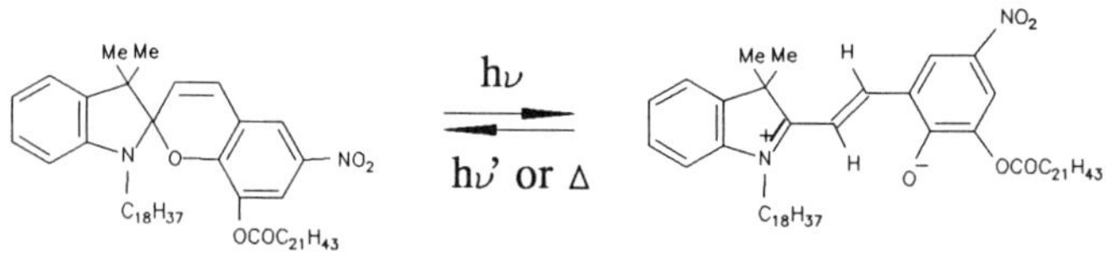

Figure 6. Photochromic switching of a spiropyran.

Many are only strongly photochromic in solid solution, the stability of the coloured form being dependent upon the method by which the solution is formed. Bleaching is often faster in solvent cast films than in films formed from a polymerised monomer (doped with a spiropyran) presumably because of restrictions imposed on molecular motion by the matrix in the latter case.

Thermal bleaching occurs at two rates: initial removal of the exciting radiation causes a rapid decrease in absorption followed by a much slower tailing-off, in some cases, taking several hours. This results from an initial rapid ring closure of the direct product of the photoreaction followed by ring closure of the isomerised photoproducts. These must first undergo a slow isomerisation to the *cis* form before reverting to the colourless spiropyran. In common with many photochromic systems the spiropyrans have subgroups which exhibit low fatigue resistance and undergo less than ten cycles before the photochromic properties are lost. Fatigue performance is improved by careful selection of the substituent groups [25] and some indolinospiropyrans may survive at least 30,000 bleaching cycles. Electron withdrawing groups on the indoline benzene ring reduce the resistance to fatigue whereas, conversely, donating groups enhance the effect. Substitution also affects the persistence of the open coloured form, this being important because the majority of applications of photochromic materials requires thermal stability of both states.

Three-dimensional optical memories have been disclosed: (i) Rentzepis [26] based his on an amplitude-recording medium (a spiropyran embedded in a polymer matrix) with two-photon writing/reading of information; (ii) Ando *et al.* [10,27], in their device, utilise the sharp J−aggregate bands of heat-treated spiropyran Langmuir-Blodgett films. For a multilayer film, comprising different photochromic layers with non-overlapping absorption bands, information may be recorded at different levels on the same spot by addressing each layer at a different characteristic wavelength.

Ando and co-workers [10] have reported amphiphilic spiropyrans with a half life for thermal decay 10^4 times greater than those of conventional spiropyrans. UV irradiated Langmuir-Blodgett films of 1-octadecyl-3',3'-dimethyl-6-nitro-8-[(docosanoyloxy)methyl]-spiro-[2H-1-benzopyran-2,2'-indoline], display a broad absorption profile (λ_{max} 583 nm) with a half life for thermal decay of 20 hours. In contrast, when irradiated at temperatures above 35°C aggregate formation causes the absorption profile to sharpen and shift to longer wavelengths (λ_{max} 618 nm; half width at half maximum, HWHM 25 nm), the half-life increasing to a few thousand hours at room temperature. The stability and sharpness have important implications for three-dimensional multifrequency optical data storage [27]. However, to store several bits per pixel requires components with sharp non-overlapping absorption bands.

1.6 CHARGE TRANSFER

Few compounds utilise charge transfer for their photochromism, the best known example being AgCl $\rightleftharpoons$ Ag° + Cl˙. Organic equivalents include the copper and silver salts of TCNQ (7,7,8,8-tetracyano-p-quinodimethane) and its derivatives [28], which switch according to the following equation.

$$[Ag^+TCNQ^-]_n \underset{\Delta}{\overset{h\nu}{\rightleftharpoons}} [Ag^+TCNQ^-]_{n-x} + xAg° + xTCNQ°$$

The 4,4'-bipyridinium salts, although better known for their electrochromic behaviour, are also photochromic. They undergo a one-electron reduction to a coloured radical cation but back charge transfer is usually rapid. Nonetheless, Nagamura et al. [29] have disclosed persistent (and reversible) colouration by using tetrakis[3,5-bis(trifluoromethyl)phenyl]borate as the counterion. The lifetime of the coloured form is dependent upon temperature but at 0°C some elastic polymer films (see Fig. 7) do not manifest thermal decay within 24 hours [30]. The stability of this switched state, as with the Ag⁺TCNQ⁻ films mentioned above, is attributable to structural changes within the film, in particular, to changes in the orientation and/or separation of the interacting moieties.

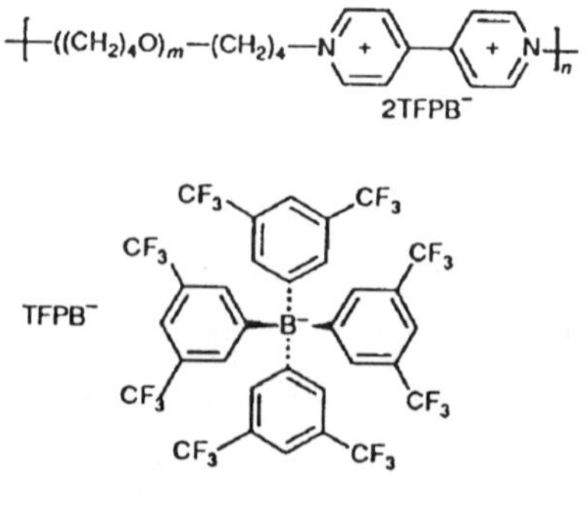

Figure 7. Molecular structure of an elastic polymer of 4,4'-bipyridinium with tetrakis[3,5-bis(trifluoromethyl)phenyl]borate anions.

Other materials which may be classified under this heading include an unusual series of π-bridged zwitterions (*e.g.*, see Fig. 8). They bleach when irradiated at visible wavelengths [31-33]. In the dark, the solution process is reversible but with erratic colour return, whereas Langmuir-Blodgett films show no recolouration. The mechanism has not been elucidated but may involve charge transfer as the alternative explanations are less plausible: (i) *trans-to-cis* isomerisation is feasible but the low-energy visible transition is inconsistent with those normally associated with a sterically hindered *cis* form and, in addition, the switched form is unlikely to persist; (ii) photodimerisation of the ethylene bridge may occur but reversion to the unswitched form, as occurs in solution, would again be unlikely. However, erratic behaviour in solution may result from more than one mechanism.

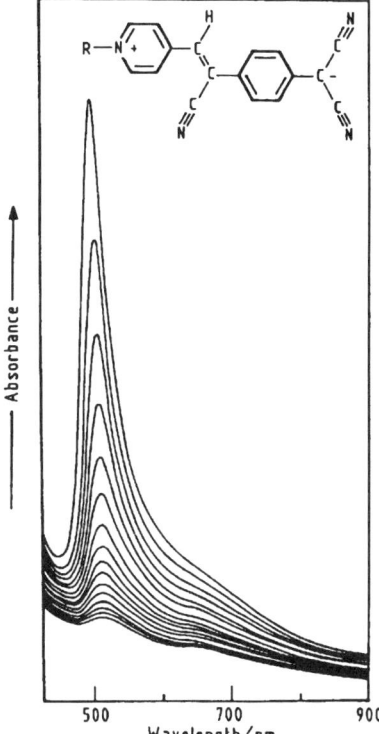

Figure 8. Langmuir-Blodgett film spectrum of Z-ß-(1-decyl-4-pyridinium)-α-cyano-4-styryldicyanomethanide showing progressive photobleaching.

The photobleaching of zwitterionic Langmuir-Blodgett films was at first attributed to *intra*molecular electron transfer between the negatively charged dicyanomethanide group and the positively charged heterocycle followed by a rearrangement to the neutral quinonoid structure, the stability of this form being explained by changes to the planarity of the molecule [33]. However, for the pyridinium zwitterion (see Fig. 8) the charge transfer band at 495 nm has recently been reassigned as an *inter*molecular transition and now, as a less contentious explanation, the photobleaching may be attributed to orientational changes which weaken the intermolecular interaction [34]. This is supported by the fact that substitution with bulky groups suppresses the absorbance of the charge transfer band.

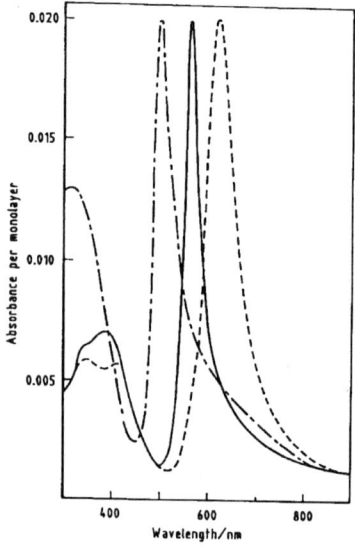

Figure 9. Langmuir-Blodgett film spectra of Z-ß-(1-decyl-4-pyridinium)-α-cyano-4-styryl-dicyanomethanide (λ_{max} 495 nm), Z-ß-(1-hexa-decyl-4-quinolinium)-α-cyano-4-styryldicyano-methanide (λ_{max} 565 nm) and Z-ß-(1-octyl-4-quinolinium)-α-cyano-4-styryldicyanomethanide (λ_{max} 614 nm).

A useful feature of the Langmuir-Blodgett film spectrum (Fig. 8) is the narrowness of the photobleachable charge transfer band (HWHM 27 nm) and this is characteristic of other zwitterions in the series. Of particular interest are the quinolinium analogues whose film spectra are dependent upon the quaternary group (see Fig. 9); the long-chained quinolinium zwitterions (pentadecyl and above) exhibit a sharp *inter*molecular charge transfer band at 565 nm with a 22 nm HWHM whereas for shorter chain lengths the band is bathochromically shifted to 614 nm with a 37 nm HWHM. This variation is attributed to different molecular tilts.

At wavelengths which correspond to the maxima of the pyridinium and quinolinium zwitterions (Fig. 9) the absorption bands only partially overlap. Thus, as with Ando's spiropyrans these materials also have the potential to be used as components of a multifrequency memory, albeit limited to a *write once read many* system.

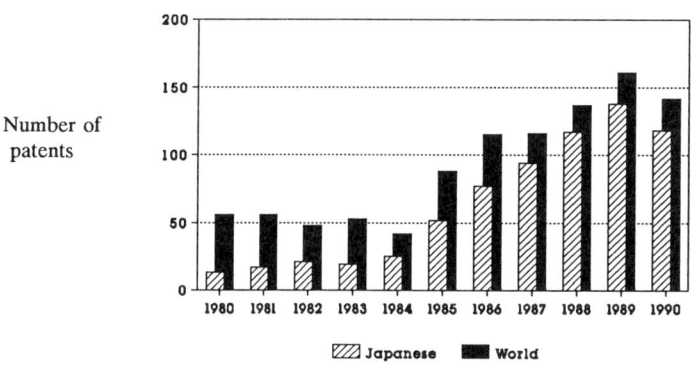

Figure 10. Growth of photochromic patents since 1980.

1.7 OPTICAL DATA STORAGE

The market for *direct read after write* (DRAW) rewritable discs for optical data storage is estimated to be approximately US$3 billion in the 1990's [35]. This has resulted in recent increased research activity in this field, particularly in Japan (see Fig. 10), and many different approaches of achieving reversibility in optical recording media are being followed.

Photochromic materials have potential as optical data storage media (as outlined in the previous sections) avoiding some of the disadvantages of other systems. However, they present their own set of problems, in particular, fatigue and an instability of the switched form. Information may be recorded by either colouring or bleaching, the latter being preferable due to the availability of suitable lasers in the visible region, but this depends on the absorption profile of the compound coinciding with the laser wavelength. Reading the recorded information may be carried out *either* by an absorptive readout (the beam being attenuated by, for example, the presence of a coloured read form) *or* by phase readout where the read beam detects a change in refractive index between the colourless and coloured areas. This technique has the advantage of using a read wavelength which is not necessarily absorbed by the photochromic and, thus, should not switch the recorded areas.

For use as erasable memory materials few photochromics have demonstrated the reliability (thermal stability and low fatigue) to make them commercially viable but, with further improvement, some clearly have potential. However, to compete they need to out-perform other data storage media. Probably the most advanced of these, in terms of a marketable product, uses changes in polarisation of a plane polarised laser beam caused by changes in magnetic orientation of the recording media (magneto optic effect). In its blank state the disc is formed from randomly oriented magnetic particles in a polymer base. Upon irradiation with a solid state IR laser the polymer softens and allows the magnetic particles to align with the field of a magnet held below the disc. The aligned area is then locked by the polymer cooling. Reading may be carried out using a laser of the same frequency but of reduced power to avoid reheating; changes in polarisation, brought about by the magnetic medium, may be recorded as intensity changes by using a polariser.

Magneto optical devices provide storage densities of 10^8 bits/cm^2 with transfer rates of *ca.* 0.5 Mbyte/sec [36]. It will be an uphill battle for photochromics to match this performance.

2. ELECTROCHROMISM

Electrochromic materials are analogous to the photochromics and have been described by Silver [37] as chemical chameleons. They undergo reversible electric field induced reactions (oxidation or reduction) accompanied by colour changes and have uses as antiglare mirrors, smart windows with controllable absorption characteristics, displays and optical filters. There are numerous electrochromic materials but probably the most important example is tungsten trioxide which turns blue when switched [38]. The process depends on an electric field induced diffusion of group I elements, particularly hydrogen, into polycrystalline films of tungsten oxide and is described by:

$$WO_3 + xH^+ + xe \rightleftharpoons H_xWO_3$$

In a practical form the WO_3 may be evaporated or sputtered on a transparent electrode such as ITO glass, a solid electrolyte deposited on the oxide (*e.g.* by spinning or spraying from solution) and the cell completed by a counter electrode which may be evaporated or held against the electrolyte. With an appropriate voltage bias ions diffuse through the oxide layer and react with it to form the tungsten oxide bronze. The system may be cycled between yellow (WO_3) and blue (H_xWO_3) without undue fatigue and, as both states are stable, the colour persists when the current is switched off.

Other examples include V_2O_5 and Nb_2O_5 which undergo cathodic colouration and IrO_2 (anodic colouration) [38] as well as several organic compounds. Of these, the lanthanide bis-phthalocyanines have attracted much recent attention; they undergo a one-electron oxidation to $[M(pc^.)_2]^+$, a red form, and reduction to $[M(pc)_2]^-$ or $[M(pc)_2]^{2-}$ which are blue and purple respectively with 10^7 oxidation/reduction cycles being achievable [36]. However, of the organics, the best known examples are the viologens, *e.g.* the N,N'-diheptyl-4,4'-bipyridinium and N,N'-bis(*p*-cyanophenyl)-4,4'-bipyridinium salts (reviewed in [39]). They undergo a one-electron reduction to the highly coloured radical monocation which, being less soluble in aqueous solution, deposits at the cathode. Colouration is fast (10 to 100 ms) with $> 10^5$ reduction/oxidation cycles being achievable but the solution cell has the disadvantage of a limited temperature range.

3. CONDUCTIVE CRYSTALS

Most homomolecular organic crystals are electrical insulators; the molecules are held together by van der Waals forces and the interaction is too weak for there to be any appreciable π–overlap except at high pressure (*e.g.* iodanil: σ_{rt} 10^{-12} S cm^{-1} at 1 bar; 20 S cm^{-1} at 500 kbar [40]). Conductivity enhancement is achieved by turning to the charge transfer complexes which may be classified into separate groups based upon the degree of charge transfer. Non-bonding complexes, of which there are many, have low conductivities, $\sigma_{rt} < 10^{-8}$ S cm^{-1}, and their structures are characterised by mixed stacks in which the donor and acceptor moieties alternate, whereas dative complexes have intermediate to high conductivities of $10^{-8} < \sigma_{rt} < 10^{4}$ S cm^{-1}. In this category, the less conductive examples exhibit complete charge transfer and/or mixed stacking whereas for metallic behaviour the criteria are non-integral charge transfer and segregated stacking with a uniform spacing and favourable π-overlap (Fig. 11).

$D^{\circ} \ A^{\circ} \ D^{\circ} \ A^{\circ}$	$D^{+} \ A^{-} \qquad D^{+} \ A^{-}$	$D^{\delta+} \ A^{\delta-} \ D^{\delta+} \ A^{\delta-}$
$A^{\circ} \ D^{\circ} \ A^{\circ} \ D^{\circ}$	$A^{-} \ D^{+} \qquad D^{+} \ A^{-}$	$D^{\delta+} \ A^{\delta-} \ D^{\delta+} \ A^{\delta-}$
$D^{\circ} \ A^{\circ} \ D^{\circ} \ A^{\circ}$	$D^{+} \ A^{-} \qquad D^{+} \ A^{-}$	$D^{\delta+} \ A^{\delta-} \ D^{\delta+} \ A^{\delta-}$
$A^{\circ} \ D^{\circ} \ A^{\circ} \ D^{\circ}$	$A^{-} \ D^{+} \qquad D^{+} \ A^{-}$	$D^{\delta+} \ A^{\delta-} \ D^{\delta+} \ A^{\delta-}$
$D^{\circ} \ A^{\circ} \ D^{\circ} \ A^{\circ}$	$D^{-} \ A^{+} \qquad D^{+} \ A^{-}$	$D^{\delta+} \ A^{\delta-} \ D^{\delta+} \ A^{\delta-}$
$A^{\circ} \ D^{\circ} \ A^{\circ} \ D^{\circ}$	$A^{-} \ D^{+} \qquad D^{+} \ A^{-}$	$D^{\delta+} \ A^{\delta-} \ D^{\delta+} \ A^{\delta-}$
Low–σ	Intermediate–σ	High–σ

Figure 11. Stacking and electrical characteristics of charge transfer complexes.

The first highly conductive organic salt was reported three decades ago and now there are numerous examples of synthetic metals and superconductors (for reviews see [41-43]). However, recent advances in this field may be seen to stem from the synthesis of tetrathiafulvalene [44]; it forms nonstoichiometric salts with inorganic anions and stoichiometric charge transfer complexes with organic acceptors such as TCNQ. The electrical properties of TTF–TCNQ and its derivatives [45-57] are summarised in Table 1.

Table 1. CHARGE TRANSFER COMPLEXES OF TTF-TCNQ AND 1:1 CONGENERS.

Donor	Acceptor	$\sigma_{rt}/S\ cm^{-1}$	T_{max}/K	σ_{max}/σ_{rt}	Ref.
TTF	TCNQ	200-920	59	3-150	[45]
TTF	TCNQ-Me	200-500	210	1.2	[46]
TTF	TCNQ-Et$_2$	200	210	1.4	[47]
TTF	TCNQ-ClMe	250	210	1.2	[48]
TTF	TCNQ-BrMe	435	225	1.1	[48]
TTF	TCNQ-IMe	185	225	1.3	[48]
TTF	TCNQ-Br$_2$	295	275	1.0	[48]
DMTTF	TCNQ	50	50	25	[49]
TMTTF	TCNQ	350	60	15	[49]
TMTTF	TCNQ-Me$_2$	120	80	3	[50]
HMTTF	TCNQ	500	80	3.5	[51]
tTTF	TCNQ	200-400	120	2.5	[52]
MEtTTF	TCNQ	90	165	1.6	[53]
DBTTF	TCNQ-Cl$_2$	40	260	1.1	[54]
TSF	TCNQ	800	40	12	[55]
DEDMTSF	TCNQ	500	55	9	[56]
TMTSF	TCNQ	1200	65	7	[56]
HMTSF	TCNQ	2000	75	3.5	[57]

TTF tetrathiafulvalene; DMTTF dimethyltetrathiafulvalene; TMTTF tetramethyltetrathiafulvalene; HMTTF hexamethylenetetrathiafulvalene; tTTF trimethylenetetrathiafulvalene; MEtTTF methylethyltrimethylenetetrathiafulvalene; DBTTF dibenztetrathiafulvalene; TSF tetraselenafulvalene; DEDMTSF diethyldimethyltetraselenafulvalene; TMTSF tetramethyltetraselenafulvalene; HMTSF, hexamethylenetetraselenafulvalene.

These complexes display conductivity maxima and, with the exception of HMTSF-TCNQ, undergo metal-to-insulator transitions at lower temperatures. The loss of metallic behaviour is often attributed to a Peierls transition [58] (a distortion of the lattice with localisation of the carriers) and this has been substantiated for TTF-TCNQ by X—ray and neutron scattering (reviewed in [59]). However, for other TCNQ complexes an alternative explanation which may apply is that solvent, trapped during crystal growth, induces the transitions as it contracts upon freezing [60]. Five of the TTF-TCNQ complexes listed in Table 1, as well as many metallic salts (ca. 30 % of the total), including the now classical N-methylphenazinium$^+$(TCNQ)$^-$ and quinolinium$^+$(TCNQ)$_2^-$ salts [61,62], have conductivity maxima between 210 and 230 K. Many semiconducting salts also have transitions in this narrow temperature range [60]. A common feature of these is that acetonitrile is used in the synthesis; it melts at 229 K and undergoes a solid state transition at 214 K. This distinctive double transition has been found in the DSC traces of TCNQ salts, and electrical anomalies have been suppressed by reducing the trapped solvent concentration [60].

We have so far considered TTF-TCNQ and substituted derivatives but there are many different classes of organic metals. Most undergo metal-to-insulator transition but a few remain highly conductive to low-temperatures; examples include (2,5-dimethyl-N,N'-dicyanoquinoneimine)$_2$Cu [63], nickel phthalocyanine iodide [64] and the dimercaptoisotrithione salts, NMe$_4$[Ni(dmit)$_2$] (T$_c$ 5 K) and TTF[Ni(dmit)$_2$] (T$_c$ 1.6 K) which superconduct at 7 kbar [65]. Their molecular structures are diverse and the properties are strongly influenced by the size and geometry of the counterion.

Among the organic conductors the low-temperature superconducting examples are of most interest (see [41-43]). Developments in this field have been rapid; the first ambient pressure example, TMTSF$_2$ClO$_4$ (T$_c$ 1.4 K), was discovered in 1981 and a decade later there are ca. 40 known examples. These achievements were initially overshadowed by the discovery by Bednortz and Müller in 1986 of

Table 2. AMBIENT PRESSURE ORGANIC SUPERCONDUCTORS.

Salt[*)	T_c/K	Ref.	Structure
ß(BEDT-TTF)$_2$I$_3$	1.4	[76]	
θ(BEDT-TTF)$_2$I$_3$	3.6	[77]	
κ(BEDT-TTF)$_2$I$_3$	3.6	[78]	
γ(BEDT-TTF)$_3$(I$_3$)$_{2\frac{1}{2}}$	2.5	[79]	
ß(BEDT-TTF)$_2$IBr$_2$	2.8	[80]	
ß(BEDT-TTF)$_2$AuI$_2$	5.0	[81]	
κ(BEDT-TTF)$_2$Au(CN)$_2$·H$_2$O	5.0	[82]	
κ(BEDT-TTF)$_2$Cu(NCS)$_2$	10.4	[83]	
κ(BEDT-TTF)$_2$Cu[N(CN)$_2$Br	11.6	[68]	
ß$_m$(BEDO-TTF)$_3$Cu$_2$(SCN)$_3$	1.1	[84]	
(DMET)$_2$I$_3$	0.47	[85]	
(DMET)$_2$IBr$_2$	0.59	[85]	
(DMET)$_2$AuCl$_2$	0.83	[86]	
κ(DMET)$_2$AuBr$_2$	1.9	[87]	
κ(MDT-TTF)$_2$AuI$_2$	4.5	[88]	
(TMTSF)$_2$ClO$_4$	1.4	[89]	
K$_3$C$_{60}$	18	[73]	
Rb$_x$C$_{60}$	30	[75]	

[*) ß, γ, κ and θ denote the crystal phase.

the high T_c ceramic superconductors [66]; in this field the highest verifiable T_c is currently 125 K for $Tl_2Ba_2Ca_2Cu_3O_{10+y}$ (reviewed by Grant [67]). However, the recent discovery of the superconducting buckminster fullerenes has once again opened up the field.

The different classes of superconducting salts are listed in Table 2. Of these the BEDT-TTF series provides the largest number of representative examples, the highest ambient pressure T_c being 11.6 K for κ(BEDT-TTF)$_2$Cu[N(CN)$_2$]Br [68]. The Cu[N(CN)$_2$]Cl analogue has a higher T_c of 12.8 K but at a critical pressure of 0.3 kbar [69]. These materials have a higher dimensionality than the organic metals (Table 1) and the restriction of uniform stacking is removed. Their properties were extensively reviewed by Williams et al. in 1987 [42] and developments since then have been updated in a separate article [43].

In this field it is now likely that attention will focus on the hollow-cage buckminster fullerene (C_{60}). Reported by Krätschmer et al. [70] in 1990 it may be synthesised by resistive heating of graphite in an inert atmosphere. Its properties have been summarised by Diederich and Whetton [71]; it is a strong electron acceptor, the reduction potential in THF being −0.18 V vs. Ag/AgCl. The C_{60} films, doped with alkali metal vapour, are highly conductive (e.g. σ_{rt} 500 S cm^{-1} for K_xC_{60} [72]) comparable with the TTF complexes in Table 1. At low temperatures the samples superconduct; potassium doped samples have an onset T_c of 18 K (Hebard et al. [73] confirmed by Wang et al. [74]). Hoiczer et al. [75] have reported 19.3 K and have disclosed a stoichiometric compound (K_3C_{60}). The onset T_c of the rubidium doped buckminster fullerene (Rb_xC_{60}) is higher still and is reported as 28.6 K [71] and 30 K [75] (higher than those of Nb_3Sn and Nb_3Ge).

These very recent preliminary findings clearly open the way for a series of carbon-based superconductors and are likely to trigger similar feverish activity as observed after the initial discovery of the high T_c ceramic superconductors in 1986.

4. NONLINEAR MATERIALS

4.1 MOLECULAR RECTIFIERS

The quest for an ever decreasing size of electronic components combined with the possibility that molecules may be designed to operate as self-contained devices is the driving force behind molecular scale electronics. Carter has advanced the idea and has proposed some molecular analogues of conventional electronic switches, gates and connecting [1]. The simplest of these, the molecular *pn* junction, was first suggested in 1974 [90]. However, fabrication of a molecular rectifier device comprising an organic monolayer (2 to 3 nm thick, pinhole-free and sandwiched between metal electrodes) requires an extremely careful technique and, if successful, it would be difficult to assess whether a Schottky contact or the molecule is responsible for the observed *I-V* behaviour.

The idea, originated by Aviram and Ratner [90] and developed by Metzger [91], requires a nonlinear material; the molecule proposed in the original paper was never synthesised but consisted of an electron-donor (TTF) linked *via* a bicyclo[2.2.2]octane bridge to an electron-acceptor (TCNQ). It was suggested that under forward electrical bias the steps would be (i) resonant (elastic) electron tunnelling from anode to acceptor and from donor to cathode; (ii) reversion to the neutral form *via* intramolecular inelastic tunnelling.

$$M_1^\frown |\,D\text{-}\sigma\text{-}A^\frown|\,M_2 \;\rightarrow\; M_1\,|\,D^+\text{-}\sigma\text{-}A^-\,|\,M_2 \;\rightarrow\; M_1^\frown|\,D\text{-}\sigma\text{-}A^\frown|\,M_2$$

With opposite bias the mechanism would lead to the reverse polarity zwitterion and Aviram and Ratner postulated that their material would rectify because the barrier to $TTF^+\text{-}\sigma\text{-}TCNQ^-$ is low relative to $TTF^-\text{-}\sigma\text{-}TCNQ^+$. Alternatively, in this direction, the first step may involve intramolecular charge transfer to $TTF^+\text{-}\sigma\text{-}TCNQ^-$ (uphill) followed by tunnelling to and from the electrodes.

In such a device it is necessary to closely match the work electrode functions with the donor ionisation energy and acceptor electron affinity; the flexible σ−bridge should be short enough to prevent bending of donor over acceptor;

the molecule should be capable of forming a pinhole-free Langmuir-Blodgett monolayer. Following these criteria representative D–σ–A materials have been synthesised by Metzger [91] with a –NH–(C=O)–O–CH₂CH₂–O– bridge linking the donor (*e.g.* *p*-alkoxybenzene or N,N-dialkylaniline) and acceptor (*e.g.* 5-bromo-TCNQ) parts.

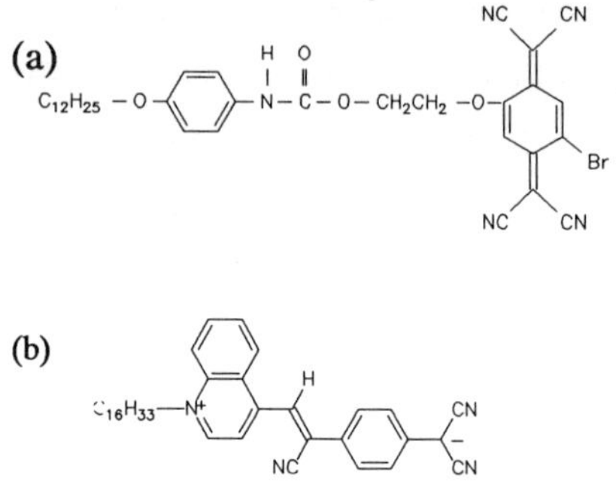

(a)

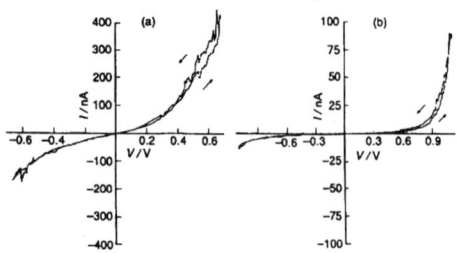

(b)

Figure 12. Molecular structures of (a) DDOP-C-BHTCNQ and (b) $C_{16}H_{33}$-Q3CNQ.

Recently, two examples of rectifying devices have been reported [92,93], one based on a σ-bridged system (DDOP-C-BHTCNQ, Fig. 12a), first synthesised by Metzger [94], and the other on a photobleachable π-bridged zwitterion ($C_{16}H_{33}$-Q3CNQ, Fig. 12b). The monolayer films were deposited on platinum and, to avoid damage during overlay, evaporated silver-coated magnesium was used as the top electrode. The *I-V* curves of the $Mg \vert (C_{16}H_{33}$-Q3CNQ$) \vert Pt$ structures are shown in Fig. 13; the asymmetric behaviour is as expected for molecular rectification but the reactive Mg electrode also adds the possibility of Schottky contact behaviour. The mechanism remains to be resolved.

Figure 13. *I-V* plots for $Mg \vert (C_{16}H_{33}$-Q3CNQ$) \vert Pt$ structures with (a) one and (b) three monolayers.

4.2 FREQUENCY DOUBLERS

In recent years there has been considerable interest in advanced nonlinear optical materials and the fabrication of non-centrosymmetric structures for second harmonic generation (SHG). The material requirements (inorganic as well as organic) are discussed at length in Chapter 4 and, thus, they will not be duplicated here. Instead, the fabrication of low-loss Langmuir-Blodgett film structures for SHG will be outlined. The molecular criteria are similar to those discussed in the previous section for the Aviram and Ratner rectifier except that a π-electron-bridge (e.g. $-CH=CH-$, $-N=N-$ or $-C_6H_4-$ or combinations or multiples thereof) is necessary as the link between the donor and acceptor parts. A further requirement is for the packing to be non-centrosymmetric.

The packing may be controlled in monolayer structures whereas in multilayer films the stacking arrangement is usually head-to-head and tail-to-tail (Y-type). In most cases the films provide a small SH intensity when the number of layers is odd but the signal diminishes when the number is even. In contrast, SHG from non-centrosymmetric structures should increase quadratically with the number of layers but a lack of long range order usually depresses the signal.

Table 3. SECOND ORDER NONLINEAR OPTICAL COEFFICIENTS.

Dye	$\chi^{(2)}$ (pm/V)*	Film type	Ref.
DCANP	16	Y-type (herringbone)	[95]
C12PPy	40	Interleaved	[96]
Hemicyanine	100	Molecular Zip	[97]
$C_{16}H_{33}$-Q3CNQ	180	Z-type	[98]

*Obtained using Nd:YAG laser, λ 1064 nm.

26

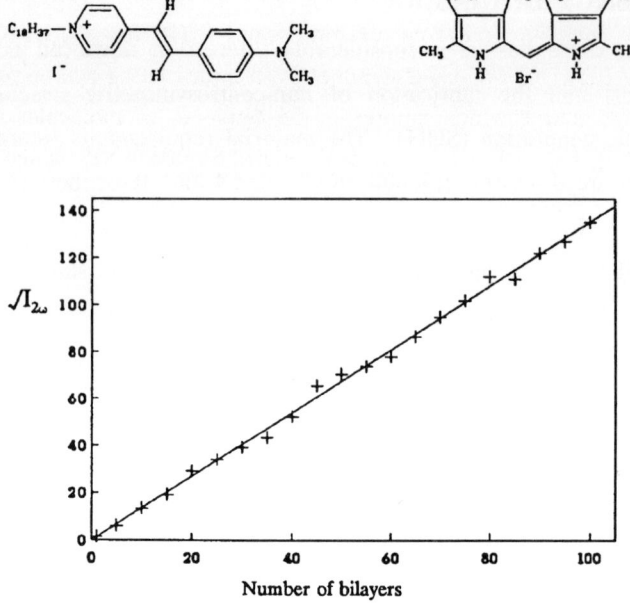

Figure 14. Square root of the SH intensity *vs.* the number of bilayers of a hemicyanine dye and two-legged spacer.

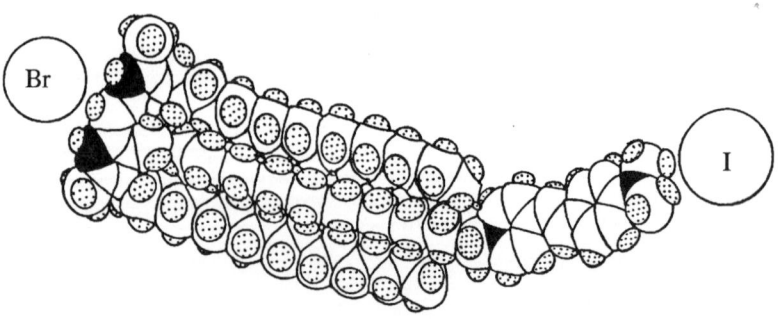

Figure 15. Interlocking (*Molecular Zip*) structure of the hemicyanine dye and two-legged spacer shown above.

Quadratic SHG enhancement from micron thick Langmuir-Blodgett films has been reported for the following examples only: (i) the $C_{16}H_{33}$-Q3CNQ zwitterion (Fig. 12b), its non-centrosymmetric structure being stabilised by the terminal negative charge which suppresses the tendency of molecules to stack head-to-head [98]; (ii) 2-docosylamino-5-nitropyridine (DCANP) which has an unusual non-centrosymmetric herringbone Y-type packing arrangement [95]; (iii) 5-(p-dodecyloxyphenyl)pyrazine-2-carboxylic acid (C12PPy) interleaved with arachidic acid [96]; (iv) a polymeric azo dye interleaved with tricosenoic acid [99]; (v) a hemicyanine dye interleaved with a two-legged spacer (Fig. 14) [97]. Their properties are summarised in Table 3.

To overcome the limitations of Y-type film forming materials attempts have been made to fabricate non-centrosymmetric Langmuir-Blodgett structures by interleaving an optically nonlinear dye with a spacer, generally a fatty acid [100]. However, with the exception of C12PPy and azo dye above, the SHG falls short of the expected quadratic dependence for films comprising more than a few bilayers. A recently reported approach, uses a two-legged dipyrrylmethene spacer (see Fig. 15), the spacer being designed to fasten the bilayer structure *via* insertion of the hydrophobic tail of the optically nonlinear material between the legs of the spacer (*a "Molecular Zip"*). Ordered low-loss films have been fabricated by this technique [97, 100] and, unlike many other Langmuir-Blodgett structures, the signals have shown no sign of deterioration even after one year.

5. LIQUID CRYSTALS

When a crystalline solid is heated or added to solvent, it is usual to think of it passing directly into a molten or dissolved state with no residual ordering of the molecules. In practice, however, it is common for compounds to pass through intermediate states (*mesophases*) having less order than a crystal, but more than the familiar isotropic liquid state. Compounds in which the molecules have a near-spherical form frequently gain rotational freedom while maintaining a rigid three dimensional lattice structure, thereby forming *plastic crystal* phases. Correspondingly, for molecules with a highly anisotropic shape, the rigid three-dimensional lattice of the crystal can break down while orientational order of the molecules is retained. Such phases, which combine birefringence and other anisotropies typical of a non-cubic crystal with some capacity for viscous flow, are termed *liquid crystal* phases.

The liquid crystal phases share the feature of orientational molecular order in a phase which lacks three-dimensional long-range translational order of the molecules. A wide variety of liquid crystal phases is known, these being distinguished by various degrees of short-range or low-dimensional translational order, by the shape (rod or disc-like, or more complex) of the constituent molecules, and by the mode of formation (melting or solution) of the phase. The complete picture is, therefore, complex but the best studied liquid crystal phases and those of the greatest technological interest are formed by melting compounds with rod-like molecular form. Such compounds include esters of cholesterol which provided the first recognised liquid crystal phases to be studied [101].

A wide range of rod or lath shaped (*calamitic*) liquid crystals are known. A selection is illustrated in Table 4, the phases exhibited being conveniently referred to by single letter abbreviations. The nematic phase (N; Fig. 16a) is the least ordered liquid crystal phase and the one most commonly used in electro-optic applications. It lacks all long-range translational order and forms a fluid oily liquid with a viscosity characteristically lower than the other liquid crystal phases. Because of its simplicity and technical importance, the nematic phase has been the subject of the most intense experimental and theoretical study and provides much of the basis for understanding the other liquid crystal phases.

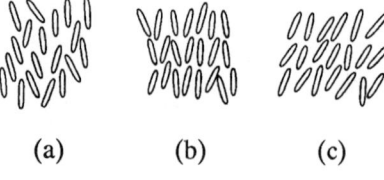

(a)   (b)   (c)

Figure 16. Schematic structure of (a) nematic, (b) smectic A and (c) smectic C liquid crystal phases.

Table 4. SOME THERMOTROPIC CALAMITIC LIQUID CRYSTALS.

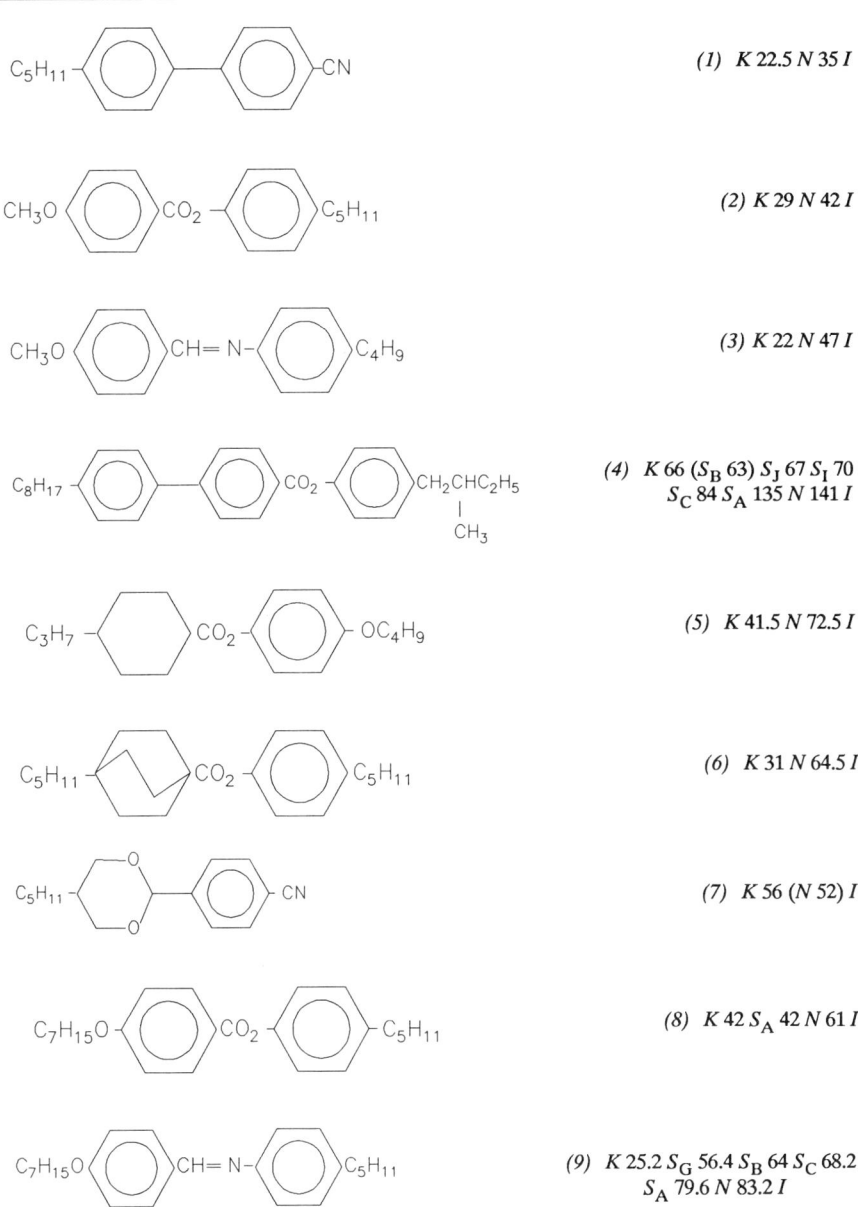

(1) K 22.5 N 35 I

(2) K 29 N 42 I

(3) K 22 N 47 I

(4) K 66 (S_B 63) S_J 67 S_I 70
S_C 84 S_A 135 N 141 I

(5) K 41.5 N 72.5 I

(6) K 31 N 64.5 I

(7) K 56 (N 52) I

(8) K 42 S_A 42 N 61 I

(9) K 25.2 S_G 56.4 S_B 64 S_C 68.2
S_A 79.6 N 83.2 I

Table 4. Continued.

C_3H_7 — (cyclohexyl) — C_2H_4 — (phenyl) — (phenyl) — C_3H_7

(10) K 67 S 119 N 144 I

C_3H_7 — (cyclohexyl) — C_2H_4 — (phenyl) — (phenyl, F) — C_3H_7

(11) K 40 N 108 I

$C_{10}H_{21}O$ — (phenyl) — $CH=N$ — (phenyl) — $CH=CH$ — $CO_2CH_2CHC_2H_5$ | CH_3

(12) K 76 S_C^* 92 S_A 117 I

$C_8H_{17}O$ — (phenyl) — (phenyl) — CO_2 — (phenyl, F) — C_5H_{11}

(13) K 49 $(S_I$ 30$)$ S_C 121 S_A 128 N 164 I

$C_8H_{17}O$ — (phenyl) — CO_2 — (phenyl) — $CO_2CH_2CHC_2H_5$ | CH_3

(14) K 35 $(S_C^*$ 31$)$ S_A 58 I

$C_5H_{11}CO_2$ — (phenyl) — (pyrimidine) — C_7H_{15}

(15) K 48 $(S_A$ 44$)$ N 52.5 I

$C_8H_{17}CO_2$ — (cholesteryl)

(16) K 80.5 $(S_A$ 77.5$)$ Ch 92 I

5.1 THE NEMATIC PHASE

The nematic phase has only orientational long-range ordering of the constituent molecules. Viewed over a macroscopic volume, the long molecular axes are found to be preferentially aligned along a unique direction, which can be influenced by surface forces and by imposed electrical and magnetic fields. The locally favoured alignment direction is known as the *director* and in common nematic phases it coincides with the unique optic axis of a phase which behaves as a uniaxial crystal. The nematic phase is highly disordered by thermal agitation and each molecule can be characterised by an angle θ between the molecular long axis and director. (For this purpose, the molecule is supposed to show cylindrical symmetry, reflecting the symmetry of the nematic phase. The difficulty of defining a long axis for a real molecule is also laid to one side.) The ordering of the phase is, therefore, described by the distribution of θ values for its constituent molecules. Experimentally, it is found that no distinction exists between the two directions along the director, for example, a nematic phase responds in an identical manner to an electric field in a specified direction regardless of whether its polarity is positive or negative. Taking this additional symmetry into account, it is convenient to be able to quantify the degree of order in a nematic phase by a scalar, and this is provided [102] by the order parameter (S), given by:

$$S = \tfrac{1}{2} <3\cos^2\theta - 1> \qquad (1)$$

This parameter describes the degree of order of the molecular axes along the director subject to the symmetry constraints described above: it has a value of unity in a hypothetical perfectly ordered nematic liquid in which all molecules are parallel, and zero in the isotropic phase. It may also be regarded as the first term in the description of the complete distribution of θ values in terms of the even-order Legendre polynomials; higher order terms in this expansion, as well as the microscopic biaxiality which results from the non-cylindrical form of real molecules, can be investigated by spectroscopic methods [103].

The scalar order parameter S is related solely to the mean angle between each molecule and the director, *i.e.* to the disorder in the alignment direction. Because of this, it can be studied from properties of the phase which reflect the anisotropy of the liquid crystal molecules. Many properties are anisotropic and the diamagnetic susceptibility of a nematic liquid crystal is particularly suitable for this purpose. The susceptibility is dependent chiefly on the freedom of an induced electron current to flow in response to an applied magnetic field [104]. Structural sub-units which make a large contribution to such an induced current include aromatic rings and triple-bonded groups such as acetylene and cyano functions, but the relative orientation of the magnetic field and the symmetry axis of the π-bonded group is also crucial. The measured susceptibility of, for example, a crystal in which these groups are rigidly held in position will vary strongly according to the orientation of the sample. In a nematic liquid crystal, the magnetic field interacts with the sample in such a way that for most materials (those which have a positive diamagnetic anisotropy, *i.e.* $\chi_\parallel > \chi_\perp$) the director tends to align parallel to the field. This means that for such materials, the experimental geometry illustrated in Fig. 17a is automatically obtained, and that in Fig. 17b cannot be achieved. The measured susceptibility now depends on the different diamagnetic components measured along the molecular axes, and on the orientation of each molecule in the sample relative to the director, *i.e.* on the order parameter S. Expressing the relationship in the more usual and useful form gives [105]:

$$S = (\chi_\parallel - \chi_\perp)/(\chi_l - \chi_t) \qquad (2)$$

The quantities χ_l and χ_t are values of susceptibility measured on a perfectly ordered system (*e.g.* in an oriented solid crystal) respectively along and transverse to the molecular long axes. In many cases, suitable measurements on the solid are not available and an alternative procedure must be used. The assumption that S will tend to unity at zero Kelvin provides such an alternative and a procedure for extrapolation from measurements made at accessible temperatures has been given [106].

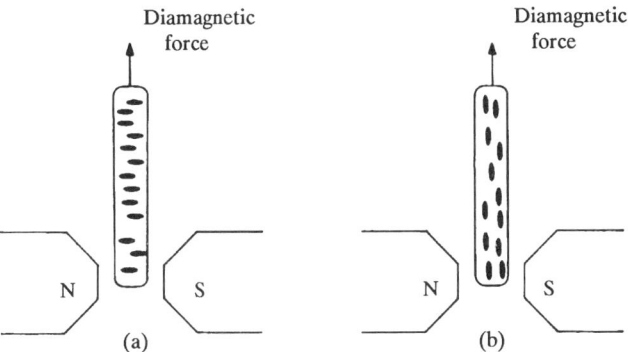

Figure 17. Geometry for the measurement of (a) $\chi_\parallel$ and (b) $\chi_\perp$. (b) is experimentally inaccessible for reasons described in the text.

The suitability of the magnetic susceptibility for the measurement of S derives from the fact that it is small and the field in the sample is essentially identical with that imposed externally. In the case of, for example, dielectric constant measurements this is not so: the local field around a molecule is distorted by the response of adjacent molecules to the applied field. If the form of the local field distortion is assumed, S can be estimated through such measurements. The convenience and accuracy of refractive index measurement has led to this being a standard route to the measurement of order in thermotropic nematic systems.

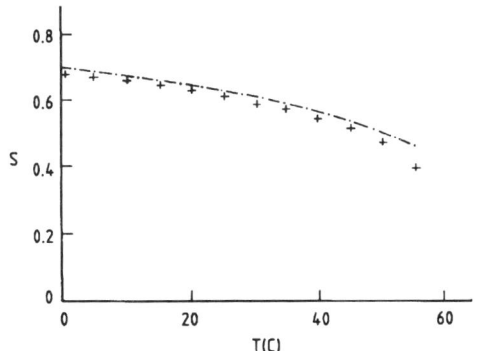

Figure 18. Variation of the order parameter with temperature for the cyanobiphenyl based mixture E7, which has a Nematic-Isotropic transition at 60°C (+) and the Maier-Saupe prediction (line).

The variation of the order parameter with temperature is different for each material studied, but in general it follows a smooth curve until close to the clearing temperature before dropping abruptly to zero (see Fig. 18). A semi-quantitative agreement with experimentally determined values of S can be achieved from a remarkably simple theoretical treatment of nematic ordering. The Maier-Saupe theory [107,108] regards liquid crystal molecules as rigid ellipses which interact only *via* long range dispersion forces. Each molecule responds to and contributes to an anisotropic potential which provides the driving force for the establishment of an orientationally ordered phase. The result is a self-consistency equation whose solution provides the variation of order parameter with temperature.

The Maier-Saupe theory is successful in many of its predictions but remains unsatisfactory from a conceptual point of view. The reason for its success has been pointed out [109]: the short range interactions including steric and permanent dipolar forces must be comparatively very large in absolute magnitude but be substantially isotropic in the contribution they make to the molecular force field. It is the remaining weaker forces that dominate the anisotropy of the intermolecular field. Attempts to account for orientational ordering based on pure steric interactions of rod-like molecules also lead to the prediction of a stable nematic phase [110]. The detailed predictions of the variation of order with temperature and of the resulting density of the phase are rather unsatisfactory. The application of modern computational techniques is allowing progress in the use of more complex and realistic intermolecular potentials in the modelling of liquid crystalline ordering [111].

Although the theoretical treatment of nematic phases in terms of dispersion type forces works well for some physical parameters of the LC phase, it does not provide a very enlightening insight into the relationship between chemical structure and mesophase stability. In particular, a naive consideration of the theory would lead one to expect that nematic phase stability should correlate well with molecular polarisability anisotropy and, therefore, that conjugated aromatic groups should form the best structural units for designing liquid crystal molecules. Inspection of Table 4 will show that this is not necessarily the case, a point which will be expanded on below.

The discussion of nematic orientational order above has been framed in terms of rod-like molecules but may be applied, virtually unchanged, to thermotropic discotic phases. The director now lies along the unique *short* molecular axis and it is usual for the optical and (in the absence of a strong permanent dipole moment) dielectric anisotropies to be negative [112]. Furthermore, a similar discussion may be applied to amphiphilic nematic phases in which the anisotropic units are not molecules but rod or disc-like micelles. The geometrical form of the micelle may be deduced from the diamagnetic anisotropy of the phase [113]; the micelles are believed to be polydisperse and may undergo a transition to lamellar or hexatic phases. Biaxial nematic phases have been observed in micellar systems [114]; this behaviour has been described more recently in thermotropic nematic systems [115,116].

The nematic phase has a low viscosity comparable with isotropic liquids. (Nematic phases are also formed by some polymeric systems and in these the viscosity is of course rather high). Nematic liquids are distinguished from ordinary fluids by the intense scattering of light observed from the phase. This light scattering has its origin in thermally driven fluctuations in the orientation of the director, causing corresponding local changes in the effective refractive index of the medium. The abrupt disappearance of this scattering at the liquid crystal to isotropic transition leads to its alternative name of *clearing point*.

The nematic is usually easy to recognise among LC phases because of its fluidity. When definite identification is required, it may be obtained by placing a small sample of the material in question between a microscope slide and cover slip to create a thin film of liquid (*ca.* 10 μm thick). When observed under a microscope, between crossed polarisers, the various liquid crystal phases have characteristic appearances [117] (*textures*) resulting from their orientational anchoring to the glass surfaces, deformation under flow and the presence of defects in the LC structure. In the case of the nematic phase, a fluid birefringent texture containing broad black brushes and hair-like defect lines is seen, termed a Schlieren texture. Diffraction studies are also readily capable of identifying a nematic phase, on the basis of the complete lack of long range translational order.

5.2 THERMOTROPIC SMECTIC PHASES

Smectic phases are usually formed by calamitic thermotropic liquid crystals at lower temperatures than the nematic phase. There are, however, examples of nematic phases in dipolar materials which underlie smectic phases as well as compounds and mixtures which show a Smectic − Nematic − Smectic sequence on heating or cooling [118,119]. The underlying nematic phase in these cases is referred to as a *re-entrant* phase. Smectic phases, in addition to the orientational order of the nematic phase, have a one-dimensional translational ordering into layers. A number of smectic phases, denoted S_A, S_B etc (Table 5), can be distinguished by their possession of additional structural features. The more ordered phases are formally disordered solids and are not presently of technological importance. Only the true liquid crystal phases are considered here.

Smectic liquid crystal phases usually show a significantly higher apparent viscosity than nematic materials of similar chemical constitution. The different smectic phases are identified either by their characteristic textures and lack of complete miscibility when observed under a polarising microscope or by X-ray diffraction studies [120].

5.2.1 SMECTIC A AND SMECTIC C PHASES

The smectic A (S_A) phase is the least ordered of the thermotropic smectic phases. The molecules are arranged in disordered layers, each layer having a liquid-like freedom of motion of its constituent molecules in two dimensions. The layers are, moreover, far from being as well defined as the layer planes of molecules in a crystalline solid. Study of S_A phases by X-ray diffraction reveals that the variation in density along an axis perpendicular to the layer planes is sinusoidal, *i.e.* the 'layers' are diffuse and molecular motion from one layer to another is rather easy, despite the existence of an energetic barrier. This mirrors the orientational order, where molecules have a preferred alignment but are nevertheless able to rotate about their centres of gravity. In the S_A phase, the orientational director is perpendicular to the layer planes, termed an *orthogonal* phase. In the smectic C (S_C) phase, by contrast, the molecules are tilted away from this direction at an angle up to about 35°. The tilt direction is correlated between molecules within each layer and from one layer to another. The order in the S_A and S_C phases is otherwise identical.

Table 5. TYPICAL MICROSCOPIC TEXTURES OF SMECTIC LIQUID CRYSTAL PHASES

Phase	Textures	
S_A	Focal conic,	Orthogonal
S_C	Focal conic,	Schlieren
S_B^H	Mosaic,	Orthogonal
S_F	Mosaic,	Schlieren
S_I	Mosaic,	Schlieren

5.2.2 HEXATIC SMECTIC PHASES

In the smectic B phase, there is once again ordering of orientationally aligned molecules into layers. Within the layers there is an unusual form of ordering termed *hexatic*. The molecules are arranged in a hexagonal array but the translational order is short-range only. The orientation of the hexagonal net is, however, maintained over a long range and unlike the ordering of molecular positions is correlated between layers.

Two tilted analogues of the S_B phase exist in which the tilt direction is constrained to point either toward one face of the hexagonal lattice (S_F) or toward one apex (S_I).

5.3 DISCOTIC COLUMNAR PHASES

A close analogy can be drawn between the calamitic smectic phases in which rod-like molecules associate to form two dimensional layers, and the columnar phases in which disc-like molecules stack into essentially one dimensional columns. These columns then tend to pack into a hexagonal arrangement, or a rectangular one if the molecules are tilted relative to the axis of the columns. Three rectangular phases can be distinguished by X-ray; they are identified by their various space groups. Phases in which the molecules are regularly spaced within the columns, and in which the centres of gravity of the molecules are correlated in position between columns are once again formally solid phases, although they may be highly disordered. The known columnar phases [121] are summarised in Table 6. Like the smectic phases, the columnar phases have

---

Table 6. ORDER IN COLUMNAR DISCOTIC PHASES

Translational order	Hexagonal	Rectangular	Oblique
Liquid	D_{hd}	$D_{rd(P2_1/a)}$, $D_{rd(P2/a)}$, $D_{rd(2/m)}$	$D_{ob,d}$
Crystal	D_{ho}		

characteristic microscopic textures, but diffraction studies remain the most useful structural probe.

5.4 AMPHIPHILIC PHASES

Apart from the amphiphilic nematic phases, referred to above, there exist at least three fundamental classes of mesophases formed in surfactant solutions. Lamellar phases are formed as analogues of the thermotropic smectic phases by association of surfactant molecules into infinite two dimensional sheets. These sheets are normally bilayer structures which present the hydrophilic heads of the surfactant molecules to the polar solvent phase. In common with the other amphiphilic phases, an inverse phase is possible if the solvent is non-polar. The spacing between lamellae in these phases can be surprisingly large (sufficiently large for optical interference effects to be observed in a few cases [122]).

In cases where tubular micelles of essentially infinite length are formed, these can pack into a hexagonal arrangement reminiscent of the discotic columnar phases. Both normal and inverse hexagonal phases are known.

Two distinct types of cubic amphiphilic phases are known. The first of these results from the close-packing of spherical micelles, while the other is a bicontinuous phase [123] in which a polar and non-polar fluid interpenetrate one another, separated by a curved amphiphile sheet which forms a minimal surface.

The formation of the different amphiphilic phases relies heavily on the stability of a micelle or other structure of appropriate shape. This in turn depends chiefly on the ratio of effective sizes of the polar head and hydrophobic tails of the surfactant molecules [124]. The situation is frequently complex: the phases are often best developed in mixed solvent systems where the solvent polarity, surfactant concentration and temperature must all be considered as variables. This reflects the fact that the molecular aggregates which form the basis of the surfactant LC phases are not permanent rigid bodies like molecules of thermotropic phases. Rather, they exist in dynamic equilibrium with each other and with the solvent system associated with them.

5.5 MISCELLANEOUS LIQUID CRYSTAL PHASES

The liquid crystal phases described above represent the main classes of structures which have been investigated. Increasingly in recent years, novel molecular architectures have been investigated which extend the range of LC materials by introducing new structural features or by crossing the dividing lines between phase types which were implicit in the above discussion. Space allows only a brief mention of the more important developments here.

5.5.1 POLYMER LIQUID CRYSTALS

Any of the mesogenic entities described above may be incorporated into a polymer liquid crystal. The fundamental architectures available are the incorporation of the mesogen into the polymeric chain itself, or to attach the substituents laterally onto the polymer chain by more or less flexible linking groups (Fig. 19). The former main chain LC polymers must either contain flexible links or lateral substituents to reduce their melting points, or have sufficient polarity to dissolve in polar solvents to form their LC phases. Important examples of the latter lyotropic rod LC phases are formed by certain cellulose derivatives [125], polypeptides [126], and DNA fragments. Extensive work on synthetic main chain LC polymers, both lyotropic and thermotropic, has derived from their importance as structural polymers. When such materials are extruded or spun from their liquid crystal phases, the polymer chains are efficiently aligned in the direction of shear resulting in a very high modulus fibre. Poly-1,4-phenylene terephthalamide (Kevlar) spun from sulphuric acid or phosphoric acid solution is probably the best known example [127].

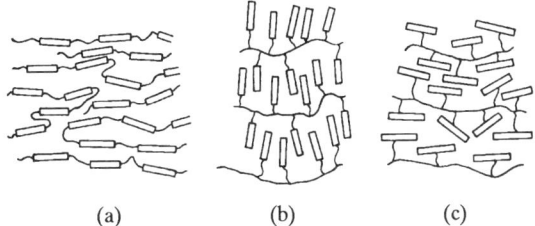

Figure 19. Polymer liquid crystal structures: (a) main chain jointed rod; (b) side chain with terminal attachment; (c) side chain with lateral attachment.

(a)   (b)   (c)

Liquid crystal side-chain polymers have been thoroughly explored using polysiloxane, polyacrylate and to a lesser degree other backbones [128]. Calamitic mesogenic groups may be attached at a lateral or terminal position, and discotic and other groups may also be used. The side chain polymers are of great interest for some technological applications because their physical properties mirror those of the parent mesogen, and they can in principle provide self supporting films [129]. The high viscosity inherent in polymeric materials means that molecular alignment processes are usually rather slow in such materials.

5.5.2 NON-CLASSICAL MOLECULAR ARCHITECTURES

The thermotropic liquid crystals described above are derived from essentially rod or disc shaped molecules. Many examples are now known of LC's formed by compounds whose molecular form deviates markedly from either of these shapes. Compounds based on highly branched terminal groups, two rods joined laterally or end to end by a flexible chain, and rod/disc combinations have all been examined. Progressive exploration of structures intermediate between rod and disc has provided the first examples of thermotropic biaxial phases. The reader is referred to an excellent review for work up to 1988 in this field [130].

5.5.3 AMPHITROPIC PHASES

Compounds which form LC phases either by melting or solution have been known for a considerable time but the design of such amphitropic materials by

42

substitution of hydrophilic groups onto classic rod or disc like mesogen molecules is a recent development. The full scope of this approach, together with the properties and value of the resulting materials, remain to be established.

5.6 OPTICALLY ACTIVE LIQUID CRYSTAL PHASES

The presence of optically active species in a liquid crystalline material can lead to the induction of a helical structure in the phase. The chiral counterpart of the nematic phase is the chiral nematic or cholesteric phase. In this phase, a helical twist axis is developed along some direction at right angles to the local director orientation. Progressing along this helical axis, the director orientation is seen to rotate at a uniform rate with distance travelled (Fig. 20a). Although, for convenience, the molecules in the helical structure are shown in a column in the figure, it should be emphasised that no such ordering into columns or layers exists in the cholesteric phase; the local order is identical to that in the nematic phase, and the nematic phase may be considered a special case of the cholesteric phase in which the pitch length is infinite. The physical properties of cholesteric materials become different from those of nematic substances when the pitch length is short enough to be comparable with the distance scale of the experiment. The striking optical properties of short-pitch chiral nematic phases are discussed further below. Lyotropic chiral nematic phases may be formed in an analogous manner to thermotropic ones [131,132], and chiral nematic discotic phases are also well established [133].

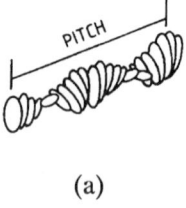

(a)

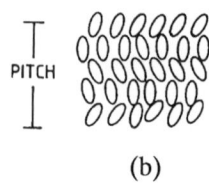

(b)

Figure 20. Schematic structures of (a) the chiral nematic phase and (b) chiral S_C phase.

Orthogonal smectic phases generally show no change in their structure when chiral materials are included, but tilted smectic phases can show a helical twist axis orthogonal to the smectic layer planes (Fig. 20b). The chiral S_C phase has been most thoroughly explored and pitch lengths in the sub-micron region, comparable with cholesteric phases, are easily obtained [134].

In short-pitch chiral nematic compounds and mixtures in the immediate vicinity of their clearing points, the so-called *blue phases* can often be observed [135]. At least three such phases are known, two having a cubic structure and the remaining one of an unknown nature. These phases appear to arise from frustration of the tendency of the LC structure to twist uniformly along different axes simultaneously. Their name comes from the opalescent reflection of light (not always blue) observed from bulk samples of the phases.

5.7 CHEMICAL STRUCTURE AND STABILITY

The investigation of the correlation of liquid crystal phase stability with chemical structure is a well-developed science [136]. Nevertheless, very few definitive rules exist to predict phenomena in this area. Furthermore, there is essentially no useful theory to give guidance. Rather, a body of empirical observations allows quite accurate predictions to be made about structure/property correlations only where rather closely analogous materials are available to support the argument. Furthermore, the available body of knowledge is heavily concentrated in the field of thermotropic calamitic phases because these materials have the longest history, are generally convenient to synthesise, and in the last two decades have assumed commercial importance through their use in electro-optic displays.

5.7.1 THERMOTROPIC CALAMITIC PHASES

Many of the typical liquid crystal phases of this type are derived from two phenylene rings joined by a linking group which may be a single bond, or a more complex group such as ester, azoxy, or azomethine. Although such a structure clearly forms a good basis for the formation of a liquid crystalline structure, the very large number of these compounds described in the literature stems largely from the convenience of their synthesis. Recent systematic exploration of liquid crystalline compounds has been stimulated by their applications in electro-optic displays, and it has been demonstrated   that aromatic, alicyclic and heterocyclic ring structures may all be used in the design of liquid crystal materials. In order to obtain high liquid crystal phase stability, it has been found that aromatic rings and other groups of high polarisability or polarity (cyano or nitro groups, double or triply-bonded linking groups *etc*) should be directly linked, without intervening σ-bonded groups which would interrupt the electronic conjugation of the structure. The point is well illustrated by the compounds in Table 7.

The standard molecular architecture for calamitic liquid crystals consists of a rigid core with terminal substituent groups at each end. These terminal groups may be flexible alkyl or alkoxy chains which serve to reduce the crystalline melting point as well as increasing the mesophase stability, or a polar group such as cyano or halogen. A list of the relative efficiencies of terminal groups in stabilising LC phases has been compiled for nematic phases:

$$Ph > CN > OCH_3 > Cl \sim CH_3 > F > H > CF_3$$

In smectic phases, a different order of stabilisation is found; it is dependent on the nature of the rigid molecular core and on the type of smectic phase being considered. Alkyl and alkoxy chains present the simplest pattern; usually LC molecules substituted with short chains tend to show predominantly nematic phases. As the chain length increases, smectic phase stability increases at the expense of the nematic phase and smectic polymorphism may arise (Table 4; compounds 2, 8; 3, 9). Long alkyl or alkoxy chains may lead to the nematic phase being totally replaced by the smectic. Branching of the alkyl chain may often increase the stability of smectic phases relative to the nematic as well as

Table 7. MESOPHASE STABILITY AND ELECTRONIC CONJUGATION.

(17) K 47.2 N 79.2 I

(18) K 59.6 (N 15) I

(19) K 62 N 100 I

(20) K 113 (N 50) I

(21) K 69 (N 33.5) I

(22) K 63 (N -210) I

increasing the relative stability of tilted smectic phases over orthogonal ones [137]. As an homologous series is ascended, there is usually an odd/even alternation in clearing points which is most marked in the shorter chain members; odd numbered alkyl chains (CH_3, C_3H_7 etc) and even numbered alkoxy chains (C_2H_5O etc) give higher clearing points.

Liquid crystal molecules which lack any strongly dipolar group frequently provide rather wide smectic phase ranges. The more highly ordered smectic phases often predominate. The stability of the nematic phase can be increased in these compounds and in general by lateral substitution onto the core. The clearing point is almost always reduced by lateral substitution, but the temperature of the $S-N$ transition is reduced to a greater degree than the $N-I$. Of the available substituents, fluorine alone produces a small effect on the clearing point but has a significant effect on the smectic phase [138]. Such lateral fluorine substitution may also affect the nature of the smectic phase which is formed by favouring tilted phase stability [139].

The liquid crystalline polymorphism of a series of compounds is also strongly influenced by the nature of linking groups in the core. It is well established, for example, that Schiff's base derivatives show a rich smectic polymorphism when analogous ester compounds show only S_A or S_A and S_C phases (Table 4, compounds 8,9). The stability of smectic phases is increased markedly by the presence of certain dipolar terminal groups substituted onto the core, such as $-CO_2R$.

The requirement of mechanical rigidity in a molecule in order for a LC phase to form means that in thermotropic mesogens, at least two ring structures are normally needed. There are, however, a number of exceptions to this principle. The 4-alkoxy benzoic acids [140], among other related series, form nematic and smectic phases in which strongly hydrogen bonded dimers form the mesogenic species. Sufficient rigidity can be obtained in non-cyclic structures to allow LC phase formation if conjugated double bonds are present in the molecule [141,142], but the products have poor stability.

A series of thermotropic LC's of particular historical importance is provided by the esters and other derivatives of cholesterol. This sterol does in fact provide a conventional rod-like skeleton on which to base a LC structure, but until quite recently was unusual because of its largely saturated ring structure. Since the early 1970's, a large number of synthetic saturated ring mesogens have been prepared and the behaviour of sterol derivatives can be viewed in context.

A tabulation of a very large number of liquid crystalline materials, together with their transition temperatures is available [143,144]. It provides ample examples of the principles outlined above, together with many of which space precludes elaboration.

5.7.2 THERMOTROPIC DISCOTIC MATERIALS

Just as calamitic liquid crystals have a 'standard architecture' comprising a rigid core with two polar or flexible end groups, discotic LC's commonly have a rigid core with several such groups substituted onto it. There is considerable diversity available in the structures of discotic compounds and their synthesis has been far less exhaustively explored than is the case for calamitic systems. A fundamental point arises from the rotational symmetry of the 'discs', or the number of substituent groups around the rigid discotic core. In this respect, there is great flexibility available and mesogens with 3, 4, 6 and 8 substituents are known. As in calamitic LC's, aromatic core structures provided the first discotic phases (and many of those discovered subsequently) but alicyclic compounds such as esters of *scyllo*-inositol also form stable mesophases. One of the principal difficulties is that the high molecular weights of the compounds encourage high melting points, so that most discotic phases are formed substantially above room temperature. A selection of representative discotic compounds is listed in Table 8; the total number of known discotic mesogens is still comparatively small and correlation between structure and phase sequence is still being elucidated [121].

48

Table 8. SOME THERMOTROPIC DISCOTIC MESOGENS.

(23) K 63 D$_{rd}$ 86 I

(24) K 152 D$_{rd}$ 168 N$_D$ 244 I

(25) K 96 D 147 I

(26) K 107 (D' 95) D" 127 I

5.8 PHYSICAL PROPERTIES OF LIQUID CRYSTALS

The orientational order found in liquid crystals profoundly affects the physical properties of the phases. In most cases, the main result is the introduction of anisotropy to the properties similar to that found in truly crystalline solids. The unique feature of the LC phases is the combination of this typical anisotropy with the ability to distort the optic axes of the phase on a very short distance scale without destruction of the sample that would result from such a distortion of a solid. Coupling between the anisotropic properties of the phase and its alignment direction is also quite general, so that the alignment can be changed by imposition of electric, magnetic, optical and shear fields *etc*. In this section the more important properties are considered in turn, concentrating on the nematic phase which provides the best studied, simplest and technologically most important case. Some properties specific to other phases are also discussed briefly.

5.8.1 ALIGNMENT PHENOMENA

Two quite different phenomena are to be noted under this heading, each of which has its own importance; these are respectively the use of solid surfaces to control the director alignment of a LC phase, and the behaviour of non-mesogenic solute molecules in a LC solvent.

It has been implicit in what has gone before that LC phases interact with solid surfaces to give a more-or-less fixed director pattern. It is this which is responsible for stabilising the characteristic microscopic textures of the phases, for example. The interaction can be controlled by providing a properly uniform solid substrate and there are important special cases [145]. *Homeotropic alignment* is an alignment of the director perpendicular to the substrate. It can be achieved by treating the surface with a variety of surface active materials such as lecithin, silicones *etc*, and some liquid crystals align homeotropically on a clean glass surface. *Planar* or *homogenous* alignment has the director lying in the plane of the substrate. It is easily obtained by applying a thin polymer layer to the surface. Suitable polymers include nylon 6-6, polyvinyl alcohol, and various polyimides. Subsequent mechanical rubbing of the coated surface induces a preferred alignment direction along which the director will orient itself. Most practical surface alignment treatments produce *strong anchoring*, *i.e.* the director of the LC remains aligned at the surface for all practical applied fields and alignment away from the surface changes smoothly, over a characteristic field dependent distance, from this orientation to that determined by an applied field.

Solutes in a liquid crystalline medium generally take up the orientational order of the solvent to a greater or lesser degree. The effect is strongest in the case of rod shaped solutes (or, presumably, disc shaped solutes in a discotic LC). The order parameter of the solute can be similar to, or even higher than, that of the solvent. The effect is important in the case of dissolved dyes, in which the optical absorption can be varied by changing the director orientation of the LC [146]. The alignment of solutes is also utilised in NMR investigations in LC solvents.

5.8.2 BIREFRINGENCE

Common nematic liquids behave as uniaxial crystals; they possess two principal refractive indices denoted n_e or n_o according to whether the plane of polarisation of the light is parallel or perpendicular to the director and the birefringence is defined as

$$\Delta n = n_e - n_o \tag{3}$$

For calamitic LC's the birefringence is positive, *i.e.* $n_e > n_o$. In discotic materials this situation is reversed [112]. The magnitude of the birefringence is of crucial importance for many practical applications of LC's because it determines the strength of interaction of the material with light.

The origin of the refractive index of any medium lies in the electric polarisation which results from the electric field associated with a light ray. In organic compounds this depends mainly on the degree of π-bonding to be found within the molecule. Practical birefringence values in the range of 0.03 to 0.4 can be achieved in calamitic thermotropic LC's though not necessarily in stable materials or those with a wide mesophase range. Lyotropic phases typically have a low birefringence below *ca.* 0.01. The refractive indices can be related to the orientational order parameter:

$$S = K(n_e^2 - n_o^2)/(\overline{n}^2 - 1) \tag{4}$$

$$\overline{n}^2 = \frac{1}{3}(n_e^2 - n_o^2) \tag{5}$$

In an intense optical field, the LC director tends to rotate to bring the axis having the larger refractive index parallel to the polarisation axis of the light. The resultant change in effective index represents a source of non-linear optical activity in nematic LC's which is slow (*ca.* 100 ms) but provides a very large index change under modest power levels [147]. The effect may in practice be dominated by thermal effects.

5.8.3 THE OPTICS OF TWISTED LIQUID CRYSTAL LAYERS

Liquid crystal layers in which the optic axis is twisted arise in two important circumstances: the naturally twisted chiral nematic and smectic phases, and many display devices where chirality and/or surface forces may impose the twisted structure. The cases differ chiefly in the pitch length of the twisted structure, which can be comparable with the wavelength of light in the spontaneously twisted phase but is much longer in display devices. In the phenomena described below, it is assumed that light is propagating parallel to the helical axis of the LC phase unless it is specifically stated otherwise.

A nematic layer with a 90° twist was first examined in the early part of this century [148]. The behaviour was unique; light entering the LC layer polarised along (or perpendicular to) the local alignment direction emerged from the layer with its plane of polarisation rotated through 90° as if it followed the twist in the LC. This simple and ideal behaviour occurs at the Mauguin limit, for which the condition is that $d\Delta n >> \lambda$, where d is the layer thickness and λ is the wavelength of light. At smaller values of $d\Delta n$, more complex behaviour results, with elliptically polarised light propagating through the cell and emerging with a rotation which need not match the twist angle. A similar result to that found in the Mauguin limit is, however, obtained [149] for specific values of the parameter $u = 2d\Delta n/\lambda$.

When the helical pitch length is similar to the wavelength of light, a quite different effect is seen. A very strong rotation of incident polarised light is still observed, which changes sign on passing the critical wavelength given by

$$P = \lambda / \overline{n} \tag{6}$$

Within a wavelength band centred on this wavelength and of width

$$\Delta\lambda = \Delta n\lambda / \overline{n} \tag{7}$$

light having the same sense of circular polarisation as the helical sense of twist in the LC layer is reflected, while the opposite sense of circular polarisation is transmitted unchanged through the LC. The reflected light does not undergo the reversal of polarisation common when reflection of circular polarised light occurs from a solid surface. The chiral LC layer therefore appears coloured in reflection when viewed against a dark background. The apparent colour may be changed by adjusting the pitch length of the LC. It also changes according to the angle of view:

$$\lambda = \overline{n}P\cos\theta \tag{8}$$

This angular variation of wavelength is the same as that observed in Bragg reflection, which the phenomenon resembles [150,151].

The temperature dependent change in pitch of a chiral nematic LC in the vicinity of an underlying smectic phase is widely used in temperature sensors [152].

Chiral smectic phases show broadly similar optical properties to those described for chiral nematic materials. The tilted smectic phases are formally biaxial but the observed biaxiality is rather small.

5.8.4 DIELECTRIC ANISOTROPY

The dielectric constant of an organic material depends on the polarisability of its constituent molecules and on their dipole moments. In the absence of any strong dipole moment in the molecules of a thermotropic phase, the polarisability anisotropy mirrors the molecular shape, giving a small positive anisotropy for calamitic phases and a modest negative anisotropy for discotic ones. Most LC materials, however, do have an appreciable dipole moment which contributes to the dielectric anisotropy and frequently dominates the polarisability term. The dipole contribution is of the form:

$$\Delta\varepsilon = -K\mu^2(1 - 3\cos^2\beta)S \qquad (9)$$

where β is the angle between the molecular long axis and the dipole. Substitution of a dipolar group onto the terminal position of a calamitic LC therefore induces a positive anisotropy. Lateral substitution of a dipole of the same strength gives a negative anisotropy of just half the magnitude. In practice, strongly terminal dipolar LC's usually have an appreciably lower $\Delta\varepsilon$ than would be expected from the above expression. Evidence is available that in these cases there is local antiparallel pairing of dipoles which reduces their effective magnitude [153]. The degree of this local antiferroelectric order cannot be readily predicted, but can be influenced by molecular structure and is strongly dependent on composition in mixtures. The preparation of negative anisotropy LC's is hindered by the relative inefficiency of lateral dipoles, and by the tendency for lateral dipolar substituents to reduce the stability of the phases of interest and to increase the viscosity of the product to an unreasonable level. Values in the range of −20 to +50 can be achieved.

5.8.5 FERROELECTRIC LIQUID CRYSTALS

The nematic phase is strictly non-polarised in its undistorted state with an equal number of molecular dipole moments pointing each way along the director, which itself is an axis of cylindrical symmetry. (A bulk polarisation can arise in a distorted nematic, the flexoelectric effect). It was argued first, purely on symmetry grounds, that chiral tilted smectic phases should show a ferroelectric polarisation and this was confirmed by the synthesis of DOBAMBC (Table 4, compound 12). The ferroelectric S_C^* phase has been extensively explored as a result of the discovery of a fast electro-optic effect in thin cells containing such materials [154].

The ferroelectric polarisation arises from a breaking of the symmetry in the packing of molecules into smectic layers when the chiral centre interacts with the tilt. This interaction causes a preferred orientation of a dipole moment disposed lateral to the molecular long axis, and results in a polarisation being observed in the plane of the smectic layers and perpendicular to the tilt direction. In chiral smectic C phases, however, as previously described, a twisted structure develops which causes the polarisation vector to sum to zero within a macroscopic volume. The twisted structure can be unwound by an electric field, by surface forces or by adding a different compound of opposite twist-sense to give a ferroelectric domain which can respond to the polarity of an applied field. The coupling of the polarisation to the director allows optical switching effects to be obtained. Associated with the ferroelectric polarisation, these phases are expected to show piezo and pyroelectric effects and these have indeed been observed.

5.8.6 ELASTIC BEHAVIOUR

Liquid crystal phases are true liquids which cannot sustain a static shear force. Nevertheless, they are able to transmit elastic torque forces and deformation of a liquid crystal from its lowest energy state, in which the director is everywhere parallel, is opposed by an elastic restoring force. It is in nematic phases that the elastic theory is best established and simplest [102,104].

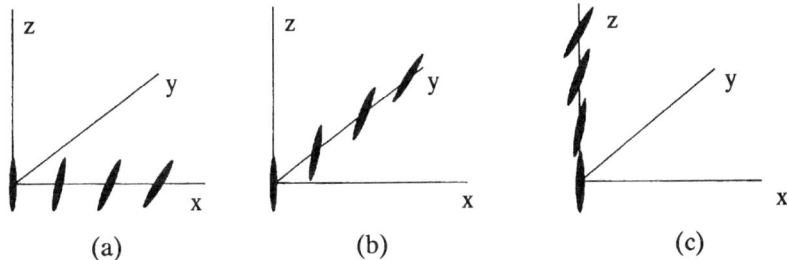

Figure 21. Elementary deformations of splay (a), twist (b), and bend (c) in a nematic liquid crystal.

All deformations of the nematic director can be expressed in terms of the three basic operations of splay, twist and bend which are mediated by the elastic constants K_{11}, K_{22} and K_{33} respectively (Fig. 21). The energy density associated with an arbitrary deformation can be written:

$$g = \tfrac{1}{2}[K_{11}(\nabla . n)^2 + K_{22}(n.\nabla \times n)^2 + K_{33}(n.\nabla n)^2]$$ (10)

The elastic constants have dimensions of force and magnitudes of the order of $10^{-11}N$, corresponding to a rather weak elastic response by comparison with solids. Viewed against the available dielectric anisotropy of common materials, this means that substantial deformation of the director of a nematic liquid can be achieved by an applied potential of around 1 volt. The absolute values of elastic constants decline as the temperature is raised, only very approximately in line the prediction of theory which suggests that $K_{ii} = kS^2$. The elastic constants can be correlated with molecular structure to some degree; for example, compounds containing a C_2H_4 link often have large K_{ii}, and long alkyl chains as terminal groups give a small ratio K_{33}/K_{11}.

The elastic behaviour of S_A liquid crystals is dominated by one effect: any deformation which would alter the inter-layer spacing of the smectic phase requires an input of energy akin to the compression of a solid. A splay deformation of the director is permitted by this constraint, but twist and bend cannot be sustained easily. Most real deformations which would be driven by imposed fields involve a combination of these elementary distortions and are therefore ruled out on energetic grounds. In practice most field induced deformations of S_A materials require high field strengths and proceed through the generation of disclinations (*i.e.* by breakup of the crystal structure). The elastic behaviour of tilted smectic phases and especially of chiral tilted smectic phases is the subject of continuing theoretical investigation.

The elastic constants of nematic phases may all be measured by studying the director deformation under an applied electric or magnetic field and complete sets of data, together with their temperature variation, are available for many compounds.

5.8.7 VISCOSITY IN NEMATIC LIQUID CRYSTALS

Like other physical properties, the viscosity of a nematic LC is anisotropic, and there is coupling between viscosity and director. The director configuration can therefore be altered by an applied shear. The situation for viscosity is a little more complex than for most other properties, however, and *five* independent viscosities are required to characterise a nematic liquid [104]. Three of these viscosities are the anisotropic counterparts of the classic shear viscosity of ordinary liquids. In each case, the director is disposed in a different direction with respect to the shear (Fig. 22). The remaining two viscosities have no counterpart in isotropic liquids. One, denoted $\eta_{1,2}$ provides a subtle correction to the viscosity corresponding to a geometry intermediate between those shown in Figs. 7b and 7c. The remaining coefficient is a viscosity corresponding to a situation in which no flow occurs, but the director rotates (in other words, each molecule rotates about its own centre of gravity). It is usually assigned the symbol γ. Although the five viscosities are all, in principle, amenable to measurement either by using a strong magnetic field to fix the director in a flow cell in the required direction, or by using a rotating magnetic field to induce director rotation, the experimental difficulties are considerable and rather few complete sets of reliable data are available.

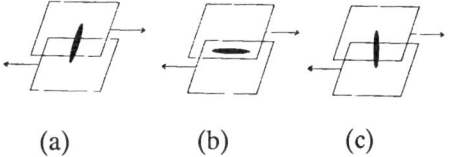

Figure 22. Orientation of LC director and shear to measure η_1 (a), η_2 (b), and η_3 (c).

(a)   (b)   (c)

Only an overview of some few important aspects of liquid crystal science has been presented in the above paragraphs. The area remains one where rapid progress is being made both in the synthesis of new materials and in understanding the complex interactions and behaviour of the products.

Liquid crystals are in some ways an anomalous inclusion in the field of molecular electronics, because the peculiar properties of the phases derive not from the properties of the constituent molecules but from interactions between them. Nevertheless, these compounds provide examples of self-assembly and organisation, and a mature example of molecular engineering applied to the electronics industry. Above all their properties are fascinating and promise a rich source of new phenomena yet to be uncovered.

REFERENCES

1. *Molecular Electronic Devices*, ed. F.L.Carter, Marcel Dekker, New York (1982).

2. *Photochromism*, eds. H.Dürr and H.Bouas-Laurent, Elsevier (1990).

3. H.Dürr, *Angew Chem. Int. Ed. Engl.*, **28**, 413 (1989).

4. M.Emmelius, G.Pawlowski and H.W.Vollmann, *Angew Chem. Int. Ed. Engl.*, **28**, 1445 (1989).

5. Z.F.Liu, K.Hashimoto and A.Fujishima, *Nature*, **347**, 658 (1990).

6. H.H.Wasserman and P.M.Keehn, *J. Am. Chem. Soc*, **89**, 2770 (1967).

7. A.Castellan, J.M.Lacoste and H.Bouas-Laurent, *J. Chem. Soc., Perkin Trans.*, **2**, 411 (1979); A.Castellan, J.P.Desvergne and H.Bouas-Laurent, *Nouv. J. Chim.*, **3**, 231 (1979).

8. M.Usui, T.Nishiwaki, K.Ando and M.Hida, *Chem. Lett.*, 1561 (1984).

9. H.G.Heller, *IEE Proc.*, **130**, 209 (1983).

58

10. E.Ando, J.Miyazaki, K.Morimoto, H.Nakahara and K.Fukuda, *Int. Symp. on Future Electronic Devices-Bioelectronic and Molecular Electronic Devices*, Tokyo (1985) 47; E.Ando, J.Miyazaki and K.Morimoto, *Thin Solid Films*, 133, 21 (1985); E.Ando, J.Hibino, T.Hashida and K.Morimoto, *ibid.*, 160 , 279 (1988).

11. G.J.Ashwell, *Thin Solid Films*, 186, 155 (1990).

12. A.J.Henry, *J. Chem. Soc.*, 1156 (1946); D.Gegiou, K.A.Muszkat and E. Fischer, *J. Am. Chem. Soc.*, 90, 3907 (1968); E.Fischer, *Chem. Unserer Zeit*, 9, 85 (1975).

13. D.G.Whitten, *J. Am. Chem. Soc.*, 96, 594 (1974); F.H.Quina and D.G. Whitten, *ibid.*, 99, 877 (1977); W.F.Mooney *et al.*, *ibid.*, 106, 5659 (1984); L.Collins-Gold, D.Mobius and D.G. Whitten, *Langmuir*, 2, 191 (1981).

14. S.Shinkai *et. al.*, *J. Am. Chem. Soc.*, 104, 1960 (1982); Y.Kawabata *et. al.*, *Chem. Lett.*, 1933 (1986); M.Tanaka *et. al.*, *ibid.*, 1307 (1987); A.Yabe *et al.*, *Thin Solid Films*, 160, 33 (1988).

15. H.Stobbe, Annalen, 380, 1, (1911).

16. A.Santiago and R.S.Becker, *J. Am. Chem. Soc.*, 90, 3654 (1968).

17. P.J.Darcy, R.J.Hart and H.G.Heller, *J. Chem. Soc., Perkin Trans I*, 571 (1978).

18. C.Trundle, PhD Thesis (Wales), 1980.

19. W.J.Thomlinson, E.A.Chandross, R.L.Fork, C.A.Pryde and A.A.Lamola, *Appl. Opt.*, 11, 533 (1972).

20. E.Hadjoudis, I.Moustakali-Mauridis and J.Xexakis, *Isr. J. Chem.*, 18, 202 (1979); I. Moustakali-Mauridis, E.Hadjoudis and A.Mavridis, *Acta Crystallogr.*, B36, 1126 (1980); E. Hadjoudis, *J. Photochem.*, 17, 355 (1981); E.Hadjoudis, M.Vitorakis and I.Moustakali-Mauridis, *Mol. Cryst. Liq. Cryst.*, 137, 1 (1986).

21. E.Hadjoudis and E.Hayon, *J. Phys. Chem.*, 74, 3184 (1970); T.Rosenfield, M.Ottolenghi and A.Y.Meyer, *Mol. Photochem.*, 5, 39 (1973); P.Jacques, J.P.Biava, A.Gourso and J.Faure, *J. Chim. Phys. Phys. Chim. Biol.*, 76, 566 (1979); P.Jacques, *ibid.*, 79, 352 (1982); J.W.Lewis and J.W.Sandorfy, *Can. J. Chem.*, 60, 1738 (1982); R.S.Becker, C. Lenoble and A.Zein, *J. Phys. Chem.*, 91, 3509 & 3517 (1987).

22. S.Kawamura, T.Tsytsui, S.Saito, Y.Muaro and K.Kima, *J. Am. Chem. Soc.*, **110**, 509 (1988).

23. S.M.Binder and N.W.Lord, *J. Chem. Phys.*, **36**, 540 (1961); K.Maeda, A. Chinone and T.Hayashi, *Bull. Chem. Soc. Jp.*, **43**, 1431 (1970).

24. *Photochromism*, ed. G.H.Brown, Wiley Interscience (1971).

25. A.Karube, M.Yamazaki, H.Matsuoka and S.Suzuki, *Chem. Lett.*, 691 (1983).

26. P.M.Rentzepis, *US Pat.* 07 342 978 (1989); D.A.Parthenopoulos and P.M.Rentzepis, *Science*, **245**, 843 (1989).

27. J.Miyazaki, E.Ando, K.Yoshino and K.Morimoto, *Eur. Pat. Appl.* EP 193931 A2 (1986); *US Pat.* 4737427 A (1988).

28. R.S.Potember, R.C.Hoffman and T.O.Poehler, *Johns Hopkins APL Technical Digest*, **7**, 129 (1986) and references cited therein.

29. T.Nagamura and K.Sakai, *J. Chem. Soc., Chem. Commun.*, 810 (1986); T. Nagamura, K.Sakai and T.Ogawa, *ibid.*, 1035 (1988); T.Nagamura, Y.Isoda, K.Sakai and T.Ogawa, *ibid.*, 703 (1990); T.Nagamura, K.Sakai and T. Ogawa, *Thin Solid Films*, **179**, 375 (1989); T.Nagamura and K.Sakai, *Ber. Bunsenges. Phys. Chem.*, **93**, 1432 (1989).

30. T.Nagamura and Y.Isoda, *J. Chem. Soc., Chem. Commun.*, 72 (1991).

31. G.J.Ashwell, *UK Pat. Appl.* 9007230.7 (1990); *Eur. Pat. Appl.* 9030347.4 (1990); *Jpn. Pat. Appl.* 84760/90 (1990).

32. G.J.Ashwell, E.J.C.Dawnay and A.P.Kuczynski, *J. Chem. Soc., Chem. Commun.*, 1355 (1990).

33. G.J.Ashwell, E.J.C.Dawnay, A.P.Kuczynski and M.Szablewski, *Mat. Res. Soc. Symp. Proc.*, **173**, 507 (1990).

34. G.J.Ashwell and M.Szablewski, *Proc. Roy. Soc. London, Series A*, submitted for publication (1991).

35. J.Hecht, *Electronics Times*, p. 12, 23rd July 1987.

36. F.J.A.M.Greidanus and W.B.Zeper, *Mat. Res. Soc. Bull.*, **XV**, 31 (1990).

37. J.Silver, *New Scientist*, 48 (1989); J.Silver, P.Lukes, S.D.Howe and B. Howlin, *J. Mater. Chem.*, **1**, 29 (1991) and references cited therein.

38. K.Bange and T.Gambke, *Adv. Mat.*, **2**, 10 (1990).

39. L.A.Summers, *The Bipyridinium Herbicides*, Academic Press, London (1980) and references cited therein.

60

40. A.Anodera, I.Shirotani, H.Inokuchi and N.Kawai, *Chem. Phys. Lett.*, **25**, 296 (1974).

41. *The Physics and Chemistry of Organic Superconductors*, G.Saito and S. Kagoshima eds., Springer-Verlag, Berlin (1990).

42. J.M.Williams *et al.*, in *Progress in Inorganic Chemistry*, S.J.Lippard ed., pp.51-218, Wiley, New York (1987).

43. J.M.Williams *et al.*, *Science*, **252**, 1501 (1991).

44. F.Wudl, G.M.Smith and E.J.Hufnagel, *J. Chem. Soc., Chem. Commun.*, 1453 (1970).

45. G.A.Thomas *et al.*, *Phys. Rev. B*, **13**, 5105 (1976).

46. C.S.Jacobsen, J.R.Andersen, K.Bechgaard and C.Berg, *Sol. St. Commun.*, **19**, 1209 (1976).

47. A.J.Schultz, G.D.Stucky, R.Craven, M.J.Schaffman and M.B.Salamon, *J. Am. Chem. Soc.*, **16**, 155 (1977).

48. R.C.Wheland and J.L.Gillson, *J. Am. Chem. Soc.*, **98**, 3916 (1976).

49. A.N.Bloch, T.F.Carruthers, T.O.Poehler and D.O.Cowan, page 47 of *Chemistry and Physics of One-dimensional Metals*, ed. H.J.Keller ed., *NATO ASI Series B*, Vol. 25, Plenum Press, New York (1977).

50. C.S.Jacobsen, K.Mortensen, J.R.Andersen and K.Bechgaard, *Phys. Rev. B*, **18**, 905 (1978).

51. P.Delhaes *et al.*, *J. Phys. Lett.*, **38**, 233 (1977); R.M.Friend, D.Jérome, J.M. Fabre, L.Giral and K.Bechgaard, *J. Phys. C*, **11**, 263 (1978).

52. G.Keryer *et al.*, *Sol. St. Commun.*, **26**, 541 (1978).

53. G.Keryer *et al.*, *Lect. Notes in Phys.*, **95**, 65 (1978).

54. C.S.Jacobsen, H.J.Pedersen, K.Mortensen and K.Bechgaard, *J. Phys. C*, **13**, 3411 (1980).

55. E.M.Engler and V.V.Patel, *J. Am. Chem. Soc.*, **96**, 7376 (1974).

56. C.S.Jacobsen, K.Mortensen, J.R.Andersen and K.Bechgaard, *Phys. Rev. B*, **18**, 905 (1978).

57. A.N.Bloch, D.O.Cowan, K.Bechgaard, R.E.Pyle and R.H.Banks, *Phys. Rev. Lett.*, **34**, 1561 (1975).

58. R.E.Peierls, *Quantum Chemistry of Solids*, Oxford Univ. Press, London (1955).

59. S.Kagoshima, *Jpn. J. Appl. Phys.*, **20**, 1617 (1981).

60. G.J.Ashwell, I.M.Sandy, A.T.Chyla and G.H.Cross, *Synth. Met.*, **19**, 463 (1987).

61. A.J.Epstein, S.Etemad, A.F.Garito and A.J.Heeger, *Phys. Rev. B*, **5**, 952 (1972); L.B. Coleman, J.A.Cohen, A.F.Garito and A.J.Heeger, *ibid.*, **7**, 2122 (1973).

62. V.Walatka and J.H.Perlstein, *Mol. Cryst. Liq. Cryst.*, **15**, 269 (1971); K. Holczer, G.Mihaly, A.Janossy, G.Grüner and M.Kertesz, *J. Phys. C*, **11**, 4707 (1978).

63. J.U.V.Schultz *et al.*, in *Organic and Inorganic Low-dimensional Crystalline Materials*, eds. P.Delhaes and M.Drilon, *NATO ASI Ser. B*, **168**, pp. 297-300, Plenum Press, New York (1987).

64. J.Martinsen, R.L.Greene, S.M.Palmer and B.M.Hoffman, *J. Am. Chem. Soc.*, **105**, 677 (1983).

65. L.Brossard, M.Ribault, M.Bousseau, L.Valade and P.Cassoux, *C. R. Acad. Sci. Ser. 2*, **302**, 205 (1986); L.Brossard, M.Ribault, L.Valade and P. Cassoux, *J. Phys. (Paris)*, **50**, 1521 (1989).

66. J.G.Bednortz and K.A.Müller, *Z. Phys. B*, **64**, 189 (1986).

67. P.M.Grant, *Adv. Mat.*, **2**, 232 (1990).

68. A.M.Kini *et al.*, *Inorg. Chem.*, **29**, 2555 (1990).

69. J.M.Williams *et al.*, *Inorg. Chem.*, **29**, 2372 (1990).

70. W.Krätschmer, L.D.Lamb, K.Fostiropoulos and D.R.Huffman, *Nature (London)*, **347**, 354 (1990).

71. F.Diederich and R.L.Whetton, *Angew. Chem. Int. Ed. Engl.*, **30**, 678 (1991).

72. R.C.Haddon *et al.*, *Nature (London)*, **350**, 320 (1991).

73. A.F.Hebard *et al.*, *ibid.*, **350**, 600 (1991).

74. H.H.Wang *et al.*, in note added in proof of ref. 4.

75. K.Hoiczer *et al.*, *Science*, **252**, 1154 (1991).

76. E.B.Yagubskii *et. al.*, *JETP Lett.*, **39**, 12 (1984).

77. H.Kobayashi *et al.*, *Chem. Lett.*, 789 (1986).

78. R.Kato *et al.*, *ibid.*, 507 (1986).

79. R.P.Shibaeva, V.F.Kaminskii and E.B.Yagubskii, *Mol. Cryst. Liq. Cryst.*, **119**, 361 (1985).

62

80. J.M.Williams *et al.*, *Synth. Met.*, **23**, 3839 (1984).

81. H.H.Wang *et al.*, *Inorg. Chem.*, **24**, 2465 (1985).

82. H.Mori, I.Hirabayashi, S.Tanaka, T.Mori and H.Inokuchi, *Sol. St. Commun.*, **76**, 35 (1990).

83. H.Urayama *et al.*, *Chem. Lett.*, 55 (1988).

84. M.A.Beno *et al.*, *Inorg. Chem.*, **29** 1599 (1990).

85. K.Kikuchi *et al.*, *J. Phys. Soc. Jpn.*, **56**, 3436 (1987).

86. K.Kikuchi *et al.*, *ibid.*, **56**, 4241 (1987).

87. K.Kikuchi *et al.*, *Sol. St. Commun.*, **66**, 405 (1988).

88. G.C.Papavassiliou *et al.*, *Synth. Met.*, **27**, B379 (1988).

89. K.Bechgaard *et al.*, *J. Am. Chem. Soc.*, **103**, 2440 (1981).

90. A.Aviram and M.A.Ratner, *Chem. Phys. Lett.*, **29**, 277 (1974).

91. R.M.Metzger *et al.*, *Langmuir*, **4**, 298 (1988) and references cited therein.

92. N.J.Geddes, J.R.Sambles, D.J.Jarvis, W.G.Parker and D.J.Sandman, *Appl. Phys. Lett.*, **56**, 1916 (1990); D.J.Sandman *et al.*, *Synth. Met.*, **41-43**, 1415 (1991).

93. G.J.Ashwell, J.R.Sambles, A.S.Martin, W.G.Parker and M.Szablewski, *J. Chem. Soc., Chem. Commun.*, 1374 (1990).

94. R.M.Metzger *et al.*, *J. Mol. Electron.*, **2**, 119 (1986).

95. G.Decher, B.Tieke, C.Bosshard and P.Günter, *J. Chem. Soc., Chem. Commun.*, 933 (1988); C.Bosshard *et al.*, *Appl. Phys. Lett.*, **56**, 1204 (1990).

96. M.Era *et al.*, *Jpn. J. Appl. Phys.*, **29**, 2261 (1990).

97. G.J.Ashwell, *UK Pat. Appl.*, 9025832.8 (1990); G.J.Ashwell, E.J.C. Dawnay, A.P.Kuczynski and P.J.Martin, *SPIE - Int. Soc. Opt. Eng.*, **1361**, 589 (1991).

98. G.J.Ashwell *et al.*, *J. Chem. Soc., Faraday Trans.*, **86**, 1117 (1990).

99. T.L.Penner, N.J.Armstrong, C.S.Willand, J.S.Schildkraut and D.R. Robello, *Proc. SPIE Conference on Nonlinear Optical Properties of Organic Materials IV*, San Diego (1991).

100. G.J.Ashwell, *Adv. Mat.*, in press.

101. F.Reinitzer, *Monatsh Chem*, **9**, 421 (1888).

102. G.Vertogen and W.H.deJeu, *Thermotropic Liquid Crystals, Fundamentals*, Springer-Verlag (1988).

103. L.M.Blinov, *Electro-optical and Magneto-optical Properties of Liquid Crystals*, Wiley (1983).

104. W.H. de Jeu, *Physical Properties of Liquid Crystalline Materials*, Gordon and Breach (1980).

105. A.Saupe and W.Maier, *Z. Naturforsch.*, **16a**, 816 (1961).

106. I.Haller, *Prog. Solid State Chem.*, **10**, 103 (1975).

107. W.Maier and A.Saupe, *Z. Naturforsch.*, **14a**, 882 (1959).

108. W.Maier and A.Saupe, *ibid.*, **15a**, 287 (1960).

109. G.Luckhurst and C.Zannoni, *Nature*, **287**, 412 (1977).

110. L.Onsager, *Ann. NY Acad. Sci.*, **51** 627 (1949).

111. G.Luckhurst and P.Simpson, *Mol. Phys.*, **47**, 251 (1982).

112. G.Heppke *et al.*, *Mol. Cryst. Liq. Cryst. Lett.*, **6(3)**, 71 (1988).

113. J.Charvolin, A.Levelut and E.Samulski, *J. de Phys. Lett.*, **40**, L-587 (1979).

114. L.J.Yu and A.Saupe, *Phys. Rev. Lett.*, **45**, 1000 (1980).

115. S.Chandrasekhar *et. al.*, *Pramana*, **30**, L491 (1988).

116. J.Malthete *et al.*, *C. R. Acad. Sci. Paris*, **t303**, 1073 (1986).

117. G.W.Gray and J.W.Goodby, *Smectic Liquid Crystals - Textures and Structures*, Leonard Hill (1984).

118. G.Sigaud *et al.*, *Mol. Cryst. Liq. Cryst.*, **69**, 81 (1981).

119. D.Demus *et al.*, *ibid.*, **56**, 289 (1980).

120. A.Leadbetter in *Thermotropic Liquid Crystals*, ed. G.Gray, J.Wiley (1987).

121. C.Destrade *et al.*, *Mol. Cryst. Liq. Cryst.*, **106**, 121 (1984).

122. N.Satoh and K.Tsuji, *J. Phys. Chem.*, **91**, 6629 (1987).

123. G.Tiddy and M Walsh, *Stud. Phys. Theor. Chem.*, **26**, 151 (1983).

124. H.Hoffmann, *Ber. Bunsenges. Phys. Chem.*, **88**, 1078 (1984).

125. R.Werbowyj and D.Gray, *Macromolecules*, **13**, 69 (1980).

126. C.Robinson, *Tetrahedron*, **13**, 219 (1961).

127. T.Bair, P.Morgan and F.Killian, *Macromolecules*, **10**, 1396 (1977).

128. C.McArdle (ed.), *Side Chain Liquid Crystal Polymers*, Blackie (1989).

129. H.Finkelmann, *Angew. Chem. Int. Ed. Engl.*, **26**, 816 (1987).

130. D.Demus, *Liquid Crystals*, **5**, 75 (1989).

131. K.Radley and A.Saupe, *Molec. Phys.*, **35**, 1405 (1978).

132. M.Boidart, A.Hochapfel and P.Peretti, *Mol. Cryst. Liq. Cryst.*, **172**, 147

64

(1989).

133. J.Malthete *et al.*, *Mol. Cryst. Liq. Cryst. Lett.*, **64**, 233 (1981).

134. G.Gray and D.McDonnell, *Mol. Cryst. Liq. Cryst.*, **37**, 189 (1976).

135. H.Stegemeyer *et al.*, *Wiss. Beit. Martin Luther Univ. Halle. Witt.*, **52**, 64 (1986).

136. K.Toyne, in *Thermotropic Liquid Crystals* (ed. G Gray), Wiley (1987).

137. D.Coates, *Liq. Cryst.*, **2**, 63 (1987).

138. P.Balkwill *et al.*, *Mol. Cryst. Liq. Cryst.*, **123**, 1 (1985).

139. M.Chambers *et al.*, *Liq. Cryst.*, **5**, 153. (1989).

140 B.Jones, *J. Chem. Soc.*, 1874 (1935).

141. K.Markau and W Maier, *Chem. Ber.*, **95**, 889 (1962).

142. J.Malthete, *Adv. Mat.*, **2**, 150 (1990).

143. D.Demus, H.Demus and H.Zaschke, *Flussige Kristalle in Tabellen*, Vol. 1, VEB Deutscher Verlag fur Grundstoffindustrie (1976).

144. D.Demus and H.Zaschke, *Flussige Kristalle in Tabellen*, Vol. 2, VEB Deutscher Verlag fur Grundstoffindustrie (1984).

145. J.Cognard, *Mol. Cryst. Liq. Cryst.*, **78**, 1, (1981).

146. A.Ivashchenko and V.Rumyantsev, *Mol. Cryst. Liq. Cryst.*, **150A**, 1 (1987).

147. I.Khoo and Y Shen, *Opt. Eng.*, **24**, 579 (1985).

148. C.Mauguin, *Bull. Soc. Fr. Mineral Crystallog.*, 34, 71 (1911).

149. C.Gooch and H Tarry, *J. Phys. D*, **8**, 1575 (1975).

150. D.Berreman and T.Scheffer, *Phys. Rev. Lett.*, **25**, 577 (1970).

151. H.Gleeson and H.Coles, *Mol. Cryst. Liq. Cryst.*, **170**, 9 (1989).

152. D.McDonnell, in *Thermotropic Liquid Crystals*, Wiley (1987).

153. K.Toriyama and D.Dunmur, *Mol. Cryst. Liq. Cryst.*, **139**, 123 (1986).

154. N.Clark and S.Lagerwall, *Appl. Phys. Lett.*, **36**, 899 (1980).

CHAPTER 2
Conjugated Polymeric Conductors
M. F. Rubner

1. INTRODUCTION AND BACKGROUND

Generally, when one considers the electrical properties of the various classes of materials it is tacitly assumed that inorganic metals and semiconductors are best suited for applications requiring materials with high electrical conductivities whereas polymers are better exploited in applications that demand excellent insulating properties. In fact, these latter materials have traditionally been utilized in the electronics industry as inactive packaging and insulating materials whose sole purpose is to isolate and/or encapsulate (for environmental protection) the various electronically active components of an electrical circuit or device. This rather narrow view of the role of polymers in the electronics industry, however, is rapidly changing as new polymeric materials with a full range of electrical and optical properties have suddenly become available. A new class of polymers known as intrinsically conductive polymers or electroactive polymers has recently emerged which exhibit electrical and optical properties previously only found in inorganic systems. Although still in their early stages of development, the potential utility of these polymers in device schemes requiring specialized electroresponsive materials could be quite significant.

The majority of these new polymeric conductors are conjugated polymers characterized by highly delocalized electronic states. In the mid 1970s, it was discovered [1] that these π-electron rich materials were highly susceptible to chemical or electrochemical oxidation which, in turn, could be used to modify their electrical and optical properties over a tremendous range. The main focus of the early research on polymeric conductors was on the electrical conductivity of these new materials which stimulated the discovery of many new types of conjugated polymers and led to the development of our current understanding of the mechanisms of charge storage in these systems. Once the basic structural features required to obtain highly conductive materials were identified, the focus of this research shifted to the development of highly conductive polymers with good environmental stability and more acceptable processing attributes. These latter two parameters became important as soon as it was realized that most of the conjugated polymers of interest were essentially unstable in air and not capable of being processed into scientifically or technologically useful forms. Most recently, the nonoxidized forms of these conjugated polymers have also been under investigation due to their exceptionally large nonlinear optical susceptibilities [2] that could prove useful in the development of new materials for the emerging area of nonlinear optics (photonics). Thus, current research efforts are focussed on the development of stable, processible materials that exhibit high electrical conductivities in the oxidized state or large nonlinear optical susceptibilities in the neutral state.

The aim of this chapter is to provide the reader with a basic understanding of the principles and concepts underlying the physics and chemistry of conjugated polymeric conductors. In keeping with the theme of this book, an emphasis has been placed on the potential role of conjugated polymers in the area of molecular electronics. A number of issues related to the development of conjugated polymer based electronic and optical devices have been addressed throughout the chapter. Clearly it is not possible to cover all of the many exciting developments that have occurred in the field of conducting polymers since their initial discovery. The reader is therefore referred to the extensive literature on conducting polymers and the many highly informative reviews [3-14] covering many different aspects of these materials for additional information.

2. CONDUCTING POLYMERS - BASIC PRINCIPLES AND STRUCTURES

2.1 DELOCALIZED ELECTRONIC STATES OF CONJUGATED POLYMERS

According to band theory, the two basic requirements for electronic conduction in a material are: (1) a continuous system of a large number of strongly interacting atomic orbitals - which leads to the formation of band-like electronic states and (2) the presence of an insufficient number of electrons to fill these bands. In the case of traditional conductors such as inorganic semiconductors or metals, the atomic orbitals of each atom overlap with each other in the solid state in such a way as to create a number of continuous energy bands comprised of closely spaced energy states (see Figure 1). The electrons provided by each orbital are said to be delocalized throughout the entire array of atoms. Each of these energy bands is separated by a range of forbidden energies from which the electrons are excluded,called the band gap. The highest occupied energy band is referred to as the valence band whereas the lowest unoccupied band is called the conduction band. When the valence band is partially filled, the electrons present in this band are free to move under the application of an applied electric field,resulting in metal-like conduction. In the case of a completely filled valence band, conduction can only occur when electrons are promoted across the forbidden energy gap separating the valence band from the conduction band thereby leaving a vacancy or hole in the valence band and adding free electrons to the empty conduction band. The strength of the interaction between the overlapping orbitals determines the extent of delocalization that is possible for a given system which in turn establishes the spread of energy levels available to the electrons within each band (the bandwidth). The greater the degree of electron delocalization, the larger the bandwidth, and the higher the mobility of the carriers within the band.

It is clear from the above discussion that electronic conduction in the solid state is best facilitated by the presence of a continuous system of overlapping atomic orbitals which establishes the delocalized electronic states that are needed for facile movement of electrons in an electric field. For polymers however, it is often the case that the molecular orbitals responsible for bonding the carbon atoms of the chain together are sp^3 hybridized sigma orbitals which are spatially not suited for

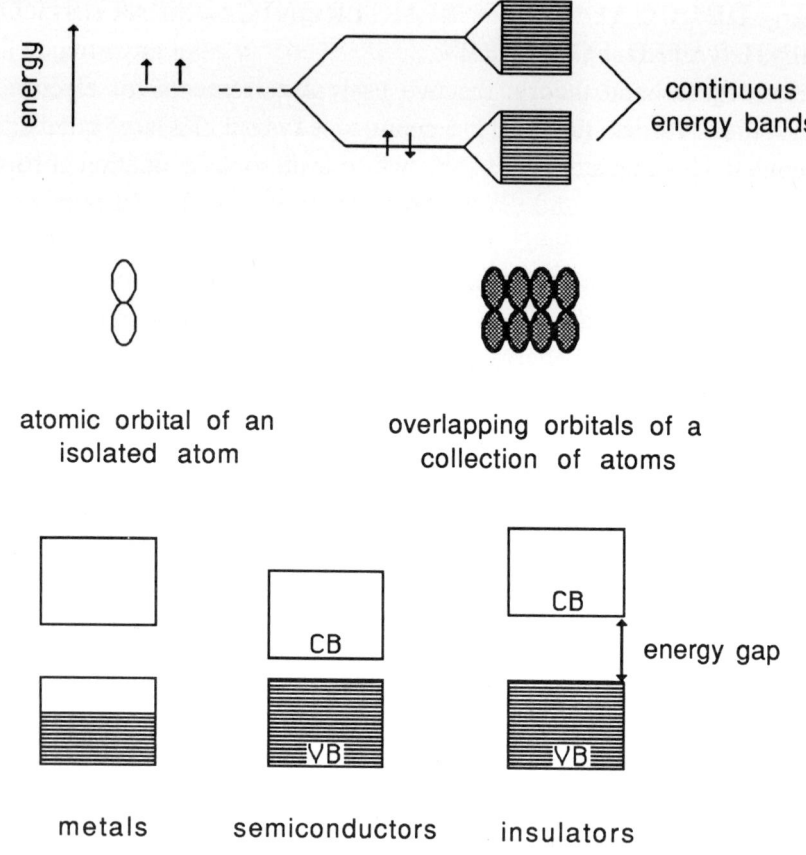

Figure 1. Formation of band-like electronic states via the overlap of atomic orbitals. The bottom portion of the figure displays the highest occupied bands for the three basic types of materials.

extensive overlapping and thereby preclude the possibility of significant electron delocalization. In other words, the electrons involved in bonding are strongly localized between the carbon atoms that they hold together and cannot easily contribute to the conduction process. In fact, any attempt to excite these electrons to higher energy states where conduction may be possible will most assuredly risk altering the molecular structure of the polymer. In terms of a simple band theory, this type of bonding produces a large energy gap between the highest occupied energy band (which is completely filled) and the lowest unoccupied energy band imparting to the material the characteristics of a good electrical insulator. This is illustrated in Figure 2 for the familiar polymer polyethylene (shown in its fully extended all trans conformation) which is known to be an excellent insulator.

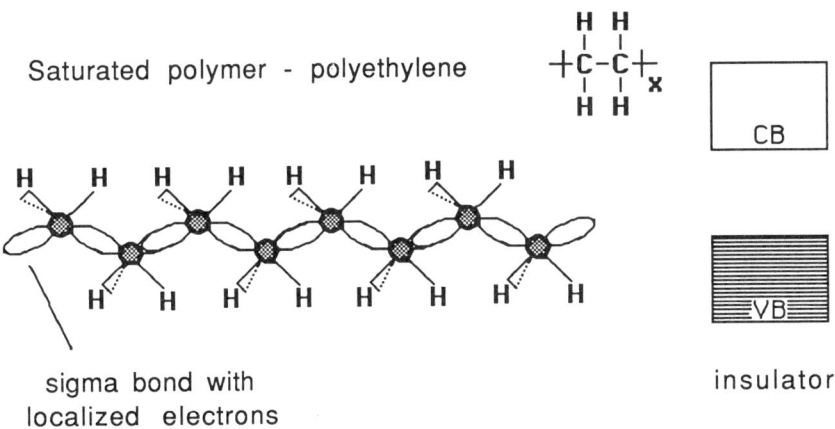

Saturated polymer - polyethylene

sigma bond with
localized electrons

insulator

Figure 2. Saturated backbone structure of the electrically insulating polymer polyethylene.

The creation of an electrically conducting polymer therefore requires the development of a bonding system that, in addition to establishing the molecular structure of the polymer, provides a continuous system of atomic orbitals that can approach each other closely enough to allow the formation of delocalized electronic states. This type of molecular arrangement can be found in polymers that contain conjugated unsaturated bonds. For the purposes of this discussion, a conjugated polymer can be defined as a polymer comprised of a planar or near planar sequence of alternating single and double bonds (or any other unsaturated multiple bond) along its backbone. In such a system, each carbon atom is fitted with a p_z-orbital which is oriented perpendicular to the polymer backbone and capable of overlapping with its neighbors.

70

One might anticipate that a linear polymer backbone consisting of a large number of strongly interacting coplanar p_z orbitals, each of which contributes only one electron to the resultant continuous π-system, would behave as a one-dimensional metal with a half-filled conduction band, as illustrated for the conjugated polymer polyacetylene in Figure 3.

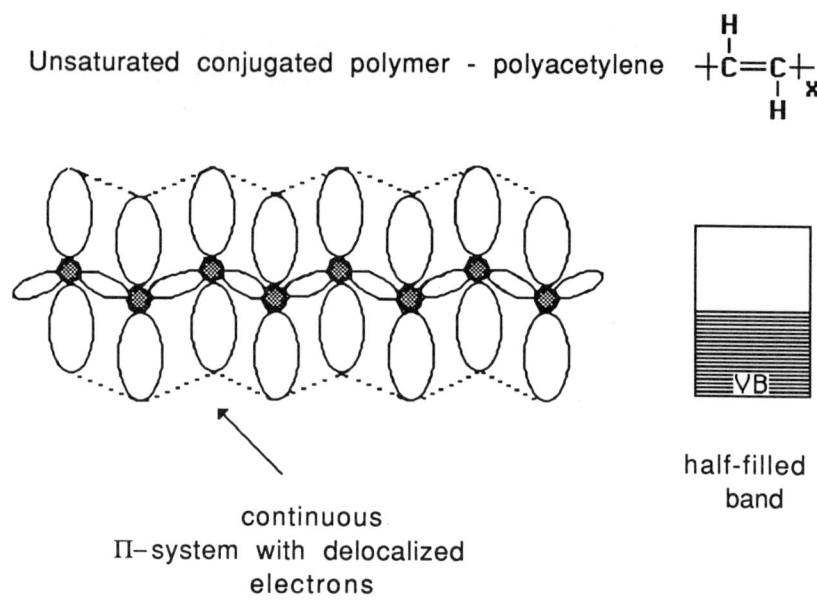

Unsaturated conjugated polymer - polyacetylene

continuous
Π–system with delocalized
electrons

half-filled
band

Figure 3. Unsaturated backbone structure of the conjugated polymer polyacetylene (pendant hydrogen atoms have been omitted for clarity). Complete delocalization (no bond alternation) would result in a one-dimensional metal with a half-filled conduction band.

This in fact would be the case if the electronic wavefunctions were completely delocalized over the entire chain and all of the backbone bond lengths were of identical length. In this case, the driving force for delocalization is provided by the energy savings associated with a resonance stabilized structure. It turns out, however, that for one-dimensional systems, the chain can more efficiently lower its energy by introducing bond alternation (alternating short and long bonds) [3] which, in effect, limits the extent of electronic delocalization that can take place along the backbone. Thus, there is a periodic modulation of the charge density along the polymer chain with the regions of space occupied by the shorter bonds carrying a greater share of the electron density. In such an arrangement, adding additional conjugated bonds to the chain will not produce a corresponding increase in the length over which the electrons are delocalized as this value is ultimately determined by bond alternation effects. For most bond alternating polymer chains, the maximum extent of delocalization is obtained when about 15-20 multiple bonds are linked together [15]. Thus, properties directly related to the extent of electron delocalization along the backbone, such as the optical band gap of the polymer, will reach their limiting values at chain lengths much shorter than the overall degree of polymerization of the chain (which in some cases can readily exceed 10^4 repeat units).

Bond alternation is a direct consequence of the strong coupling that exists between the π-electrons and the backbone skeletal vibrations (phonon modes) of the polymer which, in turn, is characteristic of many quasi one-dimensional systems [6]. If bond alternation of the backbone is viewed as a distortion of the system from its state of equal bond lengths, then it can occur only if the energy cost to distort the molecule is less than the energy that would have been needed to maintain the fully delocalized backbone. It turns out that the increase in energy associated with the distortion is more than compensated for by a decrease in electronic energy; the total energy of the system is therefore lowered when bond alternation sets in. Lattice instabilities of this type were initially proposed by Peierls [16] for one-dimensional metals and have been extensively studied in molecular charge transfer salts [6]. Thus, the low dimensionality of conjugated polymer structures dramatically influences the electronic states of these materials. As will shortly become apparent, the electronic properties of doped conjugated polymers are also strongly influenced by the coupled motions of π-electrons with lattice vibrations ie., strong electron-phonon interactions which are characteristic of one-dimensional systems.

As is illustrated in Figure 4, the effect of bond alternation and the attendant restriction it places on the extent of delocalization is to open an energy gap at the Fermi level creating a filled valence band and an empty conduction band. Thus, all conjugated polymers are large band gap semiconductors with band gaps generally in excess of about 1.5 eV. In addition, since electronic transitions between the valence band and conduction band of these polymers require energies comparable to the energy of visible light, conjugated polymers are highly colored materials, typically appearing blue or red in transmitted light. It should also be apparent that the quasi-one-dimensional nature of these materials means that, for a system of fully aligned chains, the optical and electrical properties will be highly anisotropic, i.e., strongly dependent on the direction in which the optical or electrical field is applied [14]. For example, stretched aligned conjugated polymers typically display a greater absorbance when the incident light vector is polarized along the chain direction as compared to perpendicular to the chain direction.

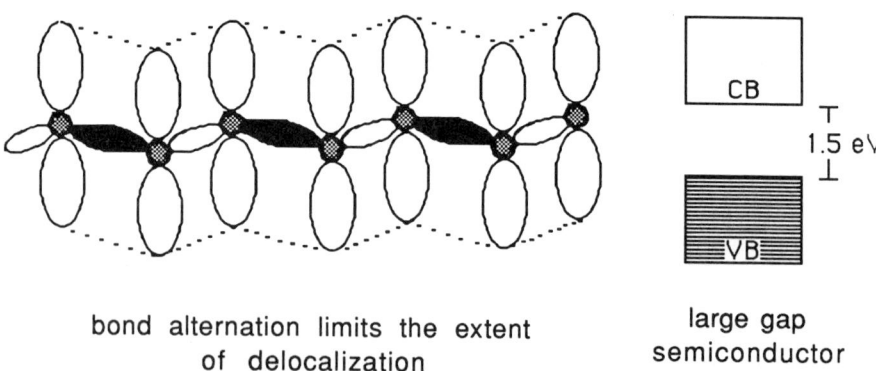

bond alternation limits the extent of delocalization

large gap semiconductor

Figure 4. Alternation of bond lengths along a conjugated polymer chain (estimated to be about 0.10 Å for polyacetylene) results in a material with the properties of a large gap semiconductor.

2.2 ELECTRONIC CONDUCTION IN POLYMERS

Figure 5 displays the repeat unit structures of a number of conjugated polymer systems that can be rendered electrically conductive by chemical oxidation or reduction. This particular list includes what would now be considered the first generation of conducting polymers as it represents the five main types of materials that were investigated in the early stages of research of electroactive polymers. It can be seen that each polymer backbone is comprised of a system of alternating single and double bonds, which, as discussed above, forms the extended π-electron conjugation needed for electron transport in these materials. In the case of poly(phenylene sulfide), it is believed that overlap of the sulfur d-orbital with the aromatic π-system of the benzene ring maintains the backbone conjugation in this polymer. Each of these polymers is electrically insulating; however, upon exposure to suitable electron acceptors (oxidizing agents) or electron donors (reducing agents), a transformation from an insulator to a conductor takes place. It was this dramatic transition from an electrically insulating state to an electrically conductive state that stimulated the tremendous interest in these materials that continues to date.

Typical oxidizing agents include I_2, AsF_5, $FeCl_3$, and $NOPF_6$, whereas chemical reduction, when possible, is best accomplished with reducing agents such as sodium naphthalide [3,6]. Upon exposure to these charge transfer agents, the conductivity of the polymer rises rapidly and eventually levels off at some maximum conductivity characteristic of the particular polymer and the type of chemical agent used to dope the polymer. The doping agents influence the level of conductivity in a number of ways including how homogeneously they are distributed throughout the polymer and whether or not they are capable of initiating side reactions that inhibit the polymer's ability to achieve high conductivity. An example of this latter effect can be found in the Br_2 doping of polyacetylene which, in addition to chemical oxidation of the polymer backbone, also produces saturated sites along the backbone by addition of Br_2 to the double bonds [6]. Strong oxidizing agents such as $NOPF_6$ are also capable of reacting with the polymer in a way that reduces the maximum conductivity level when long exposure times are utilized for doping. Note that charge transfer agents are frequently referred to as dopants in analogy to the doping process used to raise the conductivity of inorganic semiconductors such as silicon. Although, as will be seen shortly, this analogy is rather weak, its pervasive use in the literature cannot be ignored.

A curve characteristic of how the conductivity of a conjugated polymer varies with the amount of dopant incorporated in the polymer is shown in Figure 6. It should be noted that electron transfer through chemical

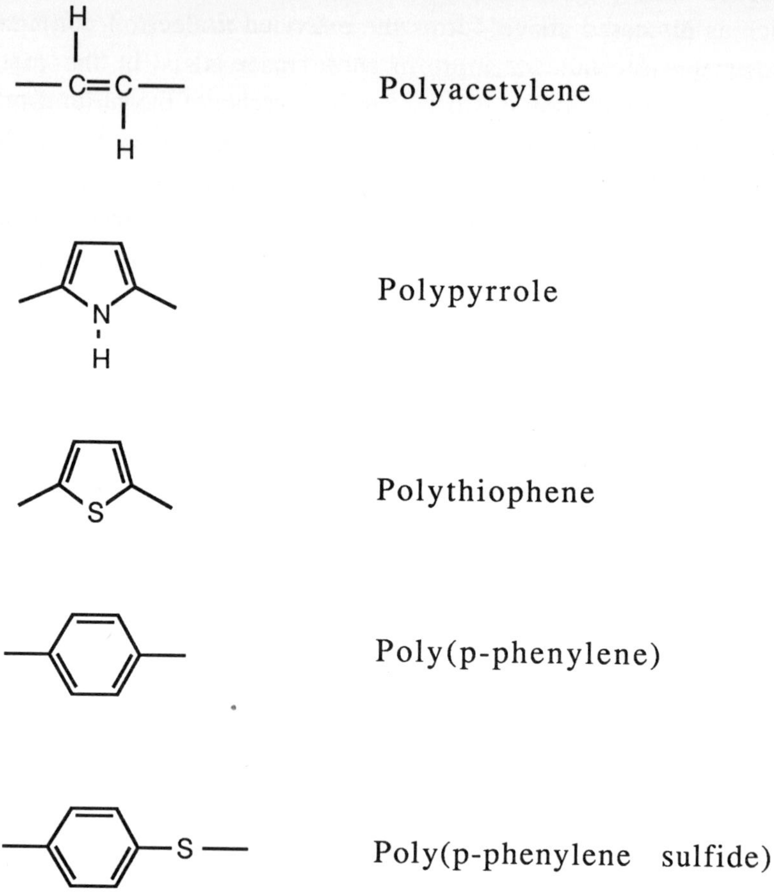

Structure	Name
$-C=C-$ (with H above and below)	Polyacetylene
(pyrrole ring)	Polypyrrole
(thiophene ring)	Polythiophene
(phenylene ring)	Poly(p-phenylene)
(phenylene-S ring)	Poly(p-phenylene sulfide)

Figure 5. Repeat unit structures of a number of conjugated polymers.

oxidation (or reduction) involves the creation of a charged polymeric backbone (polymeric carbenium ions or carbanions) and the introduction of a counterion that insures that charge neutrality is preserved. Thus, the amount of dopant incorporated in the polymer is usually expressed as the mole fraction of counterion per repeat unit. For most conducting polymers, the plateau region of this curve represents a level of conductivity between 1.0 and 1000 S/cm (ohm^{-1} cm^{-1}), although recently conductivities as high as 10^5 S/cm have been achieved in polyacetylene samples prepared with very low sp^3 defect levels [17]. In contrast to traditional semiconductor doping where doping levels in the parts per million range are typically used to enhance conductivity, most conducting polymers become highly conducting only when the doping level exceeds about 1 mole %.

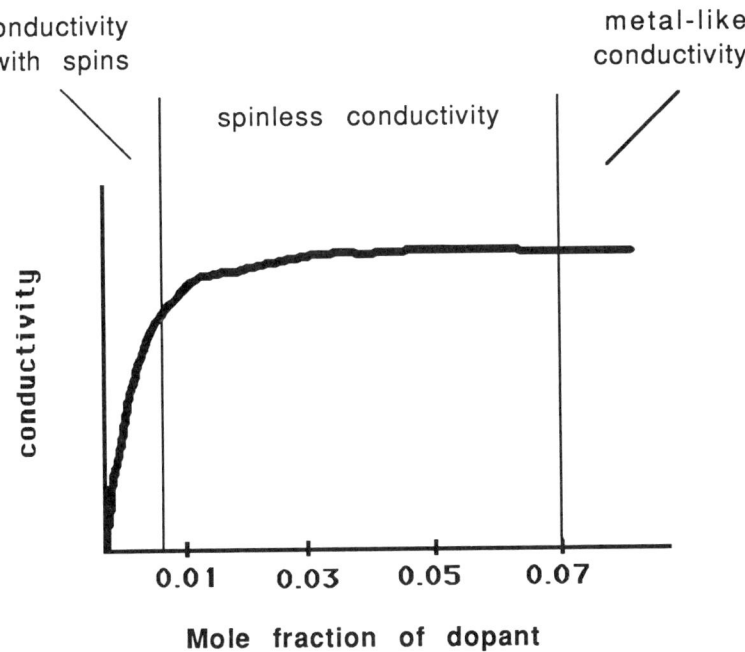

Figure 6. Curve illustrating how the conductivity of a conjugated polymer varies with dopant level. The two vertical lines on the curve represent transitions to different conduction mechanisms (see text).

Returning to the description of these materials as large band gap semiconductors with completely filled valence bands, it might be expected that oxidation of the polymer backbone simply removes electrons from the valence band thereby facilitating band-like conduction by free unpaired electrons. In this same light, reduction can be viewed as a process by which electrons are added to the empty conduction band. In both cases, we would expect conduction to take place by the movement of free spins associated with the unpaired electrons now present in either the conduction (reduction) or valence (oxidation) band. This would be analogous to inorganic semiconductor doping (such as the substitutional doping of silicon with boron) in which unpaired electrons are produced by the creation of occupied donor levels slightly below the conduction band or empty acceptor levels slightly above the valence band. At room temperature, electrons are readily promoted from the valence band to the empty impurity states (acceptor doping) or from the occupied impurity states to the conduction band (donor doping). The former process produces a p-type conductor, whereas the latter process an n-type conductor. Both processes are illustrated in Figure 7.

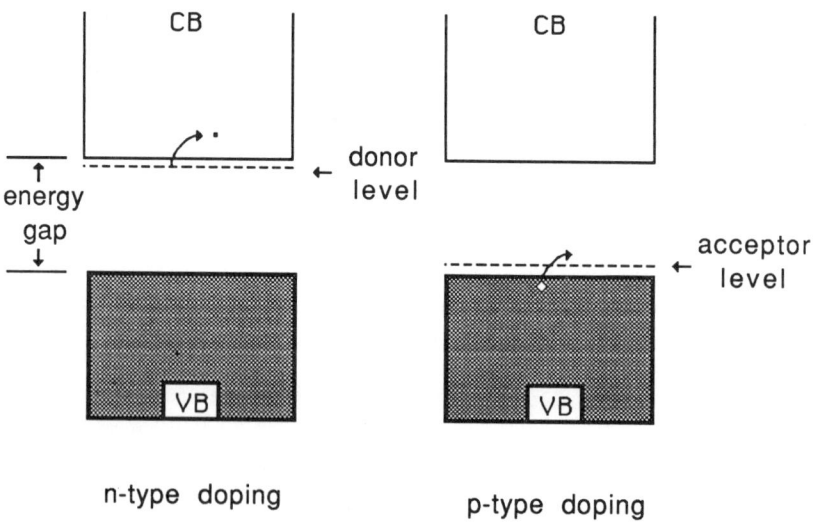

n-type doping   p-type doping

Figure 7. Substitutional doping of inorganic semiconductors either adds electrons to the conduction band (n-type doping) or removes electrons from the valence band (p-type doping).

This process of dopant induced conductivity, however, does not explain one of the most fascinating aspects of conducting polymers, namely that the concentration of free spins (as determined by electron spin resonance spectroscopy) in most conducting polymers is simply too low to account for the level of conductivity observed in these materials. Specifically, it has been found that at low dopant levels the concentration of free spins increases with the conductivity, but as doping is continued, it saturates and eventually decreases to an undetectable amount at higher doping levels [18]. Thus, in the important doping region of high conductivity, conduction occurs without the benefit of unpaired electrons (referred to as spinless conductivity). In order to understand this phenomenon, we must look more closely at how charges are stored on the polymer backbone and their influence on the electronic band structure of the polymer.

During the oxidation process, charge is transferred from the π-system of the polymer chain to the doping agent. The resultant polymeric cation and its associated anionic counterion (usually a reduced form of the doping agent) form an ionic complex which establishes the basis for p-type conduction in the polymer. For oxidation processes, it is therefore desirable to have a polymer with a small ionization potential to insure that effective charge transfer via oxidation can be readily achieved. The fate of the positive charge existing on the polymer backbone is determined by the chain's ability to support that charge in an energetically favorable manner. Delocalization of the positive charge over the entire backbone would be equivalent to removing an electron from the valence band thereby creating a partially filled band (via creation of holes) and metallic-like conductivity. The polymer chain can better accommodate this positive charge, however, if the charge becomes localized over a smaller section of the chain; the net effect being the formation of new localized electronic states with the bandgap [13]. Charge localization along the chain requires a local rearrangement of the bonding configuration in the vicinity of the charge (the polymer lattice relaxes to a new bonding geometry) and hence costs the system a gain in elastic energy. The generation of this local lattice distortion, which is needed to support the localized charge, however, also lowers the ionization energy of the distorted chain and increases its electron affinity making it more easily oxidized and better suited to accommodate the newly formed charges. Clearly, if the gain in elastic energy is offset by the reduction of the ionization energy of the chain, then charge localization will take place at the expense of complete charge delocalization via the band process [13].

A similar scenario can be easily developed for a reduction process involving the addition of electronic charge to the chain (n-type doping) and the formation of localized carbanions along the polymer backbone. Structural relaxations of the lattice surrounding the charged defect sites are possible due to the strong electron-phonon coupling that exists in quasi-one-dimensional polymeric systems. Thus, this is the second instance in which the one-dimensional nature of conducting polymers has dramatically influenced the behavior of these materials. Such distortions would not be expected in rigid 3-D structures where electrons and holes are generally found to be the dominant excitations. The specific types of charged defects formed on the polymer backbone during doping depend upon the structure of the polymer chain. Essentially two main types of conjugated polymer backbones can be identified: those with degenerate ground state structures and those with nondegenerate ground state structures.

The doping process, as illustrated in Figure 8 for the chemical oxidation of a neutral chain of polypyrrole (which is an example of a conjugated polymer with a nondegenerate ground state structure), proceeds in the following manner. First, an electron is removed from the π-system of the backbone creating a free radical (unpaired electron with spin 1/2) and a spinless positive charge (cation). The radical and cation are coupled to each other via a local bond rearrangement, which in this case takes the form of a sequence of quinoid-like rings. The quinoid-like lattice distortion is of higher energy than the remaining portion of the chain which still retains a benzoid-like bonding configuration. Thus, separation of these defects along the chain and the concomitant creation of additional high energy quinoid-like structure costs a considerable amount of energy thereby limiting the number of quinoid-like rings that can link these two bound species together. In the case of polypyrrole, the lattice distortion is believed to extend over about four pyrrole rings. This combination of a charged site coupled to a free radical via a local lattice distortion is called a polaron. A polaron can take the form of a radical cation (chemical oxidation) or a radical anion (chemical reduction). Polaron formation creates new localized electronic states in the gap, with the lower energy

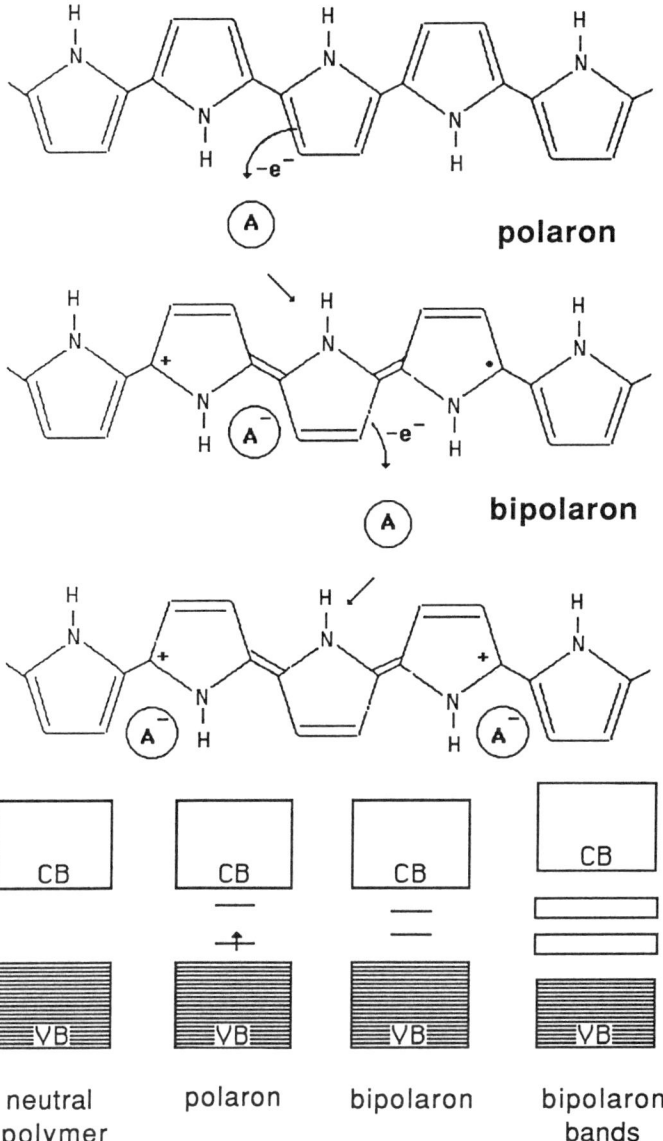

Figure 8. Oxidation of polypyrrole and the creation of polaron and bipolaron states.

states being occupied by single unpaired electrons (a polaron has spin). Theoretical studies indicate that the polaron states of polypyrrole are symmetrically located about 0.5 eV from the band edges [13].

Upon further oxidation, an electron can be removed from either the polaron or the remaining neutral portion of the chain. In the former case, the free radical of the polaron is removed and a dication is created comprised of two positive charges coupled through the lattice distortion. This new spinless defect is called a bipolaron. Removal of an additional electron from a neutral portion of the chain, on the other hand, would create two polarons. Since the formation of a bipolaron produces a larger decrease in ionization energy compared to the formation of two polarons, the former process is more thermodynamically favorable than the latter. At higher doping levels, it also becomes energetically possible for two polarons on the same chain to combine to produce a bipolaron. Thus, additional oxidation is accompanied by the elimination of polarons and the emergence of new localized bipolaron states. These new empty bipolaron states are also located symmetrically within the band gap about 0.75 eV away from the band edges in the case of polypyrrole. Continued doping of the polymer creates additional localized bipolaron states which, at high enough doping levels, can overlap to form continuous bipolaron bands. The bandgap of the polymer is also increasing during this process as the newly formed bipolaron states are created at the expense of the band edges. For conjugated polymers that can be heavily doped, it is theoretically conceivable that the upper and lower bipolaron bands will eventually merge with the conduction and valence bands respectively to produce partially filled bands and metallic-like conductivity.

In the case of conjugated polymers with degenerate ground state structures, the situation is somewhat different [19]. As illustrated in Figure 9, the initial oxidation of the conjugated polymer trans-polyacetylene also creates polarons which appear in the band picture as localized electronic states symmetrically located within the gap. Further oxidation of the polymer, in turn, creates polymeric dications as was the case for polypyrrole. However, since the ground state structure of polyacetylene is twofold degenerate, the charged cations are not bound to each other by a higher energy bonding configuration and can freely separate along the chain (assuming a chain of infinite length). In other words, the bonding configurations on either side of the charged defects only differ by a reversed orientation of the conjugated system and are

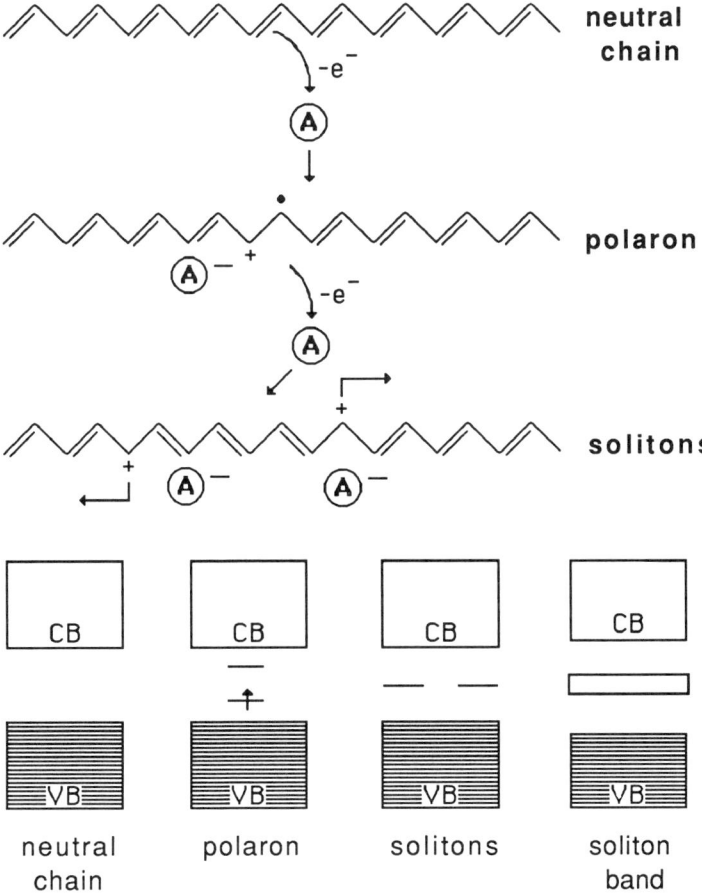

Figure 9. Oxidation of polyacetylene and the creation of polaron and soliton states.

therefore energetically equivalent resonance forms. In essence, we now have isolated, noninteracting charged defects that form domain walls separating two phases of opposite orientation but identical energy; such defects are called solitons and they can be either charged or neutral in nature. Solitons are actually believed to be delocalized over about 12 CH units with the maximum level of charge density located adjacent to the dopant counterion and the bonds in the center of the defect exhibiting a lesser amount of bond alternation than the bonds away from the defect center [19]. Soliton formation results in the creation of new localized electronic states that now appear at the middle of the energy gap as indicated in Figure 9. At high enough doping levels, the charged solitons interact with each other to form a soliton band which can eventually merge with the band edges to create true metallic conductivity.

To summarize, the excess charges created on conjugated polymer chains by oxidation or reduction are accommodated in localized electronic states located within the band gap rather than as free carriers within the existing band states. At low doping levels, the defects created by charge transfer are in the form of polarons which are comprised of unpaired electrons (free radicals) coupled to charged defect sites. Polarons create two symmetrically located defect levels in the gap and generally do not significantly enhance electrical conductivity. As the doping process continues, the free radicals are eliminated either by direct ionization of a polaron or by the reaction of two polarons existing on the same chain. For conjugated polymers exhibiting a nondegenerate ground state, the resultant charges are bound to each other in pairs via a change in bonding geometry and are called bipolarons. In the case of degenerate ground state polymers such as trans-polyacetylene, the charges, now called solitons, are not correlated along the chain and only interact with each other at very high dopant levels. Bound states such as bipolarons create defect levels symmetrically located above the valence band and below the conduction band whereas solitons result in defect levels located at the mid-gap.

The concept of localized charged defects on conjugated chains was originally proposed in a number of elegant theoretical models [19, 20, 13] devised to account for the spinless transport observed in conducting polymers. Although specific details concerning the quantitative aspects of these theories will no doubt be debated for quite some time, it is now generally well accepted that the primary mode of charge storage on conjugated polymers is via polarons, bipolarons or solitons. Experimental evidence supporting the existence of these unique electronic states has been provided by a number of different techniques and continues to accumulate. Most notably, an examination of the optical spectra of many of the doped conjugated polymers reveals the presence of electronic transitions with energies less than the gap energy (interband transition). The number of low energy optical transitions observed in various conjugated polymeric systems has also been found to correlate well with that predicted by theory; namely, three transitions when polarons are present, two transitions for bipolarons, and one transition for solitions [3]. Figure 10 displays optical spectra of a thin film of poly(3-octyl thiophene) in its neutral and oxidized state. In the neutral state, the spectrum is dominated solely by the presence of a strong interband transition centered around 500 nm. After doping to a state of high conductivity, in this case with NOPF$_6$, the interband transition decreases and two new absorption bands resulting from transitions of electrons from the valence band to the two newly created bipolaron states are clearly observed.

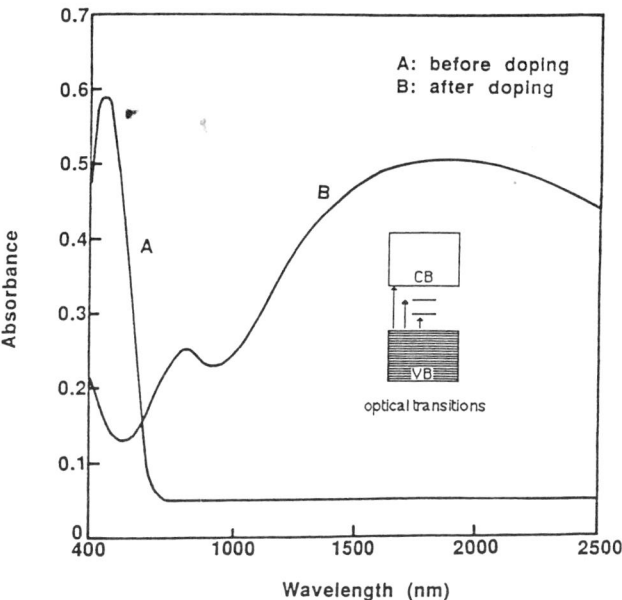

Figure 10. Optical spectra of a thin film of poly(3-octyl thiophene) A) before doping and B) after doping with NOPF$_6$. Inset shows optical transitions of a bipolaron.

2.3 TRANSPORT OF CHARGE IN CONDUCTING POLYMERS

Although solitons and bipolarons are believed to be the primary source of charge carriers in conducting polymers, the exact mechanisms by which charge is transported in these materials are still largely unknown. The problem arises when one attempts to trace the path of a charge carrier through the bulk of a polymeric matrix. Unlike the well ordered state of a crystalline metal or semiconductor lattice, most polymers are highly disordered materials comprised of both crystalline (or paracrystalline) and amorphous regions. Substantial disorder is also introduced into the polymer by the diffusion of counterions during the doping process essentially insuring that the electrically conductive form of a conjugated polymer will be dominated by the effects of disorder. This state of affairs is further complicated by the fact that many doping processes do not produce a homogeneously doped material but rather create a polymer comprised of regions of highly doped material coexisting with regions of undoped or lightly doped material [21]. It therefore becomes necessary to consider not only the transport of charges along and between chains, but also across the complex domain boundaries established by the multitude of phases that can be present in the material. The amazing fact that many conducting polymers exhibit high levels of conductivity in spite of their high degree of structural disorder continues to fuel the search for better ways to control the microstructures of these materials.

As a result of the large variations in ordering possible in conducting polymers (even within the same material) and the diverse microstructures present in these materials, it should not be too surprising to find that experiments have uncovered many different possible conduction mechanisms [22]. The large change in electrical conductivity that occurs with doping (often times more than twelve orders of magnitude) also suggests that a number of different conduction mechanisms may be needed to explain the transport behavior of these polymers. The common way to explore the mechanism of conduction in materials is to examine the temperature and frequency dependence of the conductivity over a range of many decades. These measurements, when coupled with magnetic and optical studies, provide valuable information about the nature of the charge carriers, their mobility, and the activation energies associated with charge transport.

In highly ordered materials, where band transport dominates, conduction is primarily determined by the number and mobility of the charge carriers present in the bands. The signature of true metallic conduction is the observation of a conductivity that increases (essentially linearly) with decreasing temperature and is independent of frequency up to very high frequencies. In this case, the presence of partially filled bands insures an abundance of free charge carriers,and mobility (specifically its temperature dependence) becomes the important factor in determining conduction. At low temperatures, the scattering of electrons by lattice vibrations is reduced resulting in a higher mobility and a larger conductivity. Semiconductor behavior, on the other hand, is characterized by a decreasing conductivity with decreasing temperature. The mobility of the carriers is still enhanced at lower temperatures; however, this effect is small compared to the reduction in population of charge carriers in the conduction and valence bands that occurs as the temperature is lowered and it becomes more difficult to thermally excite carriers across the band gap. In both of these cases, the conduction process is related to the mobility and density of charge carriers in extended band states.

The situation is often quite different in conducting polymers. At low doping levels, the charged defects on the polymer chains do not exist within extended states but rather are confined to localized states. As the doping level increases, these states can overlap to form soliton or bipolaron bands; however, they will be either completely filled (n-type doping) or completely empty (p-type doping). Only at very high doping levels will these latter bands merge with the conduction and valence bands to produce true metallic-like conductivity and this final state is known to occur in only a select number of conducting polymers. Thus, over much of the conducting regime of these materials, transport occurs by the movement of charge carriers between localized states or between soliton, polaron, or bipolaron states. Alternatively, in those cases where inhomogeneous doping produces metallic islands dispersed throughout an insulating matrix, conduction occurs by movement of carriers between the highly conducting domains. The net result is that transport in conducting polymers is dominated by thermally activated hopping or tunneling processes in which carriers hop across, or tunnel through, barriers created by the presence of isolated states or domains. By its very nature, charge transport of this type dictates that the conductivity will decrease with decreasing temperature with the functional dependence of the temperature dependent conductivity being highly dependent on the particular mechanism operating within the material.

A cursory review of the literature reveals the large number of transport models that have been described over the past few years. These include, for example, models based on intersoliton hopping [23], hopping between localized states assisted by lattice vibrations [24], interchain hopping of bipolarons [25], variable range hopping in three dimensions, and charging energy limited tunneling between conducting domains. These latter two mechanisms are characterized by conductivities that vary as $T^{-1/4}$ in the case of variable range hopping [26] and $T^{-1/2}$ in the case of charging energy limited tunneling [27]. This is in contrast to processes related to the thermal activation of carriers across a band gap into extended states in which the conductivity varies as T^{-1}. A detailed description of the many conduction mechanisms that have been put forth to explain transport in conducting polymers is beyond the scope of this chapter; however, the reader should be aware of the problems associated with trying to unravel this very complex puzzle.

3. MATERIALS, STABILITY, AND PROCESSIBILITY

In order to successfully exploit the unusual redox properties of conducting polymers in commercial applications, it is imperative that candidate materials exhibit good environmental stability and be amenable to a wide variety of processing techniques. Table 1 lists the current state of these latter attributes for a number of conjugated polymers along with the typical levels of conductivity reached after doping with suitable oxidizing agents. A rating of "limited" under the category of processing possibilities means that the polymer can only be processed into usable forms by specialized techniques whereas an "excellent" rating indicates that processibility can be achieved by a number of different conventional processing techniques. A polymer rated as having poor environmental stability is essentially unstable in its doped state under normal atmospheric conditions. It is not uncommon for the conductivity of the materials in this category to decrease by one order of magnitude every few months at room temperature and at much faster rates at elevated temperatures. On the other extreme are materials which exhibit good environmental stability such as polypyrrole which displays only minor changes in its conductivity even after exposure in air to temperatures as high as 200°C for extended periods of time. In general, the conducting polymers based on heterocyclic repeat units (thiophenes and pyrroles) exhibit the best overall environmental stability [28]. The environmental stability of these latter materials can be further enhanced by introducing an electron donating substituent onto the three or four position of the heterocyclic ring which has the effect of lowering the oxidation potential of the polymer and stabilizing its positively charged bipolaron states [29].

3.1 STABILITY

The stability of a conducting polymer is determined by a number of different factors which are both intrinsic and extrinsic in nature [7]. Extrinsic stability is related to the polymer's vulnerability to external environmental agents such as water vapor and oxygen. In this case, the susceptibility of the charged defect sites along the polymer chain (which in actuality are simply delocalized carbenium ions, carbanions, or free radicals) to attack by nucleophiles, electrophiles and free radicals, determines the stability of the system. When poor stability is dominated by extrinsic effects, it is possible to encapsulate the polymer in suitable barrier materials or to prepare denser morphologies both of which inhibit diffusion of the chemical agent to the active sites within the polymer and thereby improve stability. The alkyl substituted polythiophenes provide an example of a class of conducting polymers whose stability is highly sensitive to the amount of water vapor in the air [30].

Many conducting polymers, however, are also intrinsically unstable, as indicated by a substantial degradation of the conductivity over time even when the material is stored in a completely dry, oxygen free environment [28]. This latter effect is thermodynamic in origin and often times reflects irreversible chemical reactions that take place between the charged sites of

Table 1. STABILITY AND PROCESSING STATUS OF REPRESENTATIVE CONDUCTING POLYMERS.

Polymer	Conductivity ($\Omega^{-1}cm^{-1}$)	Stability (doped state)	Processing Possibilities
Polyacetylene $-HC=CH-$	$10^3 - 10^5$	poor	limited
Polyphenylene	1000	poor	limited
PPS	100	poor	excellent
PPV	1000	poor	limited
Polypyrroles	100	good	good
Polythiophenes	100	good	excellent
Polyaniline	10	good	good

the polymer chain and either the dopant counterions or the π-system of an adjacent neutral chain. The net effect of these reactions is to introduce saturated sp^3 defects along the polymer chain that break conjugation and decrease conductivity. Intrinsic instability can also take the form of a thermally driven undoping process in which conformational changes in the polymer backbone activated at elevated temperatures destabilize the charged sites created by oxidation and completely reverse the doping process. This latter process, which regenerates the undoped polymer without major structural modifications, has been observed in alkyl substituted polythiophenes [30].

In short, the stability of a conducting polymer depends on a number of factors including its susceptibility and accessibility to external chemical species, the nature and type of counterion present in the material, the reactivity of its doped sites to surrounding chains, and the flexibility and conformational states of its backbone. A great deal of progress has been made towards the development of stable conducting polymers; however, this issue continues to be of paramount importance to the successful utilization of these materials in commercial applications.

3.2 PROCESSIBILITY

Conjugated polymers have been synthesized using a wide variety of conventional polymerization routes [3] including cationic, anionic, and free radical chain growth mechanisms, coordination polymerizations, step growth polymerizations and electrochemical polymerizations. In the latter case, suitable monomers are electrochemically oxidized to create active monomeric and dimeric species capable of reacting to form a conjugated polymer backbone, which in this case is obtained in its conducting form due to subsequent electrochemical oxidation of the newly formed polymer [31]. Many of the initially prepared conducting polymers were formed as intractable, insoluble films or powders that, once synthesized, could not be further manipulated into forms with more ordered, controllable structures. In fact, the very structural attributes that give rise to the interesting electrical and optical properties of these materials, namely their rigid, planar conjugated backbones, also severely limit the ways in which the polymer can be processed. To overcome these limitations, a number of structurally modified polymers and novel processing schemes have been developed that allow substantially more control over the state of the final product. These processing schemes can be conveniently divided into four categories.

1)   The manipulation of soluble precursor polymers
2)   The manipulation of soluble conducting polymer derivatives and copolymers
3)   The in-situ polymerization of conducting polymers in insulating matrix polymers
4)   The manipulation of conducting polymers via the Langmuir-Blodgett technique

The first scheme is based on the synthesis and manipulation of a processible, nonconducting precursor polymer that, once fabricated into a suitable form using conventional polymer processing techniques, can be converted (usually by thermal treatment) into an insoluble electrically conducting polymer. The precursor polymer is usually processed in a manner that produces a state of highly aligned polymer chains which is essentially retained upon conversion to its conjugated conducting counterpart. This route has been successfully utilized to prepare highly oriented thin films and fibers of polyacetylene [32], poly(phenylene vinylene) [33, 34], poly(thienylene vinylene) [29, 35, 36], poly(2,5-dimethoxyphenylene vinylene) [37, 38, 39] and similar polymers [40, 41]. Specific examples of this route are illustrated in Figure 11. The conductivities of the resultant materials are usually highly anisotropic with maximum conductivities occurring along the stretch direction. The solubility of the precursor polymer also allows this material to be extensively purified and characterized prior to conversion to a conjugated system.

The second scheme follows the approach traditionally taken by polymer chemists when improvements in specific properties are desired; namely the synthesis of copolymers or derivatives of the parent polymer. The trick in this case is to modify the structure of the polymer in such a way as to improve the property of interest (processibility in this case) without compromising its electrical or optical properties. Attempts to improve the processibility of polyacetylene by preparing copolymers and derivatives of this material were unsuccessful because the final products, which were indeed more processible, exhibited significantly reduced electrical conductivities [42, 43]. The preparation of derivatives of the

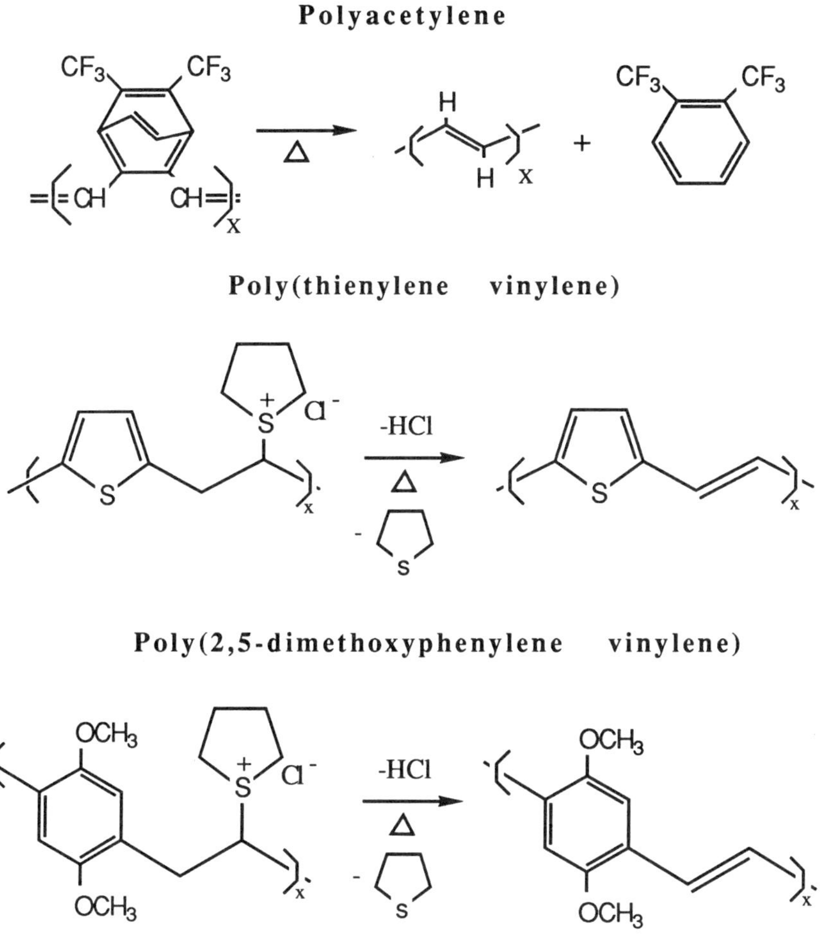
Polyacetylene
CF3
CF3
CH
CH
x
Δ
H
H
x
+
CF3
CF3
Poly(thienylene vinylene)
S
Cl
-HCl
Δ
S
x
S
S
x
Poly(2,5-dimethoxyphenylene vinylene)
OCH3
OCH3
S
Cl
-HCl
Δ
x
S
OCH3
OCH3
x

Figure 11. Soluble precursor routes to electrically conductive polymers.

polythiophenes and polypyrroles, however, has proven to be a far more productive endeavor. As illustrated in Figure 12, by simply substituting the hydrogen atom attached to the three position of the thiophene ring with an alkyl group containing at least four carbons, it is possible to obtain conjugated polythiophenes that are both solution and melt processible [44-47]. The conductivities of the doped derivatives are also comparable to the parent polymer and generally range from 1 - 200 S/cm. Thus, it is

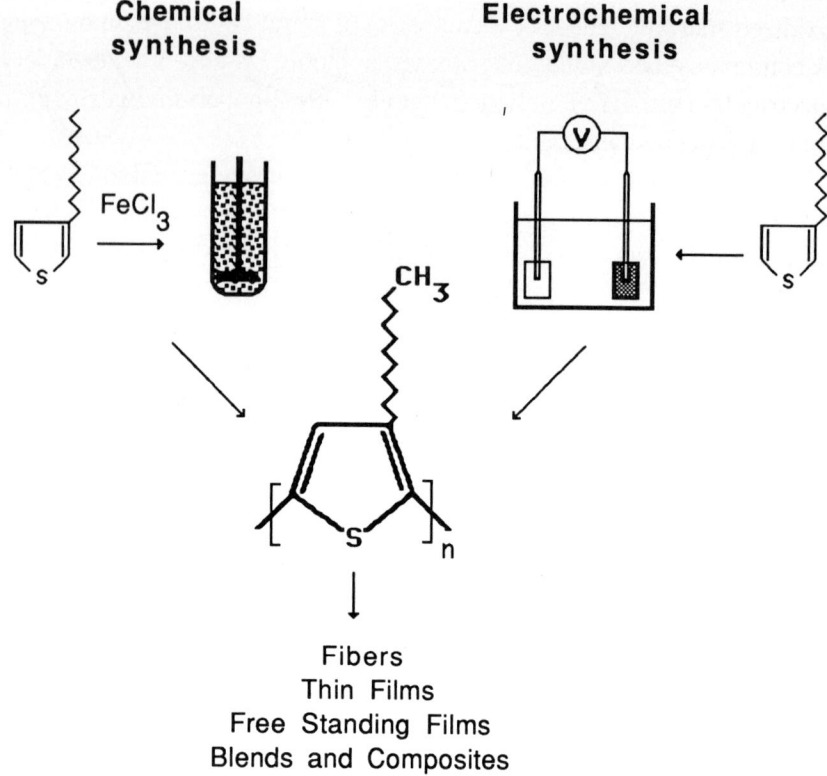

Figure 12. Melt and solution processible polythiophenes can be prepared by chemically or electrochemically polymerizing 3-alkyl substituted polythiophenes.

possible to dramatically modify the processibility of the polythiophenes without severely compromising the electrical properties. Water soluble derivatives of polythiophene [48] and polypyrrole [49] can also be prepared by placing carboxylic acid or sulfonic acid groups within the alkyl chains. These latter materials are fitted with built-in ionizable groups that can be utilized to maintain charge neutrality in the oxidized state; hence they are sometimes referred to as "self-doped" polymers. A very highly conductive derivative of polythiophene has recently been prepared [50] by replacing the hydrogen on the ring with $R = -CH_2-O-CH_2CH_2-O-$

CH_2CH_2-O-CH_3. This material is soluble in both the neutral and oxidized state and reaches conductivities of about 1000 S/cm upon doping. A continued effort along this path will no doubt produce a wide variety of electrically conductive polymers with excellent processing attributes and good environmental stability.

The third major processing scheme focuses on the growth of insoluble, intractable conjugated polymers within a preformed polymer matrix [51-55]. In this case, a processible, insulating polymer impregnated with a catalyst system is fabricated into a desired form such as a thin film or fiber. This activated polymer matrix is then exposed to the monomer, usually in the form of a gas or vapor, resulting in a blend typically comprised of an isolated or semi-continuous conjugated polymer phase dispersed throughout a continuous phase of the host polymer. Variations of this theme include electrochemical growth [56-59] of a conducting polymer within a host polymer deposited on an electrode either during the polymerization stage or prior to electrochemical polymerization. In many instances, the matrix containing the conjugated polymer serves as a processing aid for further manipulation of the order and alignment of the conducting component. For example, stretched aligned blends of polyacetylene/polybutadiene exhibit conductivities at least one order of magnitude larger than the unstretched material [53]. This enhancement in conductivity, in part, reflects a higher state of order resulting from the deformation process.

The last processing scheme relies on the ability of the Langmuir-Blodgett (LB) trough to manipulate surface active molecules into highly ordered thin films with structures and film thicknesses that are controllable at the molecular level. It has been well established that amphiphilic molecules containing hydrophilic (water attracting) groups attached to long hydrophobic (water repelling) groups will organize into well ordered monolayers at the air-water interface of an LB trough [60]. These well ordered monolayers can then be sequentially transferred onto a variety of substrates creating a multilayer structure comprised of molecular stacks (usually about 25Å thick) of highly ordered molecules. Although this technique has been mostly applied to the fabrication of insulating thin films, the recent interest in identifying new means for processing conducting polymers has lead to the development of new conducting polymer systems that can be utilized in conjunction with the LB technique. Variations of this approach include the LB manipulation of surface active monomers of conjugated polymers and the LB manipulation of mixed monolayers containing a nonsurface active conjugated polymer and a traditional surface active molecule such as stearic acid. In the former case, surface active pyrrole monomers fitted with long (> 12 carbons) alkyl groups have been manipulated into condensed monolayers and either polymerized at the air-water interface of the LB trough by adding an oxidizing agent and excess pyrrole monomer into the water subphase [61-63] or by transferring the monomer directly onto a conducting substrate and subsequently electrochemically polymerizing the resultant multilayer structure [64]. The second approach involves utilizing the spreading characteristics of stearic acid to encourage monolayer formation of soluble conjugated polymers such as the poly(3-alkyl thiophenes) that by themselves are not capable of true dispersion at the air-water interface. These mixed monolayers can be readily transferred into multilayer thin films with well defined layer structures and high electrical conductivities after doping [65]. This latter technique is generally applicable to any of the soluble conjugated polymers. LB films have also been fabricated directly from Poly(3-dodecylthiophene) [66]. In this latter case, the high rigidity of the polymer films deposited at the air-water interface required that a horizontal dipping procedure be used to transfer the polymer into LB multilayer films.

The true promise of the LB processing technique is its unique ability to allow control over the molecular architecture of conducting polymer thin films. Since the molecules are deposited one layer at a time and since layers containing different molecular species can be readily deposited onto the film, it is possible to build complex multilayer structures comprised of functionally different molecules arranged in a pattern predetermined by the operator. An example of an organic superlattice fabricated using the LB technique is illustrated in Figure 13. In this case, the film has been constructed so as to contain layers of an electrically conductive polypyrrole rich bilayer alternating with bilayers of an insulating material. The electrically conductive domains (about 90Å thick) are therefore isolated from each other by the electrically insulating domains (about 60 Å thick). The net result is a highly anisotropic film which is highly conductive within the plane of the film but essentially insulating across the thickness of the film (conductivity anisotropies as high as 10^{12} can be easily realized with this type of structure). A recent theoretical evaluation [67] of synthetic metal superlattices indicates that such novel molecular organizations, when properly designed, might exhibit unique optical and electrical properites.

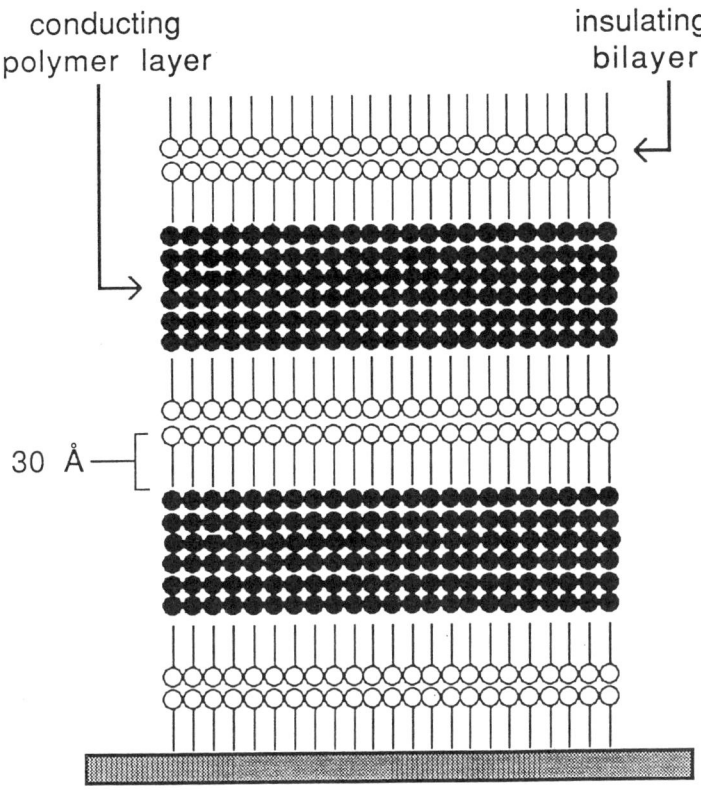

conducting
polymer layer

insulating
bilayer

30 Å

Figure 13. Polymeric superlattice comprised of layers of an electrically conductive polymer alternating with insulating bilayers.

4. TOWARDS THE OBTAINMENT OF MOLECULAR ELECTRONICS

4.1 APPLICATIONS

Thus far we have recognized that the extended π-systems of conjugated polymers make them highly susceptible to chemical or electrochemical oxidation and reduction. These charge transfer processes in turn modify the electronic states of the polymer backbone thereby dramatically altering its electrical and optical properties. Perhaps the most interesting and useful feature of this charge transfer chemistry is the fact that it can be precisely controlled and is usually completely reversible. Thus, it is possible to systematically control the electrical and optical properties of conducting polymers over a very wide range and, if desired, it is further possible to switch reversibly from a conductive state to an insulating state. It is these latter two attributes that make conjugated polymers particularly useful in the area of molecular electronics and separate them uniquely from other materials.

Table 2 displays some potential applications of conducting polymers. The table is divided into two sections. The first section focuses on applications that specifically take advantage of the reversible redox properties of these electroactive materials. The second section, on the other hand, focuses on the fact that these materials can be made electrically conductive. In the former case, each application exploits the fact that the electrical and optical properties of conducting polymers depend, in a controllable manner, on their level of oxidation or reduction. By reversibly switching between two redox states, for example, it is possible to construct electronic devices [68], electrochromic displays [69], and polymer based rechargeable batteries [70]. Controlled drug release systems [71] can also be made by systematically releasing dopant counterions that contain active drug components by electrochemically reversing the doping process. Sensors based on conducting polymers also rely on reversible changes in the optical and electrical properties of these materials that, in this case, occur in the presence of redox active chemical agents. The applications at the bottom of the table simply take advantage of the electrical conductivity of doped conjugated polymers and will not be discussed any further. Instead, the remainder of this chapter will be devoted to use of conjugated polymers in molecular electronic schemes.

Table 2. Potential applications of conducting polymers

Key Parameter ———————— Electroactivity

Molecular electronics
Electrochromic displays
Chemical and biochemical sensors
Rechargeable batteries and solid electrolytes
Drug release systems

Key Parameter ———————— Conductivity

Electrostatic materials
Conducting adhesives
Electromagnetic shielding

4.2 MOLECULAR ELECTRONICS - BASIC CONCEPTS

In the most optimistic sense, molecular electronics refers to the development of electronic and optical devices containing active elements that have the dimensions of single molecules. At this point in time, however, such molecular sized electronic devices are merely the topics of highly imaginative discussions. The difficulties associated with firstly assembling a number of single molecules into a complex device array and secondly addressing each individual element (i.e., making a contact to a single molecule) are currently insurmountable and such devices, if they are to exist at all, will only be feasible in the far future. A broader and currently more realistic view of molecular electronics centers on the utilization of organic based materials as active elements in electronic, optical, and optoelectronic devices that traditionally have exploited the properties of inorganic materials. Thus, in the case of conducting polymers, molecular electronics simply refers to the fabrication and characterization of electronic and optical devices based on electroactive polymers.

The potential for conducting polymers in the areas of electronics and photonics (nonlinear optics) is enormous. However, if this potential is to be realized, in addition to solving the problems of stability, it is equally as important to develop techniques that can be used to process conducting polymers into technologically useful forms. For many of the applications envisioned for these materials, it is essential that they exhibit flexible processing attributes that allow control over the dimensions and shape of the final form of the polymer and its molecular arrangement and organization. This latter fact is particularly important in the area of molecular electronics where the demands placed on the control over the orientation and ordering of the polymer chains will become more stringent as the active elements of optical and electronic devices based on these materials become smaller and more complex. In this light, the development of processing techniques that can be used to form highly ordered, uniform thin films with controllable thicknesses is paramount to the continued advancement of these materials in this area. A particularly interesting application of such films involves the fabrication of thin film junctions.

4.3 THIN FILM JUNCTIONS

Many solid state devices utilize junctions such as metal-semiconductor junctions (Schottky barriers), p-n heterojunctions, and metal-insulator-semiconductor (MIS) junctions as the basic building blocks of the device. When two materials with different work functions are brought together to form a junction, they establish thermal equilibrium by setting their Fermi levels equal (i.e., equilibrium is achieved when their chemical potentials are equal). The work function in this case represents the energy difference between an electron at the Fermi level and the vacuum level. The establishment of equal chemical potentials allows mobile charge carriers to lower their energy by crossing over from one material into the other. In doing so, however, they create a transition region at the junction that is significantly depleted of mobile charge carriers. The diffusion of charge carriers across the junction also creates a charge imbalance at the interface due to the many uncompensated nonmobile charges remaining within each phase. This combination of unbalanced charges of opposite sign separated by an insulating layer generates an internal electric field which eventually becomes large enough to oppose further current flow thereby limiting the size of this so-called depletion layer. At equilibrium, the depletion region therefore supports an internal barrier field caused by charge separation. Under such an arrangement, it is possible to apply an external electric field that either enhances or diminishes the effect of the internal potential barrier existing near the junction interface. When the external field is applied with a bias (forward bias) that acts to diminish the effects of the internal barrier field, carriers can acquire sufficient energy to cross the barrier and at high enough external voltages (when the external voltage exceeds the internal barrier potential) a large current will flow. If the bias is reversed, on the other hand, the barrier potential is enhanced by the external field and only a small current flows. Thus, the junction acts as a rectifier since the current flowing for a given positive external voltage is quite different from the current flowing at the same negative voltage.

As an example, consider a metal-semiconductor junction formed between a p-type semiconductor and a metal with a work function that is lower than that of the semiconductor. In this case, mobile hole carriers are depleted from the semiconductor in the vicinity of the junction thereby creating an insulating layer that supports a potential barrier between the metal and the semiconductor. The presence of this internal potential barrier means that current flow will depend on the polarity of the applied voltage. This type of junction will therefore behave as a rectifier. In contrast to this situation, a junction fabricated from a p-type semiconductor and metal with a higher work function does not exhibit rectifying behavior. In this case, it is not possible to form a depletion layer with its accompanying potential barrier and the flow of current is therefore not dependent on voltage polarity. This latter junction results in ohmic behavior; i.e., the current flow is linearly dependent on voltage regardless of whether the voltage is positive or negative. The current-voltage relationships of a rectifying junction (diode) and a ohmic junction (resistor) are shown in Figure 14. The observance of nonlinear behavior in a current-voltage plot is generally a clear signature of a rectifying junction.

If it is assumed that mobile carriers are injected across the semiconductor-metal interface, then the thermionic emission model [72] of current transport through the barrier layer can be used to derive a relationship between current and voltage in a rectifying junction. Equation

1 presents the standard relationship between current and voltage for such a diode.

$$J = J_{sat} (\exp qV/nKT - 1) \tag{1}$$

For this equation, q represents the charge of an electron (1.6×10^{-19} coulombs), V is the applied voltage, K is Boltzmann's constant, T is temperature and n is the diode quality factor which is equal to 1 for a perfect diode. J_{sat} in the thermionic emission model represents the saturation current density and is given as:

$$J_{sat} = A*T^2 \exp(-\Phi_b/KT) \tag{2}$$

In this equation, A* is the effective Richardson constant which is generally assumed to be 120 A K^{-2} cm^{-2} for an effective electron mass of one, and Φ_b is the barrier height between the metal and semiconductor.

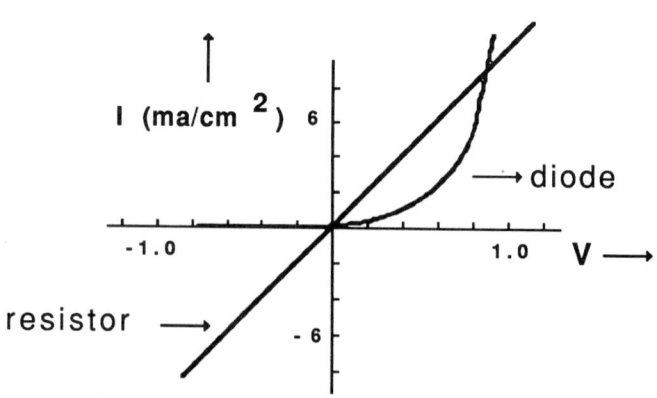

Figure 14. Current-voltage behavior of a rectifying diode and a ohmic resistor.

Capacitance-voltage measurements are also commonly used to characterize the behavior of Schottky barriers. In the simple depletion model in which the junction is represented by two layers of space charge separated by an insulating region, the junction can be modelled [73] as a circuit comprised of an ohmic resistor (with a resistance equal to that of the undepleted semiconductor) in series with a parallel circuit containing a capacitor and a resistor as illustrated in Figure 15.

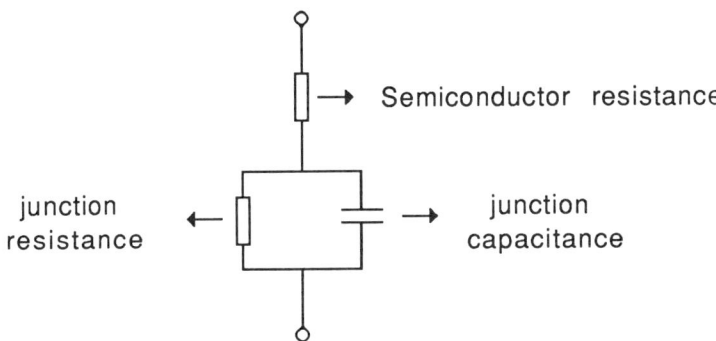

Figure 15. Model circuit used to approximate the behavior of a rectifying junction.

These latter two circuit elements represent the capacitance and resistance of the junction and are therefore dependent on the applied voltage and its bias. With a voltage bias that diminishes the internal barrier potential, the resistance of the junction becomes small and current flows readily across the junction (the capacitor is shorted-out). A reverse bias, on the other hand, will enhance the internal barrier creating a junction with a high resistance that is now dominated by its capacity. Thus, the capacitance of the diode is best measured in a reverse bias configuration.

The equation relating capacitance to voltage in a diode whose depletion region varies with voltage is

$$1/C^2 = (2/(q \varepsilon \varepsilon_0 N_a)) (V_{bi} - V - KT/q) \qquad (3)$$

where V_{bi} is the built-in voltage (internal barrier voltage), V is the applied voltage, N_a is the free carrier concentration and ε and ε_0 are the dielectric constants of the semiconductor and free space respectively. This equation predicts that a plot of $1/C^2$ versus reverse bias voltage should give a straight line. The obtainment of a linear plot also indicates that the charge density within the depletion region is uniform. The above equations can be used to extract important materials parameters from J-V and C-V curves such as the dopant concentration, the zero bias depletion width and the carrier mobility. The polarity of the J-V curves also gives information about the sign of the majority carrier in the semiconductor. A list of some of the parameters obtained from Schottky barrier junctions based on electroactive polymers can be found in the next section.

4.4 ELECTRONIC DEVICES BASED ON CONJUGATED POLYMERS

The fabrication of junctions based on conducting polymers requires sufficient processibility to allow for the deposition of homogeneous thin films onto a variety of metal and semiconductor surfaces. A number of the processing schemes described earlier have been successfully utilized to fabricate a variety of electronic devices containing conjugated polymers as an active component. For example, Schottky barrier diodes [74-83], photovoltaic cells [84-88], metal-insulator-semiconductor (MIS) diodes [89,90] and field effect transistors (FET) [91-95] have all been successfully fabricated with conducting polymers. By controlling the thickness and/or doping level of the polymer used to form the junctions of these devices, it is possible to vary the resistivity of the polymer and, hence, the device characteristics over a wide performance range. Thus, it is in principle possible to fine tune the properties of these polymer-based devices by adjusting the carrier concentration, conductivity, and the energy levels of the conjugated polymer.

4.5 SCHOTTKY BARRIER JUNCTIONS BASED ON CONJUGATED POLYMERS

Schottky barrier junctions comprised of metal-semiconductor interfaces have been fabricated with both doped conjugated polymers and undoped conjugated polymers. Heavily doped polyacetylene, for example, has been joined with a conventional semiconductor to form a junction in which the conducting polymer plays the role of a metal [77]. The more common approach, however, involves utilizing undoped or lightly doped conjugated polymers as the semiconductor in conjunction with suitable rectifying metals. In the undoped state, most conducting polymers behave as p-type semiconductors possibly due to the presence of residual impurities (such as catalyst and dopant residues) that maintain the polymer in a slightly oxidized state. Low work function metals are therefore required to form rectifying junctions with these materials. When such a junction is formed, electrons are transferred from the metal to the polymer producing an internal electric field across the junction which is supported by positive charges in the metal and excess electrons in the polymer. A typical junction configuration consists of a thin film of the conjugated polymer (typically 500 to 5000Å thick) sandwiched between the metal rectifying electrode and an ohmic collector electrode. The thin polymer film is usually deposited onto a substrate overcoated with the collector electrode using one of the processing techniques previously described followed by deposition of the top metal electrode using evaporation techniques. A forward bias in this case corresponds to a negative voltage at the blocking metal electrode relative to the ohmic contact.

Schottky diodes based on conducting polymers are generally characterized by evaluation of their J-V and C-V behavior. These data are then used in conjunction with the standard diode equations presented earlier to extract the relevant device and materials parameters of the junction. Table 3 lists a selection of such parameters obtained from metal-semiconductor junctions based on a number of different undoped conjugated polymers. Also included in this table are the work functions of the metals used to form the rectifying electrodes. In general it is found that metals with work functions greater than that of the conjugated polymer form ohmic contacts whereas metals with lower work functions form blocking contacts. This indicates that barrier formation is indeed possible with these materials and that, to a first approximation, it is related to the work function difference between the metal and semiconductor and not to the presence of surface states. It is well known that a high density of surface states can result in rectifying junctions that are independent of the work function of the metal. The former mechanism of Schottky barrier formation has been verified [76] in the case of polyacetylene using x-ray photoemission spectroscopy which provides information about changes in band bending in the polymer when different metals are deposited.

Table 3. VARIOUS DEVICE PARAMETERS OBTAINED FROM SCHOTTKY JUNCTIONS FABRICATED FROM CONJUGATED POLYMERS

Polymer	Metal	Metal Work Function* (eV)	Diode Quality (n)	Rectification Ratio	Carrier Conc. (cm^{-3})
Polyacetylene					
trans-$(CH)_x$	In	4.12	2.41		2.3×10^{17}
trans-$(CH)_x$	Pb	4.25	2.06		2.7×10^{17}
trans-$(CH)_x$	Al	4.28	1.35	100,000	9.0×10^{15}
trans-$(CH)_x$	Al	4.28	3 - 4	500	7.4×10^{16}
trans-$(CH)_x$	Au	5.10	ohmic contact		
Polythiophenes					
P(3-HT)	In	4.12	1.5 - 3.7	1000	3.0×10^{16}
P(3-MT)	In	4.12	1.5 - 2.2	1000	7.0×10^{16}
P(3-MT)	Al	4.28	1.91	10	1.0×10^{16}

* from reference 74.

As can be seen in Table 3, carrier concentrations in the undoped conjugated polymers vary from about 10^{16} -10^{17} cm^{-3}. Estimates of carrier mobilities using these data and conductivity values obtained from resistance measurements on undoped conjugated polymers range from about 10^{-4} to 10^{-5} cm^2/V s. These rather low mobilities indicate that charge is transported via either a hopping or tunneling mechanism involving movement of localized solitons, polarons, or bipolarons between chains and across domain boundaries. The low carrier mobilities of the various conjugated polymers ultimately limit the level of current that flows through the device at high forward biases.

Rectification ratios ($I_{forward}/I_{reverse}$) for Schottky diodes fabricated from conducting polymers cover a wide range with values typically around 500 to 1000. Advances in the processibility of conducting polymers, however, have resulted in dramatic improvements in the overall quality and performance of these diodes. For example, Schottky diodes fabricated with thin polyacetylene films prepared using the precursor processing route exhibit rectification ratios [90] as high as 500,000. MIS and MISFET devices fabricated from polyacetylene prepared in this manner under conditions that rigorously exclude oxygen behave like classical semiconductor devices in terms of their ability to form charge accumulation, inversion, and depletion layers at the polyacetylene-insulator interface. This is amazing considering the fact that it has been established that charge storage in these devices is via solitons and not by the formation of holes or electrons in the band states. Thus, the external electric fields imposed on the junctions of these devices serve to add or remove solitons from the junction interface. The basic physical principle underlying the operation of barrier junctions based on conjugated polymers is therefore directly related to the unique ability of these materials to reversibly store excess charges in localized defect sites such as the soliton, polaron and bipolaron states described earlier.

Although diode equations based on the thermionic emission model are a suitable starting point for the evaluation of Schottky barriers based on conjugated polymers, more often than not they represent only a basic approximation of the experimentally observed behavior of these junctions. A quick check of the diode quality of these junctions as indicated by the deviation of n from its ideal value of one, reveals that many of the conjugated polymer junctions are far from ideal. These equations also predict a very specific temperature dependence for the diode which is very rarely found to exist in devices based on conjugated polymers [74, 78]. Deviations from ideal behavior can result from a number of different sources. For example, many of the rectifying metals used to form junctions with conjugated polymers are highly susceptible to oxide formation meaning that many of these devices are actually metal-insulator-semiconductor junctions rather than pure Schottky barriers. Recombination and trapping sites in the depletion region can also cause the junction to behave in a nonideal fashion. Many of the these sites result from the high sensitivity of conjugated polymers to oxygen and other environmental agents. The observed variations in junction behavior with temperature also suggest that tunneling processes may be equally as important in these devices as the thermionic emission process. These factors coupled with the novel charge transport mechanisms active in conjugated polymers are responsible for the wide variations in behavior found in junctions fabricated from conducting polymers. Such complications also make it very difficult to compare barrier heights predicted from work function differences with those actually obtained from experiments.

4.6 FIELD EFFECT TRANSISTORS BASED ON CONJUGATED POLYMERS

The ability to reversibly vary the conductivity of conjugated polymers by the application of an applied external voltage can also be exploited to construct a number of novel field effect transistors (FET). These devices rely on the ability of a transverse electric field to modulate the flow of carriers between a source and a drain electrode. A typical configuration used to prepare a polymer-based metal-insulator-semiconductor FET or MISFET is illustrated in Figure 16.

In such a device, one is interested in controlling the flow of carriers across a polymer film which forms a continuous channel between the source electrode and the drain electrode. These latter two electrodes usually form ohmic contacts with the polymer and are typically separated from each other by about 10 μm. The junction formed between the polymer and the insulating layer/gate electrode, on the other hand, establishes a depletion layer at the insulator-polymer interface which, in turn, reduces the conductivity of the channel. By adjusting the voltage across the gate electrode, it is possible to either enhance or diminish this depletion layer thereby controlling the flow of carriers between the source and drain. In essence, the carrier concentration at the polymer-insulator interface is controlled by manipulation of the gate voltage.

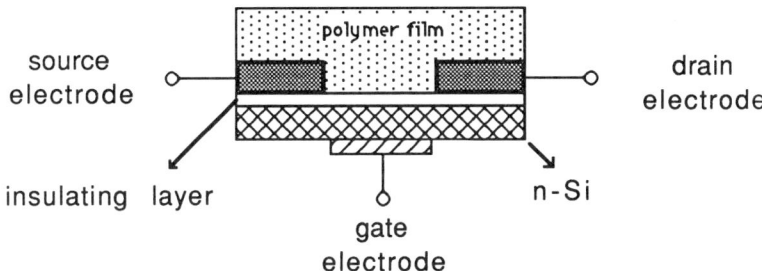

Figure 16. Typical device configuration used to prepare a conjugated polymer based MISFET.

A number of gate insulated MISFETs fabricated by depositing thin films of lightly doped p-type polythiophenes [91, 92], polypyrroles [91], or polyacetylene [90] onto SiO_2/n-Si substrates containing gold source and drain electrodes have been successfully prepared. These devices have been found to operate in the following manner. When the gate voltage is set to a positive bias or when no gate voltage is applied, the amount of current flowing between the source and drain electrodes is extremely small due the presence of a depletion layer at the polymer/insulator interface. The channel current increases significantly, however, when a negative gate voltage is applied (n-Si has a negative bias) and can be modulated by a factor of 1000 to 2000 by simply controlling the magnitude of the gate voltage. This dramatic increase in channel conductivity is believed to be due to the accumulation of positive carriers at the junction of the polymer film and SiO_2 layer that takes place under a negative bias. The accumulation layer then serves as conduction channel in the conjugated polymer. Thus, in this case, the conductivity of the conjugated polymer is varied by an external gate voltage rather than by a doping process as is normally the case.

The device characteristics of these MISETs have been found to be strongly influenced by the thickness of the polymer film and its initial doping level [91]. For the polyheterocycle based devices, very lightly doped thin films have been found to give the best overall performance. In more heavily doped films, control over the channel conductivity by variations in the gate voltage only occurs in films with thicknesses close to the estimated depletion layer thickness (about 1000Å), otherwise the high conductivity of the film away from interface will dominate the flow of current between the source and drain electrodes.

Novel microelectrochemical transistors based on the reversible redox properties of conducting polymers have also been fabricated [93, 94, 95]. By electrochemically growing conjugated polymers such as the polythiophenes, polypyrroles and polyanilines onto a substrate containing a microelectrode array comprised of many closely spaced electrodes (about 1 μm apart) it is possible to form continuous conducting channels between two or more of these electrodes. The electrodes connected by the polymer then serve as source and drain electrodes for a transistor which in this case is operated in a solution containing a suitable electrolyte. If the source electrode is then referenced to the solution as a gate, it is possible to modulate the conductivity of the connecting channel by electrochemically doping or undoping the polymer. With poly(3-methylthiophene) based transistors operated in aqueous $NaClO_4/NaH_2PO_4$ solutions, for example, it has been found that the conductivity of the connecting channel can be varied by over five orders of magnitude by changes in gate voltage [94]. Solid-state devices which operate on the same principle have also been made by utilizing solid polymeric electrolytes such as poly(ethylene oxide)/$LiCF_3SO_3$ instead of the standard electrolyte solutions [95].

5. FUTURE PROSPECTS FOR CONJUGATED POLYMERS IN MOLECULAR ELECTRONICS

The ability to controllably and reversibly add or remove charges from the backbones of conjugated polymers will continue to make these materials attractive in the area of molecular electronics. The near term possibility that electronic devices based on these materials will successfully compete with conventional electronic devices based on inorganic semiconductors such as silicon, however, is rather remote. Currently, the low carrier mobilities of conjugated polymers place severe limitations on the performance of devices fabricated from these materials. Many of the devices fabricated to date, for example, require large operating voltages to obtain acceptable current levels and can only be operated at relatively low frequencies. For those devices based on reversible electrochemical doping schemes, the operating frequency of the device is further limited by diffusion rates of ions into and out of the polymer film. Improvements in these properties can be expected as the purity, level of molecular order, and film quality of the polymers used to build these devices increases. Indeed, dramatic improvements in device characteristics have already been realized through the utilization of higher quality films with decreased film thicknesses.

Although replacement of conventional electronic and microelectronic components with devices based on conducting polymers seems unlikely at this point, these new materials offer unique characteristics that in many cases are simply not found in their inorganic counterparts. The ability to reversibly detect small quantities of various chemical and biological agents by changes in the optical or electrical properties of conjugated polymers, for example, makes these materials very attractive as sensor elements in microelectronic sensor schemes. The large variations in optical properties that occur in these materials upon formation or removal of charged defect states also suggests that devices that exploit the optical behavior of these polymers may be more useful than devices that rely on the transport of charge carriers. In this latter case, the issue of low carrier mobility is largely circumvented. Finally, it should be noted that the neutral forms of these conjugated polymers exhibit extremely large and fast (often times in the subpicosecond regime) optical nonlinearities that could prove useful in optoelectronic and all optical signal processing applications [2]. The future of conducting polymers in the area of molecular electronics clearly depends on the imagination and tenacity of the many researchers throughout the world currently involved in the development of these fascinating materials.

REFERENCES

[1]   H. Shirakawa, E. J. Louis, A. G. MacDiarmid, C. K. Chiang, A. J. Heeger, *J. Chem. Soc. Chem. Commun.* 578 (1977); C. K. Chiang, C. R. Fincher, Y. W. Park, A. J. Heeger, H. Shirakawa, E. J. Louis, S. C. Gau, A. G. MacDiarmid, *Phys. Rev. Lett.* 39, 1098 (1977); A. G. MacDiarmid and A. J. Heeger, *Synth. Met.* 1, 101 (1980).

[2] See, for example, A. J. Heeger, D. Moses, M. Sinclair, *Synth. Met.* 15, 95 (1986) and references therein.

[3] T. A. Skotheim, Ed., 'Handbook of Conducting Polymers', Vol. 1 & 2, Marcel Dekker, New York (1986).

[4] H. Gibson and J. Pochan; J. Frommer and R. R. Chance, In 'Electrical and Electronic Properties of Polymers: A State-of-the-Art Compendium', J. I. Kroschwitz, Ed., J. Wiley, New York, p.1-44 and p. 56-101 (1988).

[5] N. C. Billingham and P. D. Calvert, *Adv. Polym. Sci.*, 90, 1 (1989).

[6] J. Ferraro and J. Williams, 'Introduction to Synthetic Electrical Conductors', Academic Press, Inc., Orlando, Florida (1987).

[7] G. Baker, In 'Electronic and Photonic Applications of Polymers', M. Bowden, and S. R. Turner, Ed., American Chemical Society, Washington, DC, 271 (1988).

[8] J. C. W. Chien, 'Polyacetylene- Chemistry Physics and Material Science', Academic Press, New York (1984).

[9] R. L. Greene and G. B. Street, *Science*, 226, 651 (1984).

[10] J. Frommer, *Acc. Chem. Res.*, 19, 2 (1986).

[11] J. Reynolds, *J. Mol. Electron.* 2, 1 (1986).

[12] B. D. Malhotra, N. Kumar, S. Chandra, *Prog. Polym. Sci.*, 12, 179 (1986).

[13] J. L. Bredas and G. B. Street, *Acc. Chem. Res.*, 18, 309 (1985).

[14] R. H. Friend, *J. Mol. Electron.* 4, 37 (1988).

[15] R. H. Baughman and R. R. Chance, *J. Polym. Sci, Polym. Phys. Ed.*, 14, 2037 (1976); K. C. Schweizer, *J. Chem. Phys.*, 85, 4181 (1986).

[16] R. Peierls, 'Quantum Theory of Solids', Oxford University Press, Oxford (1955).

[17] H. Naarmann, N. Theophilou, *Synth. Met.* 22, 1 (1987).

[18] P. Bernier, In 'Handbook of Conducting Polymers', Vol. 2, p. 1099, T. A. Skotheim, Ed., Marcel Dekker, New York (1986) and references therein.

[19]   A. J. Heeger, In 'Handbook of Conducting Polymers', Vol. 2, p. 729, T. A. Skotheim, Ed., Marcel Dekker, New York (1986) and references therein.

[20]   W. P. Su, In 'Handbook of Conducting Polymers', Vol. 2, p. 757, T. A. Skotheim, Ed., Marcel Dekker, New York (1986) and references therein.

[21]   E. K. Sichel, M. F. Rubner, S. K. Tripathy, *Phys. Rev. B*, 26, 6719 (1982) and references therein.

[22]   A. Epstein, In 'Handbook of Conducting Polymers', Vol. 2, p. 1041, T. A. Skotheim, Ed., Marcel Dekker, New York (1986) and references therein.

[23]   S. Kivelson, *Phys. Rev. B*, 25, 3798 (1982).

[24]   D. Emin, In 'Handbook of Conducting Polymers', Vol. 2, p. 915, T. A. Skotheim, Ed., Marcel Dekker, New York (1986) and references therein.

[25]   R. R., Chance, J. L. Bredas, R. Silbey, *Phys. Rev. B*, 29, 4491 (1984).

[26]   N. F. Mott and E. A. Davis, 'Electronic Processes in Non-Crystalline Materials', 2nd ed., Clarendon, Oxford, (1979).

[27]   P. Sheng, B. Abeles, Y. Arie, *Phys. Rev. Lett.*, 31, 44 (1973).

[28]   M. A. Druy, M. F. Rubner, S. P. Walsh, *Synth. Met.* 13, 207 (1986).

[29]   R. L. Elsenbaumer, K.-Y. Jen, H. Eckhardt, L. Schacklette, R. Jow, In 'Electronic Properties of Conjugated Polymers", p. 400, H. Kuzmany, M. Mehring, S. Roth, Ed., Springer-Verlag, Berlin (1987).

[30]   G. Gustafsson, O. Inganas, J. O. Nilsson, B. Liedberg, *Synth. Met.*, 26, 297 (1988).

[31]   A. Diaz and J. Bargon, In 'Handbook of Conducting Polymers', Vol. 1, p. 81, T. A. Skotheim, Ed., Marcel Dekker, New York (1986) and references therein.

[32]   J. H. Edwards and W. J. Feast, *Polymer*, 21, 595 (1980); J. H. Edwards and W. J. Feast, D. C. Bott, *Polymer*, 25, 395 (1984).

[33]   I. Murase, T. Ohnishi, T. Noguchi, M. Hirooka, *Polym. Commun.*, 25, 327 (1984).

[34] F. E. Karasz, J. D. Capistran, D. R. Gagnon, R. W. Lenz, *Mol. Cryst. Liq. Cryst.* 118, 327 (1985).

[35] K. Y. Jen, M. Maxfield, L. Schacklette, R. Elsenbaumer, *J. Chem. Soc, Chem. Commun.*, 309 (1987).

[36] S. Yamada, S. Tokito, T. Tsutsui, S. Saito, *J. Chem. Soc, Chem. Commun.*, 1448 (1987).

[37] K. Y. Jen, L. W. Schacklette, R. Elsenbaumer, *Synth. Met.*, 22, 179 (1987).

[38] I. Murase, T. Ohnishi, T. Noguchi, M. Hirooka, *Polym. Commun.*, 26, 362 (1985).

[39] S. Antoun, D. R. Gagnon, F. E. Karasz, R. W. Lenz, *Polym Bull.* 15, 181 (1986).

[40] D. G. H. Ballard, A. Courtis, I. M. Shirley, S. C. Taylor, *J. Chem. Soc, Chem. Commun.*, 954 (1983).

[41] K. Y. Jen, T. R. Jow, R. Elsenbaumer, *J. Chem. Soc, Chem. Commun.*, 1113 (1987).

[42] W. Deits, P. Cukor, M. F. Rubner, H. Jopson, *Ind. Eng. Chem. Prod. Res. Dev.*, 20, 696 (1981); W. Deits, P. Cukor, M. F. Rubner, H. Jopson, *Synth. Met.*, 4, 199 (1982).

[43] J. C. W. Chien, G. Wnek, F. E. Karasz, J. A. Hirsch, *Macromolecules*, 14, 479 (1981).

[44] K. Yatsumi, S. Nakajima, M. Fujii, R.Sugimoto, *Polym. Commun.*, 28, 309 (1987).

[45] M. Sato, S. Tanaka, K. Kaeriyama, *J. Chem. Soc, Chem. Commun.*, 873 (1986).

[46] S. Hotta, S. D. D. V. Rughooputh, A. J. Heeger, F. Wudl, *Macromolecules*, 20, 212 (1987).

[47] R. L. Elsenbaumer, K. Y. Jen, R. Oboodi, *Synth. Met.* 15, 169 (1986).

[48] A. O. Patil, Y. Ikenoue, N. Basescu, N. Colaneri, J. Chen, F. Wudl, A. J. Heeger, *Synth. Met.* 20, 151 (1987); A. O. Patil, Y. Ikenoue, F. Wudl, A. J. Heeger, *J. Am. Chem. Soc.* 109, 1858 (1987).

[49] N. S. Sundaresan, S. Basak, M. Pomerantz, J. R. Reynolds, *J. Chem. Soc, Chem. Commun.*, 621 (1987).

[50]   M. R. Bryce, A. Chissel, P. Kathirgamanathaa, D. Parker, N. Smith, *J. Chem. Soc, Chem. Commun.*, 466 (1987).

[51]   G. E. Wnek, In 'Handbook of Conducting Polymers', Vol. 1, p. 205, T. A. Skotheim, Ed., Marcel Dekker, New York (1986) and references therein.

[52]   M. E. Galvin and G. E. Wnek, *Polymer*, 23, 795, (1982); M. E. Galvin and G. E. Wnek, *J. Polym. Sci Polym. Chem .Ed.*, 21, 2797, (1983).

[53]   M. F. Rubner, S. K. Tripathy, S. K. Georger, P. Cholewa, *Macromolecules*, 16, 870 (1983); E. K. Sichel and M. F. Rubner, *J. Polym. Sci Polym. Phys..Ed.*, 23, 1629, (1985).

[54]   K. I. Lee and H. Jopson, *Polym. Bull.* 10, 105 (1983).; K. I. Lee and H. Jopson, *Makromol. Chem. Rapid Commun.*, 4, 375 (1983).

[55]   A. Pron, M. Zagorska, W. Fabianowski, J. B. Raynor, S. Lefrant, *Polym. Commun.* 28 193 (1987).

[56]   S. E. Lindsey and B. Street, *Synth. Met.* , 10, 67 (1985).

[57]   O. Niwa and T. Tamamura, *J. Chem. Soc, Chem. Commun.*, 817 (1984).; O. Niwa and T. Tamamura, *Synth. Met.* 20 235 (1987).

[58]   M.-A. De Paoli, R. J. Waltman, A. F. Diaz, J. Bargon, *J. Chem. Soc, Chem. Commun.*, 1015 (1984).

[59]   J. Roncali and F. Garnier, *J. Phys. Chem.* , 92, 833 (1988).

[60]   G. G. Roberts, In 'Electronic and Photonic Applications of Polymers', M. Bowden, and S. R. Turner, Ed., American Chemical Society, Washington, DC, p. 225 (1988).

[61]   K. Hong and M. F. Rubner, *Thin Solid Films*, 160, 187 (1988).

[62]   X. Q. Yang, J. Chen, P. D. Hale, T. Inagaki, T. Skotheim, Y. Okamoto, L. Samuelson, S. Tripathy, K. Hong, M. F. Rubner, M. L. denBoer, *Synth. Met.* 28, C251 (1989); T. Skotheim, X. Q. Yang, J. Chen, P. D. Hale, T. Inagaki, L. Samuelson, S. Tripathy, I. Watanabe, K. Hong, M. F. Rubner, M. L. denBoer, Y. Okamoto, *Synth. Met.* 28, C229 (1989).

[63]   A. K. M. Rahman, L. Samuelson, D. Minehan, S. Clough, S. K. Tripathy, T. Inagaki, X. Q. Yang, T. Skotheim, Y. Okamoto, *Synth. Met.* 28, C237 (1989).

114

[64] T. Iyoda, M. Ando, T Kaneko, A. Ohtani, T. Shimidzu, K. Honda, *Tetrahedron Lett.*, 27, 5633 (1986); T. Shimidzu, T. Iyoda, M. Ando, A. Ohtani, T Kaneko, K. Honda, *Thin Solid Films*, 160, 67 (1988).

[65] I. Watanabe, K. Hong, M. F. Rubner, *J. Chem. Soc, Chem. Commun.*, 123 (1989); I. Watanabe, K. Hong, M. F. Rubner, I. H. Loh, *Synth. Met.* 28, C473 (1989).

[66] P. B. Logsdon, J. Pfleger, P. Prasad, *Synth. Met.* 26, 369 (1988).

[67] A. Saxena and J. D. Gunton, *Synth. Met.* 20, 185 (1987).

[68] R. S. Potember, R. C. Hoffmann, H. S. HU, J. E. Cocchiaro, C. V. Viands, R. A.Murphy, T. O. Poehler, *Polymer*, 28, 574 (1987)

[69] M. Gazard, In 'Handbook of Conducting Polymers', Vol. 1, p. 673, T. A. Skotheim, Ed., Marcel Dekker, New York (1986) and references therein.

[70] A. G. MacDiarmid and R. B. Kaner, In 'Handbook of Conducting Polymers', Vol. 1, p. 689, T. A. Skotheim, Ed., Marcel Dekker, New York (1986) and references therein.

[71] R. L. Blankespoor and L. L. Miller, *J. Chem. Soc, Chem. Commun.*, 90 (1985).

[72] S. M. Sze, 'Physics of Semiconductor Devices', Wiley-Interscience, New York (1969).

[73] L. Solymar and D. Walsh, 'Lectures on the Electrical Properties of Materials', Oxford University Press, Oxford (1984).

[74] J. Kanicki, In 'Handbook of Conducting Polymers', Vol. 1, p. 543, T. A. Skotheim, Ed., Marcel Dekker, New York (1986) and references therein.

[75] H. Towozawa, D. Braun, S. Phillips, A. J. Heeger, H. Kroemer, *Synth. Met.* 22, 63 (1987).

[76] J. R. Waldrop, M. Cohen, A. J. Heeger, A. J. MacDiarmid, *Appl. Phys. Lett.* 38, 53 (1981).

[77] M. Ozaki, D. L. Peebles, B. R. Weinberger, C. K. Chiang, S. C. Gau, A. J. Heeger, A. J. MacDiarmid, *Appl. Phys. Lett.* 35, 83 (1979).

[78] H. Tomozawa, D. Braun, S. D. Phillips, R. Worland, A. J. Heeger, H. Kroemer, *Synth. Met.* 28, C687 (1989).

[79] S. Glenis and A. J. Frank, *Synth. Met.* <u>28</u>, C681 (1989).
[80] S. Miyauchi, A. Fueki, Y. Sorimachi, I. Tsubata, *Synth. Met.* <u>28</u>, C691 (1989).
[81] A. El Hadri, C. Maleysson, H. Robert, *Synth. Met.* <u>28</u>, C697 (1989).
[82] S. Miyauchi, Y. Kaneko, Y. Sorimachi, I. Tsubata, *Synth. Met.* <u>28</u>, C747 (1989).
[83] F. Garnier, G. Horowitz, D. Fichou, *Synth. Met.* <u>28</u>, C705 (1989); D. Fichou, G. Horowitz, Y, Nishikitani, F. Garnier, *Synth. Met.* <u>28</u>, C723 (1989); D. Fichou, G. Horowitz, Y, Nishikitani, J. Roncali, F. Garnier, *Synth. Met.* <u>28</u>, C729 (1989).
[84] B. R. Weinberger, S. C. Gau, Z. Kiss, *Appl. Phys. Lett.* <u>38</u>, 555 (1981).
[85] J. Tsukamoto and H. Ohigashi, K. Matsumura, A. Takahashi, *Synth. Met.* <u>4</u>, 177 (1982).
[86] B. R. Weinberger, A. Akhtar, S. C. Gau, *Synth. Met.* <u>4</u>, 187 (1982).
[87] S. Glenis, G. Tourillon, F. Garnier, *Thin Solid Films*, <u>139</u>, 221 (1986); S. Glenis, G. Horowitz, G. Tourillon, F. Garnier, *Thin Solid Films*, <u>111</u>, 93 (1984).
[88] S. Glenis and G. Horowitz, In 'Electronic Properties of Conjugated Polymers", p. 423, H. Kuzmany, M. Mehring, S. Roth, Ed., Springer-Verlag, Berlin (1987).
[89] M.-S. Yun and C.-S. Huh, *Synth. Met.* <u>28</u>, C715 (1989).
[90] J. H. Burroughes, C. A. Jones, R. H. Friend, *Nature*, <u>335</u>, 137 (1988); J. H. Burroughes, C. A. Jones, R. H. Friend, *Synth. Met.* <u>28</u>, C735 (1989).
[91] A. Tsumura, H. Koezuka, T. Ando, *Synth. Met.* <u>25</u>, 11 (1988); H. Koezuka and A. Tsumura, *Synth. Met.* <u>28</u>, C753 (1989); A. Tsumura, H. Koezuka, T. Ando, *Appl. Phys. Lett.* <u>49</u>, 1210 (1986).
[92] A. Assadi, C. Svensson, M. Willander, O. Inganas, *Synth. Met.* <u>28</u>, C863 (1989).
[93] G. P. Kittlesen, H. S. White, M. S. Wrighton, *J. Am. Chem. Soc.* <u>106</u>, 7389 (1984); G. P. Kittlesen, H. S. White, M. S. Wrighton,

J. Am. Chem. Soc. 107, 7373 (1985); H. S. White, G. P. Kittlesen, M. S. Wrighton, *J. Am. Chem. Soc.* 106, 5375 (1984).

[94] J. W. Thackeray, H. S. White, M. S. Wrighton, *J. Phys. Chem.*, 89, 5133 (1985).

[95] S. Chao and M. S. Wrighton, *J. Am. Chem. Soc.*, 109, 2197 (1987).

CHAPTER 3
Langmuir-Blodgett Films
I. R. Peterson

1 INTRODUCTION
1.1 Relationship to Molecular Electronics

In the most general sense, 'Molecular Electronics' covers the use of molecular (and hence essentially organic) materials to perform a signal processing or transformation function. In many cases, such as for example liquid crystal displays and conducting polymers, the desired physical effects occur without further processing in the material synthesised by the organic chemist, as the detailed molecular arrangement either arises spontaneously or is not important. In these cases the construction of a device requires only conventional workshop techniques. There exist other types of material and phenomenon in which the detailed molecular arrangement is essential, and must be achieved using special fabrication methods. For example, for the frequency doubling of light from a pulsed laser in an organic substance, the optical scattering must be low and the molecules must be anisotropically oriented over volumes measured in cm^3. These conditions are best achieved by growing a crystal. The Langmuir–Blodgett (LB) technique is a comparable high–tech organic fabrication method, appropriate when the implementation of the function requires a high degree of molecular anisotropy in an extremely thin layer of uniform thickness.

There is a narrower sense of Molecular Electronics covering signal-processing systems in which individual molecules, or small groups of molecules in specific configurations, operate independently to give an extremely high functional density. The LB technique is important to Molecular Electronics in this sense because it provides one of the few ways of making separate electrical connection to two ends of a molecule. It is hard to conceive how the functional units might otherwise receive a constant power supply at a voltage much higher than kT/e, important for interfacing and for reliable computation, without dielectric breakdown occurring.

1.2 Elements of the LB Technique

The fabrication of an LB film involves a number of different processes, which were discovered over a period of almost fifty years. The first stage is the formation of a monomolecular layer, or monolayer, on a water surface. In some cases, such as the vegetable oils, a monolayer spreads spontaneously over any air-water interface with which the material is brought into contact. While the resulting damping of capillary ripples was known in antiquity [1,2], the phenomenon was first studied scientifically by Pockels [3,4], and then Rayleigh [5], who showed that the monolayer could be detected by the resulting reduction in surface tension, and could be manipulated by changing the area of the water surface. At maximum reduction of surface tension the area occupied per molecule is scarcely greater than its Van der Waals cross-section, so that the molecules are dense-packed and essentially perpendicular to the interface, as shown in Fig. 1a.

The spontaneous spreading initially studied by Pockels and Rayleigh is rather slow for many substances. In later work Pockels discovered a much faster and more convenient method, further developed by Devaux, Langmuir and Adam, where the substance is first dissolved in a volatile organic solvent [6]. The solution spreads rapidly over the water surface, and afterwards the solvent evaporates, leaving a pure monolayer. This method is now universally employed to prepare monolayers for the LB technique.

Pure water is not the only liquid on which monolayers can be prepared for the LB technique. Very often, specific ions are added to it to modify the monolayer properties. Moreover other fluids with high surface tension, such as ethylene glycol, glycerol and mercury [7,8], can be used. To include the possibility of materials other than water, while implying the combination of properties required, the liquid used is often referred to as the subphase.

The second basic step of the technique is to deposit the floating monolayer onto a solid substrate. This was a serendipitous discovery by Langmuir, the inventor of the tungsten-filament incandescent lamp, whose main interest at the time was trying to understand the behaviour of the oxygen monolayers influencing the emission of electrons from hot tungsten surfaces. In 1919 he discovered that by slowly withdrawing a glass slide from a monolayer-covered

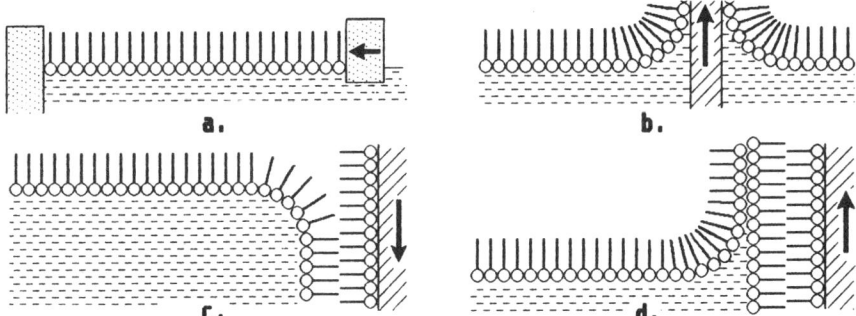

Fig. 1. Steps in the Langmuir-Blodgett technique.
a) Compression of the floating monolayer b) Transfer of first monolayer c) Subsequent downstroke d) Subsequent upstroke.

water surface as shown in Fig. 1b, the slide emerged dry and coated by the monolayer [9] .

In the early 1930's he resurrected this finding as a possible way of investigating lubrication. His assistant, Blodgett, then found that more than one monolayer could be deposited on the substrate, by repeatedly immersing and withdrawing it through the water-surface monolayer as in Figs 1c and 1d [10]. Under most conditions one monolayer was deposited on the downstroke and one on the upstroke, and the resulting film thickness was to very high accuracy equal to the total number of monolayers deposited multiplied by the thickness of a single monolayer [11]. Hence here was a way of depositing organic thin films of arbitrary but extremely precisely controlled thickness.

1.3 Ideas for Applications

Langmuir and Blodgett published many papers investigating the range of applicability of the technique for different materials and deposition conditions [12-19]. They also proposed that the resulting films could be used as accurate submicron thickness references [20], or as antireflection coatings for optical components [19,21]. However, probably because of the mechanical fragility of the films, these applications did not prove to be commercially viable.

During the following thirty years, the films remained as scientific curiosities, inspiring a few scientific studies of their structure [22-24] and physical properties [25-27], but much fewer than those concerning the water-surface monolayers from which they are made, with their wide economic relevance for detergency, lubrication, mining and the food industry. Indeed, even Gaines' 1966 book [28], which is still recognised as *the* reference work on LB films, devotes only one out of nine chapters to the deposited films.

The 'modern' period of LB film research dates from the early '70s with the work of Kuhn and his group. It is probably no coincidence that this was also the beginning of the microelectronics revolution, when the technological potential of ultraminiaturised fabrication techniques in general, and thin film deposition techniques in particular, was first appreciated. Clearly the ultimate miniaturised structures are molecules, and the existence of biological systems proves that ordered organic molecular structures can perform useful and coordinated functions. In a series of imaginative experiments, Kuhn and coworkers demonstrated that, under suitable conditions, the LB deposition process conformed to the simple picture of Fig. 1, and that it was capable of assembling molecules into a stable, well-defined functional structure. Firstly, they showed that individual monolayers could be prepared with extremely high uniformity [29]. Secondly, they showed that film molecules deposited on distinct immersion-withdrawal strokes could retain the relative position suggested by Fig. 1 over long periods of time [30-32], and even when the films are subjected to considerable perturbation by cleavage and reassembly [33,34]. Thirdly, they demonstrated that the process of exciton transfer between two different molecular species could be 'fine-tuned' by adjusting the distance between them to subnanometre precision. This work appears to have been the first concrete research towards molecular-scale functional systems, which have since been repeatedly proposed [35-37].

While progress in demonstrating a useful functional structure of this sort has been slow, many schemes have now been proposed whereby LB films can be used in conjunction with more conventional microfabrication techniques to extend the capabilities of microelectronic devices, as constituents of chemical sensors and transducers, and for optical applications. In some of these proposals the film is merely used as a spacer layer of extremely uniform thickness, for example as the gate insulator in a field effect transistor [38] or a microphone [39], or as a hyperfiltration layer [40]. In others it is used as a passive support to anchor active molecules, for example enzymes in biosensors. There are three categories where the film molecules themselves perform an active function: for example chemical, in microlithographic resists [41,42]; electronic, in high-density bulk memories [43]; or optical, in switches, modulators and nonlinear signal processing [44]. These possible future applications of the technique will be discussed in greater detail in Section 5.

1.4 The Energetics of Monolayer Formation
The only forces acting between the molecules in the monolayer and the underlying aqueous subphase are just the usual electrostatic interactions between their constituent nuclei and electrons, but their action is made somewhat less intuitively obvious by nonclassical quantum behaviour, especially of the electrons. For example in classical electrostatics, like charges repel one another. However when interactions between large numbers of organic molecules are concerned, it is possible for like materials to attract one another, and unlike to repel. In organic chemistry it is common to classify substances, especially solvents, as being 'hydrophilic'(water-loving) or 'hydrophobic' (water-fearing). Materials in the same category tend to be miscible with each other, and immiscible with members of the other category: a well-known example is the fact that water and oil do not mix. This classification has considerable explanatory power.

A physical explanation of the difference between hydrophilic and hydrophobic materials was first attempted by Hildebrand [45,46], who showed that hydrophilic materials were characterised by a high cohesive energy per unit volume, and hydrophobic materials by a small value. The square root of this value, called the solubility parameter, has proved to be a very useful concept for polymer characterisation. The solubility of a linear polymer in a solvent, or the tendency of a crosslinked polymer to swell, is very highly correlated with the difference between the solubility parameters of polymer and solvent [47]. Table 1 gives a list of solubility parameters for some common solvents and polymers.

HYDROPHOBIC			
n-Pentane	14.5	PTFE	12.7
n-Hexane	14.9	Poly(C_2ClF_3)	14.7-16.2
Cyclohexane	16.8	Polyethylene	15.8-17.1
Benzene	18.8	Polyisobutylene	16.0-16.6
Chloroform	19.0	Polybutadiene	16.6-17.6
Chlorobenzene	19.6	Polypropylene	16.8-18.8
Acetone	19.6	Polystyrene	17.4-19.0
1,2-Dichloroethane	19.8	Polyvinyl chloride	19.2-22.1
Nitrobenzene	22.7	Polymethyl methacrylate	18.6-26.2
2-Propanol	23.6	Polyethylene terephthalate	19.9-21.9
DMF	24.9	Polyaminocaprylic acid	26.0-
Methanol	29.5	Polycyanomethylacrylate	28.7-29.7
Ethylene glycol	34.8		
Formamide	39.3		
Water	47.9		
HYDROPHILIC			

Table 1. Indicative list of solubility parameters in $MPa^{1/2}$

Like many such concepts, solubility parameter theory is only approximate, and fails badly in some cases. However the anomalies can be explained in terms of the underlying molecular mechanisms which determine cohesive energy density. In organic materials, the major contributions come from Van der Waals dispersion, hydrogen bonding and dipole-dipole interactions. In one elaboration of the theory, the solubility parameter still exists, but is a vector with three components whose length is equal to the Hildebrand parameter [46]. The vector can point in completely different directions for different classes of material, and this can explain, for example, why the fluorocarbon solvents are immiscible with hydrocarbons, although the difference in solubility parameters is much smaller than that between, say, water and 2-propanol, which are miscible. However, this elaboration only serves to delimit the area of validity of the hydrophilic-hydrophobic concept, which holds for most materials encountered in the LB context.

The miscibility of two materials is very closely related to the surface tension, or energy per unit area, of their interface [48]. If this interfacial tension is negative, then there will be a tendency for the interfacial area to increase without limit until the two materials are completely mixed. Hence the adjectives hydrophilic and hydrophobic can also be applied to solid surfaces, indicating the readiness with which they are wetted by water, even though in this case subsequent mixing does not occur. Since materials of vastly different hydrophilicity have no tendency to mix, it follows that their interface has high energy.

The molecules of Fig. 1. have been drawn as 'lollipops', following a long-standing convention [49], which reflects the actual structure of many LB film-forming molecules. The 'lolly' end is a hydrophilic chemical group, typically with a strong dipole moment and capable of hydrogen bonding, like -OH, -COOH and -NH$_2$. The 'stick' end on the other hand is hydrophobic, typically consisting of a long aliphatic chain.

Such molecules, containing spatially separated hydrophilic and hydrophobic regions, are called amphiphiles. If such a molecule arrives at the air-water interface with its hydrophobic tail pointing towards the air and its hydrophilic group still in the water, the initial high-energy interface is replaced by lower energy hydrophilic-hydrophilic and hydrophobic-hydrophobic interfaces, thus lowering the total system energy. Hence the molecules at the interface are anchored, strongly oriented, and with no tendency to form a layer more than one molecule thick.

The lollipop picture is valid for a large class of LB film-forming materials, but not all. Recently, there has been considerable technological interest in films of amphiphilic polymers, whose molecules possess many hydrophilic groups and many hydrophobic tails. Because of their high molecular weight, films of these materials tend to be more stable mechanically and thermally than those of lollipop molecules. Moreover because of the dispersion of molecular weight and the many low-energy modes of deformation possible, films of polymers do not usually display any macroscopic long-range order, and this is considered to be a theoretical advantage by many researchers, although no clear-cut experimental evidence to this effect is yet available. These benefits are accompanied by very high monolayer viscosity, and maximum deposition speeds some three orders of magnitude slower than those possible with low molecular weight materials.

The first polymeric materials to be considered seriously for LB films were those formed by the polymerisation of aliphatic-chain molecules with activated vinyl groups [50]. However the repeat distance along the backbone of these materials is 0.254 nm, significantly smaller than the roughly 0.49 nm spacing between close-packed aliphatic chains, so that the configuration of such molecules as a planar, oriented monolayer is not stable. More recent work has involved alternating copolymers of a number of kinds [51–53] in which the sidechains are less densely packed along the backbone.

While polymeric amphiphiles may be considered to be nonclassical film materials, at least they are still amphiphiles. There is a further class of material which can be fabricated into films by the LB technique and which has attracted a lot of recent attention. These are the lightly-substituted aromatics, and although they possess short aliphatic side chains, they display no clearly-recognisable hydrophilic group. Of course, to assemble a layer of molecules at the water surface, distinct hydrophilic and hydrophobic moieties are not essential: these are just required for orientation and to prevent the formation of local regions in which the layer is two or more molecules thick. For some purposes, it is adequate if the resulting film is merely of some defined average thickness, rather than molecularly smooth. In these cases the molecule need only show a tendency to stabilise the water–air interface. Such materials are called surfactants. Most aromatic molecules are surfactants, including benzene, which can be seen from table 1 to be quite hydrophobic. The advantages of these materials include good thermal and mechanical stability, interesting electronic structure, strong optical effects and significant mobility of charge carriers and excitations.

1.5 Affiliated Topics

The basic amphiphilic molecular structure is found in a number of other areas of science. A detergent, for example, is also required to stabilize water interfaces, although in this case the second phase can be a variety of oily materials. Moreover the detergent is required to diffuse spontaneously to these interfaces and therefore must be appreciably soluble in water. It follows that detergent molecules are not captive at the interface, as is essential for an LB film material, but reach an equilibrium with material in solution. However, the relationship is close: a given detergent molecule can usually be converted into one suitable for LB film formation by increasing the length of the aliphatic chain and thus decreasing its solubility.

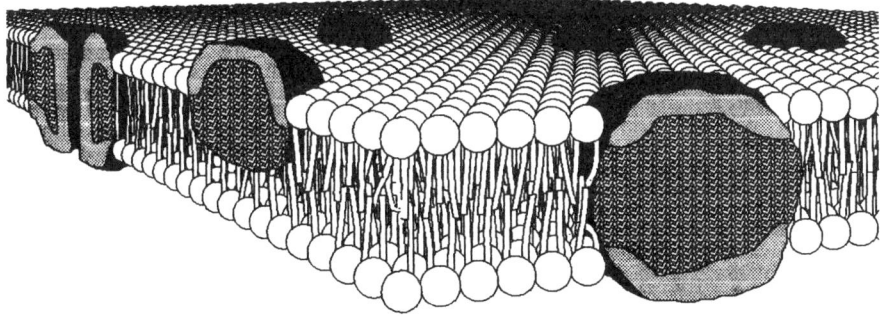

Figure 2. The Singer-Nicholson Model of the Cell Membrane Protein interior textures: ▓ , hydrophobic; ▒ , hydrophilic.

Another area in which amphiphilic molecules are routinely encountered is biochemistry. In living cells, the catalytic enzyme molecules and products of cell metabolism are in many cases not uniformly distributed within the cell, but are anchored to membranes. The essential membrane organisation is determined as shown in Fig. 2 by hydrophilic–hydrophobic interactions within a bilayer membrane of the specialised amphiphilic molecules called lipids. Like the lollipop molecules of Fig. 1 these have a hydrophilic headgroup, which is typically rather complicated and made up from glycerol, phosphate and sometimes sugar residues, but unlike the lollipop design they normally possess two aliphatic chains. LB films can be built up from biological lipids, and have been used in a number of fundamental investigations into the physical principles of biological systems.

The lipids are not the only amphiphiles found in living cells. The protein molecules shown bound to the cell membrane in Fig. 2 are attached to it by virtue of hydrophilic-hydrophobic interactions. These molecules depart radically from the lollipop design, but are nevertheless also capable of forming monolayers and LB films as was demonstrated by Langmuir and Schaeffer very shortly after the discovery of the LB technique [14,15].

Considering the possible significance of organic films of molecular thickness, it is not surprising that a number of attempts have been made to fabricate them in ways other than the LB technique. One increasingly successful approach is that of Sagiv, who employs techniques of chemisorption to attach a monolayer to a surface. In order to ensure that only a single layer of molecules is adsorbed, the top surface of the fresh monolayer must be unreactive to the chemisorbing species. However by a sequence of chemical reactions the top surface of the fresh monolayer is then converted to a reactive state, so that subsequent monolayers can be assembled on top by repeating the basic steps [54]. The resulting structure resembles an LB film in its layer structure, but the successive layers are covalently bonded, and this is claimed to result in a much greater thermal and mechanical robustness. In initial attempts, the number of molecules incorporated in each new monolayer decreased progressively, and the total film thickness stopped increasing after reaching approximately 10 nm. This appeared to be associated with increasing film disorder [55]. However more recently using purer reagents [56] and with alternative reaction and bonding schemes [57,58], it is claimed that these problems have been overcome and that uniform films of over 100 nm thickness are readily achievable.

2. EQUIPMENT

2.1 Mechanical Requirements

The basic equipment for fabricating LB films, called a preparative film balance or LB trough, consists essentially of a liquid surface of controllable area, together with mechanisms for moving a substrate through it and controlling its surface tension. The mechanical constraints on the equipment are very lax, and almost any arrangement providing for these two actions will produce films. On the other hand, the trough must be constructed so as to facilitate cleaning and reduce to a minimum the possibility of accidental contamination of the monolayer.

There exist two major schemes for obtaining a variable subphase surface. In the first scheme, part of the water surface perimeter is fixed and usually formed by the edges of the water container, or trough proper, while another part of the perimeter (the barrier) slides relative to it. These are called rigid-barrier troughs. In the second scheme, none of the perimeter is fixed, and there is no sliding motion.

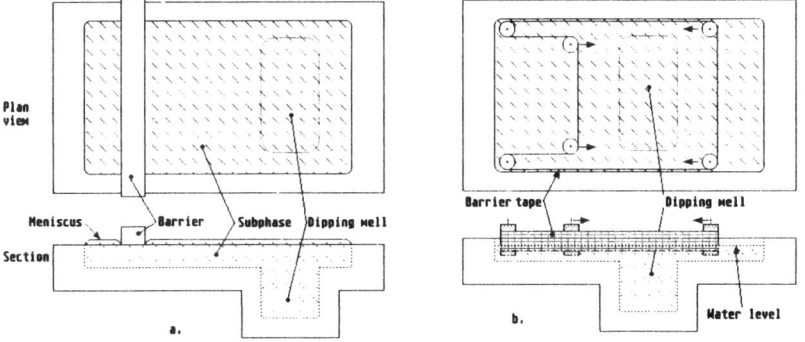

Fig. 3. Possible implementations of (a) rigid-barrier and (b) constant-perimeter troughs.

Instead, the barrier is hinged or flexed while its total length remains constant. These are called constant perimeter troughs. Each trough type has advantages and disadvantages, and commercial troughs of both types are available.

In the most common implementation of a rigid barrier trough, the barrier is a flat bar sitting horizontally on the trough edges, and only the water surface on one side of the barrier is used, as shown in Fig 3a. There must be no leakage of monolayer under the barrier, so the water level has to be higher than its underside, ie, above the edges of the trough. This is achieved by making the trough edges hydrophobic . Fig 4 shows a commercially-available trough of this type.

In the most common implementation of a constant perimeter trough, the barrier is a flexible tape mounted on a system of rollers as shown in Fig. 3b [59]. It is also possible to use a hinged diamond shape [60]. There is no direct requirement for surface hydrophobicity of either trough or barrier.

Much more important than any purely mechanical aspect of trough design are the precautions taken to avoid contamination of the monolayer. It should be possible to clean thoroughly all regions of the trough in contact with the subphase. Surface roughness and cracks of the trough walls harbour contaminants and should be avoided. Complicated assemblies forming part of the trough, for example, sensors for surface tension or surface potential, should be capable of disassembly.

2.2 Trough Materials

The materials from which the trough is made may also be prime causes of monolayer contamination. Metals of any kind have long since been excluded from the vicinity of the subphase because of the ionic contamination which they cause [6]. Glass has in the past been a popular choice, especially for constant perimeter troughs. However glass is soluble in water at the 10^{-5} M level and this has been shown to cause significant changes in the properties of some monolayers [61,62]. At the present time, polymers appear to be the most acceptable trough materials, provided that a number of precautions are taken. Those filled with plasticiser are to be avoided, because most plasticisers are surfactants and continue to diffuse out of the material for years. Condensation polymers containing ester or amide linkages should also be avoided because of their surface activity. Most adhesives are troublesome.

Many recent troughs' of both rigid-barrier and constant-perimeter types make extensive use of polytetrafluoroethylene (PTFE) for both the trough and barrier. From table 1 it is

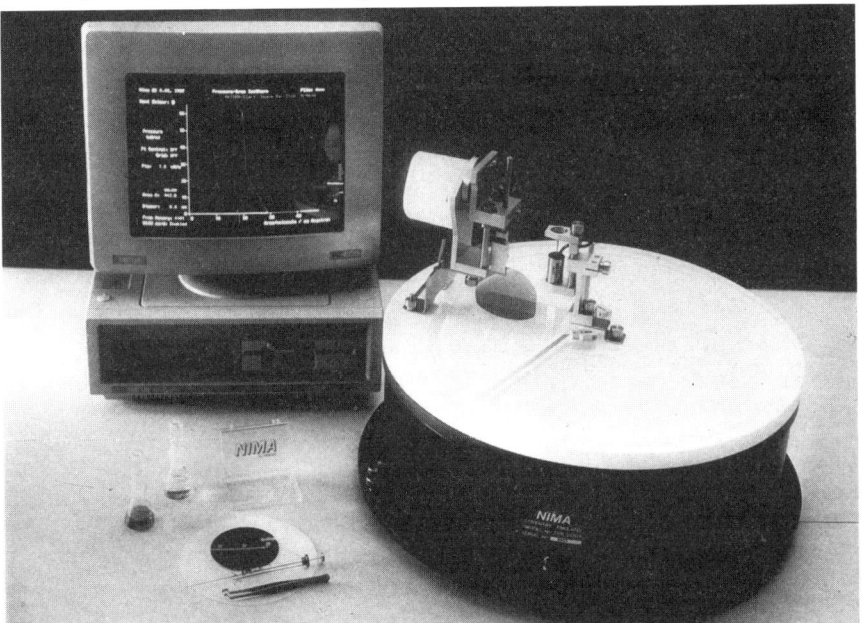

Fig. 4. A commercially available trough for the fabrication of superlattice LB films with alternating layers of two different materials. Courtesy Nima Technology.

seen to be the most hydrophobic polymer known, and hence ideal for use in a rigid-barrier trough. It is chemically essentially inert and can be subjected to some very aggressive cleaning procedures. It is readily available in the form of flexible tape, solid blocks or sheets, and may even be sprayed on to other substrates.

In the trough of Fig. 4, all parts in contact with the subphase are machined out of three pieces of bulk PTFE: one for the barrier, one for the dipping well and one to give the required subphase surface area. The dipping well and subphase area pieces are fastened together with metal screws because of the need to dismantle for cleaning occasionally. The polymer is so hydrophobic, water has little tendency to penetrate joins as long as the mating surfaces are smooth and are kept firmly fastened together.

In spite of its many advantages, PTFE is not free of problems. It is not a thermoplastic and so cannot be moulded. Most bulk forms of the material are produced by sintering the polymer powder, a procedure resulting in many small pores which can harbour contaminants. Since these pores also scatter light, the best grades in this respect are those which are most translucent. One trough manufacturer claims to be able to seal the pores by a simple treatment, but no details of the method have been revealed, and attempts by the author to duplicate it have not been completely satisfactory. It is difficult to glue, although this is not a major drawback in the LB context. PTFE components subjected to static load tend to creep, and should be reinforced mechanically by metal pieces which must of course not come into contact with the subphase: anodised aluminium is adequate. Because of the tendency to creep, a PTFE hole cannot be satisfactorily threaded to mate with a screw, but the same effect can be achieved with a captive nut or tapped insert.

The water used in an LB trough must be of the highest purity, and should ideally be changed frequently. As a consequence, it is a major item of expense in any LB laboratory. In recognition of this fact, most designs aim to minimise the working volume of water, by making the trough very shallow over most of its area. This is not possible at the point where monolayer transfer to the substrate is to take place, because it must be possible almost completely to immerse the largest substrate used. The solution, shown in both implementations of figure 3, is to have a local deepening of the trough, called the dipping well.

2.3 Surface Balances

In order to obtain uniform and repeatable deposition of an LB film, it is necessary to ensure that the monolayer conditions are well defined at the start of deposition, and remain constant as material is transferred from the water surface to the substrate. This is achieved by moving the trough barriers until the correct monolayer density is achieved. There are in principle several monolayer parameters which could be measured for this purpose, for example surface potential, water surface reflectivity, or capillary wave damping. However, in practice, the surface tension is far simpler to measure and control, and gives a more sensitive measure of monolayer compression in the range of interest for LB deposition.

The surface tension of a clean water surface at 20^0 C is 72 mN/m [63] , and this value decreases with increasing temperature and increasing density of surfactant molecules. Because many interesting phenomena occur when the monolayer tension is only slightly smaller than that of a clean surface, and because the absolute value is difficult to measure, it is not convenient to measure or specify it

134

directly. Almost universally, the degree of monolayer compression is specified by giving the difference between the surface tension of the monolayer and that of a clean water surface. This parameter is called the surface pressure. It is always positive, and usually increases with increasing surfactant density.

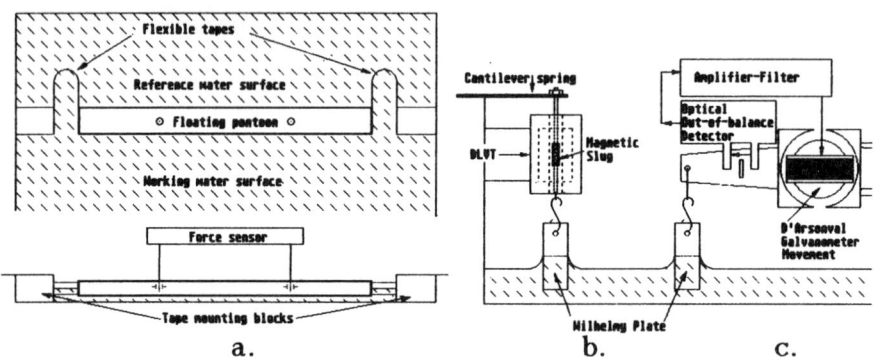

a.   b.   c.

Figure 5. Possible implementations of a) a Langmuir balance and b) and c) a Wilhelmy balance.

There are several different ways of measuring surface pressure. For most LB troughs the choice is constrained by the need for a continuous voltage representing the surface pressure to control the barrier servomechanisms, so that necking-instability methods are inappropriate. The two most popular techniques are those of Langmuir and Wilhelmy, possible implementations of which are shown in figure 5.

In a Langmuir balance [64], a floating pontoon separates a clean water surface from the monolayer, and the resultant horizontal force exerted on it is measured. The pontoon must be free to move, yet monolayer must not be able to flow around to the clean reference side. This is achieved using a floating barrier tape to connect the ends of the pontoon to the trough edges, sufficiently flexible so as not to exert any appreciable torque on the pontoon. In normal operation, the sideways force on the pontoon is equal to the difference between the surface pressures on the two sides, multiplied by the pontoon length plus half the barrier tape length.

It is fairly obvious that any bending moment around a vertical axis exerted on the pontoon by the barrier tape will cause errors in the surface pressure measurement. It is less obvious that a torque around the pontoon axis, exerted by either the barrier tape or force sensor connection, will also cause deviations from the 'correct' horizontal force. In fact the horizontal component of surface tension acting on the pontoon also depends on the angle of the water meniscus at the three-phase contact line , which is in turn determined by the equilibrium contact angle between the water meniscus and pontoon sides. The expected force is only obtained when the pontoon is free to rotate around its long axis [65].

In contrast to a Langmuir balance, a Wilhelmy balance [66] has no reference water surface, and is not connected to the trough edges. Instead, the vertical force on a plate hanging through the monolayer is measured. In this case there is no compensation for variations of the equilibrium contact angle of the water surface with the plate, which should therefore be as close to zero as possible. Many materials with zero contact angle when clean, such as platinum or mica, become coated with monolayer in an LB trough so that the balance sensitivity changes. Filter paper is highly resistant to this effect and has become the most popular choice of plate material.

In both types of surface pressure sensor it is usual to convert the force, either on the Wilhelmy plate or on the floating pontoon, to electrical form. The most widespread method involves applying the force to a spring, and measuring the resulting deformation using an LVDT (linear variable differential transformer) [67]. Alternatively the force can be converted directly to electrical form using an electrobalance [68] or a strain gauge. It is important to achieve an accuracy better than 1 mN/m, although experimental difficulties preclude determining the zero of surface pressure to better than 0.1 mN/m. In order to avoid problems due to building vibrations, mechanical resonances of the sensor should be higher than 10 Hz. The author has found that the electrobalance principle, in which the applied force is balanced against the force produced by a current flowing in a magnetic field, is more generally satisfactory than others involving stress-strain relationships in solids.

It is not always necessary to convert surface pressure into electrical form. When it is required to establish a specific surface pressure it is possible to use a piston oil, usually an unsaturated vegetable oil, which generates its equilibrium spreading pressure when present in excess. The monolayer of piston oil is separated from the monolayer under study by a thin floating thread. If variable surface pressures are required, the piston oil can be replaced by an indicator oil, usually a partially-oxidised paraffin oil, which like a spreading solution spontaneously moves to cover the whole of the available water surface, but unlike a spreading solution has no tendency to evaporate. The surface pressure developed is a function of the layer thickness, which can be deduced from the interference colour of the layer. Finally, a Langmuir balance can be used with a constant-force generator, for example a weight hanging on a thread suspended over a pulley, instead of the normal low-compliance force sensor. In this case the Langmuir balance establishes the required surface pressure rather than measuring it.

All three of these methods have been used in the past to avoid the expense of electronics and servomechanisms.

However the first two introduce unnecessary surface-active contaminants, while in the third, friction in the mechanism introduces large errors of surface tension. Due to the continuous decrease in the price of electronic components, and the increasing possibilities of computer monitoring and data processing, there is no longer any valid argument for avoiding electrical sensing of surface pressure.

2.4 Purity and Cleanliness

There exists a movie made by Langmuir demonstrating many of the techniques of handling monolayers. The action takes place in a studio, and the monolayer is prepared on the surface of tap water contained in a baking dish. When screened before an audience of LB researchers, it inevitably produces reactions of incredulity that the experiment 'worked'. It is in fact possible to produce LB films with very primitive equipment, and with such uncontrolled starting materials as tap water, but very often the films so produced are defective, and any results obtained from them nonreproducible. The quantities of material employed are so tiny that even apparently minor sources of contamination can completely change the chemical composition of the monolayer.

Perhaps the most likely source of contamination in an LB laboratory is the subphase water supply. It is true that it is easily possible to obtain water orders of magnitude purer than any organic material. However, the spread monolayer of thickness ~2 nm typically floats on water of average depth ~20 mm, a ratio of 10^7. Clearly the concentration of critical contaminants in the water must be measured in ppb (parts per 10^9). For comparison, semiconductor-grade silicon, often claimed to be the purest material available, is considered to be lightly doped at an impurity concentration of 10 ppb.

For LB work, there are two categories of critical contaminant:

ions, which can bind electrovalently to or form complexes with the hydrophilic group; and surfactants, which move preferentially to the water surface and can significantly affect the film-forming properties of the monolayer in small concentrations. While water of the required purity has in the past been prepared by distillation, using permanganate to destroy organic material, water purification equipment based on ion-exchange resins and active charcoal to remove organics is now available commercially from several sources, produces water of equal or higher quality, and is much more convenient to use.

Next in order of importance is the purity of solvents used in the laboratory. There are two main uses: for cleaning, and as a solvent for the film material during the spreading process. In the latter application, 1 mg of film material is typically dissolved in 1 g of solvent, a ratio of 10^3, so that levels of critical impurities, here essentially only surfactants, must be measured in ppm (parts per 10^6). While solvents of the required purity could in principle be prepared by distillation, in practice almost all laboratories use them as supplied commercially. It is unusual to have the surfactant concentrations in these specified, and certainly not to the ppm level. The best precaution is to buy the purest specified available grade.

Finger grease is essentially a mixture of biological lipids and is highly surface active, so that nothing which is to come into contact with the monolayer or the water surface should be touched by human hands. One way of achieving this is to avoid handling all these items, and an assortment of clean tweezers should be on hand in any LB laboratory. Another way is to wear disposable plastic gloves. Discomfort and sticking due to perspiration can be avoided by wearing cotton gloves underneath the plastic.

Finally, airborne dust is another significant source of contamination in the laboratory environment. This is particularly important when the films are intended for optical applications, because the dust particles are incorporated in the films and can be the dominant cause of light scattering. For this reason, many LB facilities are now located in electronic-grade clean rooms. A slightly cheaper solution is to locate the LB equipment in laminar flow hoods. The major problem with this latter approach is the levels of vibration from the blower motor, which can cause progressive collapse of the floating monolayer.

Any parts of the LB trough which come into contact with the monolayer or the subphase water must be regularly cleaned. However it should be realised that the impurities which give most trouble in this context are not necessarily those which must be removed from dirty clothes. Most normal cleaning agents are designed to remove hydrophobic dirt, and the active principle is usually a surface-active agent. In the present context, hydrophobic dirt is not a problem, but surface-active contamination is. This being said, some groups do use detergents to clean their troughs. The detergent chosen must form spherical micelles only at high concentration (the critical micelle concentration or CMC), and should be used at only slightly higher concentrations so that light-scattering cylindrical or lamellar micelles are not present. Prolonged rinsing using many changes of hot water is then necessary to remove the agent after use, so that this procedure can only be used on an occasional basis.

There are two other categories of agent which are useful for cleaning LB troughs: oxidising agents and solvents. Solvents are normally applied to and removed from the dry trough with the help of paper tissues. Both solvents and tissues must of course be free of surfactant impurities.

Solvent cleaning is quick and easy and should be carried out each time the trough water is changed, at least on a daily basis. The author prefers to use two solvents, firstly 2-propanol to dissolve dominantly hydrophilic impurities and remove any remaining water, then 1,1,1-trichloroethane to dissolve dominantly hydrophobic impurities. Chloroform is often used, but due to its chemical instability and its volatility, it may introduce more contamination than it removes.

Water itself can be used as the cleaning agent, but needs to be heated to $50^{\circ}C$ to achieve reasonable efficiency. The addition of methanol increases the solubility of dominantly hydrophobic impurities.

The final possibility is the use of oxidising agents. This is more time-consuming than a solvent clean but involves no risk of introducing surface-active impurities. However, the agents used are all more or less corrosive and poisonous. All these operations should be caried out in a fume cupboard taking all necessary precautions against accidental spillage.

The action of an oxidising agent on a surface-active molecule is to create many carboxylic acid groups and simultaneously to break down long aliphatic chains, so that the resultant material is readily water-soluble, with no tendency to concentrate at the water surface. For this purpose only very strong oxidising agents are adequate, for example chromic acid, fuming nitric acid, or persulphuric acid. It is necessary to ensure that the oxidising agent does not itself cause problems. For example, monolayers of carboxylate anions bind very strongly to trivalent metal ions such as Cr^{3+}, so that chromic acid would be inappropriate for work with monolayers of cadmium eicosanoate, but would not be a problem for a non-ionic material like vinyl octadecanoate. Fuming nitric acid produces no such contaminating by-products, but does produce corrosive fumes, so that this operation must be carried out in a fume cupboard and, where possible, metal parts of the trough must be detached.

3. PROCEDURES

3.1 Cleaning and Spreading

Before preparing a monolayer on a water surface, it is necessary to make sure that the water surface is free of surface-active contaminants. Naturally all efforts will have been made to ensure the cleanliness of the trough and the purity of the water added to it, but nevertheless significant quantities of contaminant will usually be found on the subphase surface. These can sometimes be detected by rapidly moving the barrier towards the surface pressure balance and looking for changes of pressure: however this is not a particularly sensitive test, as with most troughs the ratio of maximum to minimum area is less than 10, so that 10% of the area can be covered by surfactant without it being detected.

Surface-active material at the surface is removed by simply sucking it off. In almost all installations the suction is provided by a water-tap Venturi entrainment attachment. Where toxic additives such as cadmium are added to the subphase it is advisable to have a water trap in the suction line: the collected liquid must then be disposed of safely. A suction head is placed at the subphase surface in such a position that it sucks up equal quantities of air and water in a succession of small droplets and bubbles, the smaller the better, as this maximises the ratio of air-water interface to removed volume.

When material is removed from a point on the surface, the surface pressure at that point drops. The surface pressure gradient so created forces monolayer from elsewhere on the surface towards the suction head. The process is thus much more efficient if, during the cleaning process, the trough barrier is moved so as to maintain a high surface pressure. It is also much quicker if the monolayer is fluid and moves readily in response to surface pressure gradients. For very rigid monolayers, the suction head must be scanned over the trough surface to ensure that all material is removed, because this flow mechanism is then unreliable.

For spreading, the material must be dissolved in a volatile organic solvent which is immiscible with water. Trichloromethane (chloroform, $CHCl_3$) is a common choice, because in addition to it being quite volatile, it is a good solvent for a wide range of LB materials. Because of its strong molecular dipole moment, it is also itself significantly surface active, and this aids in the spreading process. There are a number of aspects of chloroform which are, however, less than ideal. Especially in the presence of light, the hydrogen atom splits off to give two chemically-reactive free radicals. For this reason chloroform is usually sold with a stabilising free radical scavenger. However, the free radicals may also react with any dissolved LB film material. Even when this does not occur, the end products of free radical attack tend to be nonvolatile and remain as contaminants in the spread monolayer. Hence, ideally, chloroform for use as a spreading solvent should be regularly distilled, stored in a refrigerator in the dark, and spreading solutions should not be kept for more than a month.

There are a number of alternatives to chloroform. Volatile paraffins such as hexane are often used: a disadvantage, due to their hydrophobicity, is the tendency of dissolved amphiphiles to form inverted micelles. Dichloromethane and 1,1,1-trichloroethane share many of the desirable properties of chloroform with much lower chemical reactivity. Benzene, much used in early work, has fallen out of favour because of its slight carcinogenicity.

It has already been stated that the spreading solvent should not be miscible with water. However, there are many reports in the literature of the use of water–miscible solvents, often as a component of mixtures [69,70]. It may be that there is no alternative to their use for certain substances where no other solvents can be found. However it should be realised that the LB film material must of necessity be water–insoluble, and when significant quantities of water mix with the spreading solvent, or when the water–soluble component diffuses out of the hydrophobic phase, the film material will precipitate, most probably before complete spreading to form a monolayer has occurred. This is incompatible with achieving molecularly uniform films.

There are two considerations limiting the maximum pressure which should be reached during spreading. Firstly, it should under no circumstances exceed the equilibrium spreading pressure (ESP) of the spreading solution. This condition can be recognised because further spreading solution dispensed onto the surface fails to dissipate and remains as small droplets visible on the water surface. If this occurs, prompt action is necessary to suck the offending droplets up and reduce the surface pressure below the ESP, otherwise when the solvent eventually evaporates, small lumps of material are left floating on the surface, and will ultimately be incorporated in the film being deposited.

3.2 Compression
The floating monolayer produced by spreading has a surface pressure less, usually much less, than the ESP of the material. However, in order to deposit an LB film, the pressure must be higher than the ESP, otherwise deposited monolayers will not

stay on the substrate when further monolayers are being deposited, but will instead tend spontaneously to spread back onto the subphase. The higher surface pressure is achieved by moving the barrier to reduce the surface area available.

As the trough area decreases during compression, the monolayer surface pressure in general rises. A plot of the surface pressure versus trough area, often normalised to area per molecule, is called an isotherm. The isotherm is very easy to measure, and at nonzero but constant compression speed has adequately reproducible features. In an LB context it requires no extra effort, as the monolayer must in any case be compressed while monitoring the surface pressure. Moreover since the shape of the isotherm is sensitive to various kinds of trough contamination, it is a quick check that everything is in order. While the rapid compression check for surface contamination is insensitive to less than 10% coverage by surfactant, the shape of the isotherm is typically sensitive to 1% contamination of the monolayer under examination, and down to 100 nM ionic subphase contamination.

When a newly-synthesised material is to be tested on an LB trough, there is of course no way of knowing whether the isotherm obtained is that of the pure material, or has been affected by impurities in some way.There is therefore a need for reference materials with known isotherms which react to contaminants in characteristic ways. Octadecanoic acid is very well characterised and falls into this category. Figure 6 shows its isotherm at room temperature. Note the straightness of the isotherm over most of the 'liquid condensed'(LC) and 'solid'(S) regions, with the gas(G)–LC and LC–S transitions extending over at most a 1 mN/m pressure range. Note also the sharpness of the LC–S transition. Curvature in the LC region, especially near the G–LC transition, and the smoothing out of the LC-S transition, are all symptoms of organic contamination of the monolayer. Ionic contamination of the subphase is typically indicated by a reduction in the LC-S transition pressure. Even if this is not the material to be deposited, the sensitivity of its isotherm as an indicator of the presence and nature of contaminants, and its ready commercial availability, mean that octadecanoic acid is an indispensable material in any LB laboratory.

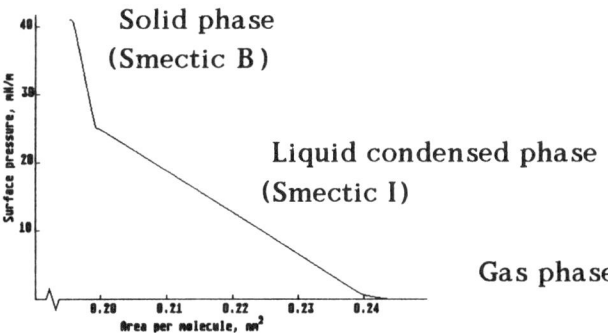

Figure 6. Isotherm of octadecanoic acid at 25 °C on a pure water subphase.

While the qualitative shape of the isotherm can be used in this way as an indication that everything is in order, its quantitative aspects are quite difficult to measure accurately. Because the water meniscus at the trough edges is curved, the total surface area is not equal to its horizontal projection as normally assumed, but can differ by up to 2% depending on the water level. Even when this variation is taken into account, the true area per molecule may itself vary. In some very careful experiments, reproducible values were only achieved with hexadecanoic acid by taking several hours to complete the compression [71]. Surface pressure measurements may also vary by about 2%, because they must in all cases be referred to the pressure of a clean water surface. A coverage of molecules in the gas phase corresponding to 1% of the surface density required for LB deposition exerts a surface pressure of 0.25 mN/m, and because of material retention on the Wilhelmy plate, or in other nooks and crannies of the trough surface, it is difficult to achieve a better reference. Hence agreement to within 2% with literature values for area per molecule and transition pressure is essentially the best which can be achieved when using the isotherm as a quick contamination check.

If the surface pressure of a monolayer ever exceeded 72 mN/m, meaning that the surface tension became negative, then it would become energetically favourable to create more surface by folding, and the resultant energy release would more than offset the increase in gravitational potential energy required. The sharp air–subphase interface would then become diffuse, and would scatter light strongly. (This is exactly what happens at the critical point of a liquid). However it has never been reported. With some materials, surface pressures above 60 mN/m can be achieved, but eventually with decreasing trough area the surface pressure suddenly starts to drop. This is the collapse point. If the surface pressure is kept below collapse, then the trough area can be cycled, and apart from hysteresis and other nonequilibrium effects, there is no loss of material from the surface. However, if the pressure is allowed to rise above the collapse point, there is an irreversible loss of film area. In many cases the collapsed film material is visible as streaks on the subphase surface.

The details of the collapse process have been studied using electron microscopy of metal-shadowed films by Ries and Kimball [72,73] and using chemical decoration techniques by Barraud et al [74]. Ries and Kimball propose the sequence of events shown in Figure 7a–d: an initial freak capillary ripple becomes unstable and develops into a bilayer sheet, which then folds over under gravity. Barraud et al are generally in agreement, but also observe the multifold structures shown in 7e.

In the presence of contaminants, the collapse pressure usually decreases, so that this aspect of the isotherm is

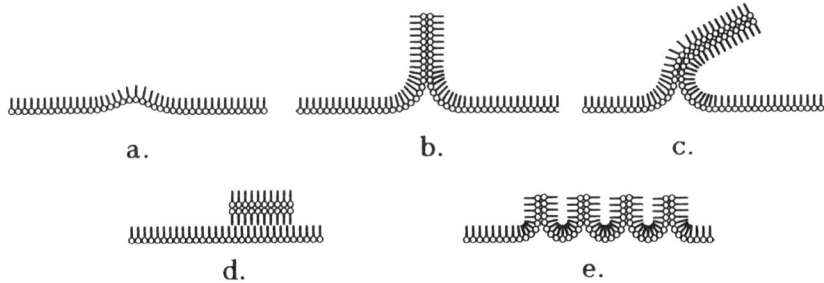

a.   b.   c.

d.   e.

Figure 7. Proposed mechanisms of film collapse: Ries and Kimball: a, initial instability; b, bilayer formation; c, gravitational instability; d, collapsed film: Barraud et al: e, observed structure.

another check of correct conditions. However, it should not be measured on monolayers destined for LB deposition. The collapsed film material remains on the subphase surface and will eventually be incorporated into the film, causing many problems. After a monolayer has collapsed, the water surface should be carefully cleaned to remove them.

Like other aspects of the isotherm, the collapse pressure is dependent on compression speed [75], and there is no satisfactory quantitative theory of the process. It is known qualitatively that the process is not homogeneous but nucleated, and accelerates with time [76]. Once a monolayer has collapsed, further measurements of the collapse pressure are always smaller.

Collapse nuclei can be produced by methods other than collapse itself. Although some of the more robust LB film materials like docosanoic (behenic) acid and 22-tricosenoic acid can be spread to pressures approaching the ESP and still produce nice uniform films, for many materials this is a recipe for disaster, as the resulting monolayer collapses readily at low pressures. In these cases the amount of material spread should be small, so that no detectable pressure is developed, and a considerable period of time – many minutes – should elapse before any compression of the monolayer is attempted.

The reasons behind these precautions have long been obscure: the time delay is often justified in terms of allowing the spreading solvent to dissipate fully, but the few scientific studies of solvent incorporation could find no evidence for it [77]. Recently, a possible mechanism for these effects has been uncovered. Although most common solvents evaporate rapidly, the choice of solvent significantly influences the long-range order in the resulting spread monolayer [78]. Moreover, allowing the monolayer to settle at low pressures before completing the transition to the ordered phase may possibly improve the stability of the film by annealing out collapse nuclei [79].

For almost all trough designs, the aspect ratio of the subphase surface changes during the compression process, so that the monolayer is sheared as well as compressed. In studies of this effect [80,81], it has been shown that the relative velocity of the monolayer at boundaries is almost always zero, just as for normal bulk fluid flow. Monolayers for LB deposition are therefore preferably ductile, with a low threshold for shear deformation and a low viscosity. In many materials, particularly the polymers and aromatic derivatives, viscous effects can cause the surface pressure near the barrier to exceed the collapse pressure even though the value at the pressure sensor, usually placed as far away from the barrier as possible, remains within acceptable limits. This is often the dominant limitation on compression speed.

3.3 Deposition

To deposit an LB film on a substrate, it is merely necessary to move a substrate smoothly downwards and upwards through the subphase surface. Many different types of substrate material are possible. Glass is a popular choice, because glass slides are readily available and films deposited on them can subsequently be examined optically [11]. For a wider transparency window it is possible to substitute fused silica or quartz [82]. Where a conductive substrate is required to allow electrical measurement of film characteristics, this is often fabricated by vacuum-depositing aluminium [83], chromium [84], tin [85], lead [86], silver [87], gold [88] or platinum [89] onto a glass slide.

Glass is by no means the only substrate which can be used. For electron microscopy, LB films have been deposited on thin polymer substrates [22] or the anodisation layer of aluminium [90]. With the recent ready availability of inexpensive and extremely smooth semiconductor wafers, particularly silicon, these have proved of value in recent studies due to their conductivity[91] and infrared transparency [92]. In fact, essentially any material with a fairly well defined surface can be used.

Substrates can, however, be divided into two categories with distinctly different behaviour towards deposition of the first layer. For hydrophilic substrates, a category which includes all metals and semiconductors with a native oxide layer, and glass as normally prepared, no monolayer is deposited as the substrate is initially immersed, so that the eventual number of layers when the substrate finally emerges into the air is odd. On the other hand hydrophobic substrates, including silicon and other semiconductors when oxide-free, silver, gold, and silanised glass, take up a monolayer on the first downstroke as well so that the final number of layers is even. Whichever is chosen, and it can be seen that for some materials like glass, either is possible depending on the surface treatment, it is essential that the surface properties are uniform.

The films cannot be deposited at arbitrarily high speeds. There is an absolute upper limit on the substrate withdrawal speed, when the water trapped between the monolayer and substrate no longer has time to drain away completely. Subsequently the water can aggregate into drops and visibly damage the film. For fatty acids, this maximum three-phase contact line velocity can be as high as 20 mm/s [17,78], but with the presence of different metal ions in the subphase and on certain substrates, it can drop to as little as 3 μm/s. Another deleterious effect of high speeds, both on the upstroke and downstroke, is mediated by monolayer viscosity, which reduces the surface pressure at the three-phase contact line, and hence the monolayer adhesion, resulting in somewhat less noticeable variations of film thickness [93]. Although further deleterious effects related to deposition speed have been claimed, there appears to be no definitive study. At maximum speed, the rate of growth of film thickness is of the order of 1 nm/s, comparable to that of vacuum techniques for thin film deposition.

The deposition of the first monolayer is especially critical and can be upset by submonolayer quantities of contaminant on the substrate surface. Many substrate cleaning procedures are commonly employed, ranging from sonication in water or solvents, the use of strong oxidising agents such as chromic acid or acidic peroxide, Soxhlet cleaning in freshly distilled and recycled solvent, and many others.

For a new or unknown film material, it is best to spread, compress and deposit as slowly as possible before its limits are established. It is also clear that all the deleterious effects are threshold phenomena. Beyond a certain point, there is no virtue in going slower. In fact for some materials there may be hazards involved in going too slow, resulting from the action of contaminants on the monolayer [61,94], increased susceptibility to meniscus instabilities leading to film striations [95], or the mixing of material in the monolayer being deposited with the underlying layer [96].

The transfer ratio is defined as the reduction of subphase surface area as a result of deposition, divided by the area of substrate coated with monolayer. Measurement of the transfer ratio is one of the quick checks of correct deposition. Under ideal conditions, the transfer ratio should lie in the range 0.95 to 1.05. Transfer ratios much less than unity usually indicate that the monolayer coverage is patchy, while ratios much greater than unity are a sign that some areas of substrate are receiving much more than one layer of molecules.

There may be difficulties in measuring the transfer ratio to high precision. The reduction in trough area is usually measured in terms of the change in barrier position at constant surface pressure assuming that the water surface is flat. However, during deposition, the line of intersection of the water surface with the substrate, the so-called meniscus line, can often be as many as 2 or 3 mm above or below the average surface level, and change according to whether the substrate has just been immersed or withdrawn. This creates inaccuracies in the measurement of true surface area which in many cases amount to 5% of the substrate area coated. In an accurate measurement of transfer ratio, it is necessary to to exclude from the calculation changes in barrier position near the upper and lower turning points which are affected in this way. A convenient way of achieving this is to plot the barrier area against substrate height: the transfer ratio away from the turning points is then the curve gradient divided by the length of the meniscus line.

3.4 Fault-finding

All the above theory is fine, but the crunch comes when the beginner deposits his first LB film following what he believes to be the recipes recommended in all the best books, and it doesn't work. This section will attempt to sketch out what general diagnostics exist, and the appropriate response to deviations from ideality.

The basic checks, which are normally carried out during each deposition sequence, are the isotherm and the transfer ratio. If graph paper is used to record these, the result in a normal laboratory is a mountain of data. The latest computer-controlled troughs allow the graphs to be displayed first on a monitor screen, and printed out only if they are of special relevance.

The significance of the isotherm is discussed in detail in Sections 3.2 and 4.1. Its shape is sensitive to very small traces of contamination, and can indicate their nature. By far the most common source of contamination is the subphase water, and rectification of a problem can be expensive. If there is some doubt, water analyses with the required ppb precision are now available. Ideally, of the ionic contaminants, only Na^+ should exceed 1 ppb (1 part in 10^9 by weight) The total oxidisable organic carbon (TOOC) should be less than 50 ppb. The pH should be 7 or slightly lower. Other possible sources of contamination are the trough, the operator's hands, and the air. Clearly, good results cannot be expected until all sources of contamination have been identified and removed.

Measurement of the transfer ratio has already been discussed in Section 3.3. Values much less than unity can usually be ascribed to poor monolayer adhesion, while those much greater than unity are typically associated with striations or monolayer collapse. When either of these unsatisfactory results is obtained, the next step is to examine the films to see exactly what they look like. The most generally applicable method involves the use of optical interference to make monolayer thickness variations of the film visible. This occurs on grey substrates like silicon, gallium arsenide or chromium when the layer thickness approaches a quarter-wavelength of visible light in the film. Silver, gold and aluminium are too reflective for this purpose, while ordinary glass is not reflective enough. At this film thickness, the reflection from the top of the film interferes destructively with that from the film-substrate interface, and is very sensitive to small changes in layer thickness. For cadmium octadecanoate maximum sensitivity is achieved at around 30 layers. Ideally no variation in interference intensity or colour should be visible, and this should be checked under a microscope as well as by the naked eye.

Striations are regions of film with greater thickness than elsewhere, extended parallel to the water meniscus, ie, perpendicular to the dipping direction. On the scale of the striation, the variation of film thickness is fairly uniform. Regions of collapsed monolayer are also thicker than elsewhere, but are not nearly so regular in thickness or orientation. Adhesion problems result in patches of film thinner than elsewhere, usually in multiples of a bilayer. Of course, the problem may very well be so severe that it is not possible to deposit 3 monolayers, let alone 30, or that film thickness variations in multilayer films are so extensive that it is not possible to determine which areas correspond to the ideal and which to deviations. In this case the film should be deposited on a substrate of silicon covered with ~80 nm of native oxide, to bring the thickness close to $\lambda/4$ for green light in the oxide. Thickness variations of 5 nm, ~2 monolayers, can then be detected by eye.

A simple, nondestructive test exists for detecting whether a layer has been deposited on a substrate or not. This is the 'breath test', and involves exhaling over it [97]. The conditions for the nucleation of water droplets are distinctly different on the LB film, and contrast is readily visible at the edge of the layer if it is there.

The problem of poor monolayer adhesion is typically much worse for the deposition of the first bilayer. The first monolayer deposits correctly, but on the second deposition stroke comes off again. Sometimes the material reappears on the water surface, resulting in an apparent transfer ratio of −1, but it can also form submerged micelles in various proportions with the water surface monolayer, resulting in a transfer ratio anywhere between −1 and +1. Poor adhesion can result from uneven or improper substrate cleaning. The available procedures are discussed in Section 3.3, and the remedies involve making the existing procedure more rigorous, or switching to another cleaning method (different materials and substrates may have different requirements). Monolayer adhesion is surface-pressure dependent, so one simple fix for this problem is to increase the pressure, if this can be done without causing film collapse.

Striations appear to be associated with visible oscillations or irregularities of the three-phase contact line or meniscus. They have been variously reported to be exacerbated by, in increasing order of correction difficulty, too slow an immersion speed, inadequate or inappropriate substrate preparation, vibrations of the dipping assembly, or contamination of the monolayer.

Collapsed film may be caused by careless compression sequences causing the collapse surface pressure to have been exceeded. When checking the isotherm with a monolayer intended for deposition, do not try to obtain the complete isotherm, but just that part below the deposition pressure, and viscous films should be compressed slowly. After a monolayer is known to have collapsed, even if only slightly, the water surface must be swept thoroughly. If this collapse has occurred at a pressure lower than the textbook value, it is almost certainly the result of monolayer contamination.

All the above problems can occur from the first monolayer on. Problems encountered only after many monolayers have been deposited are almost certainly caused by monolayer ageing and occur on the withdrawal stroke. Ageing results in a maximum speed of drainage, or three-phase contact line motion, significantly lower than that possible for the first layer [61], and correlated with increases in monolayer viscosity [80], which may be caused by a slow phase change [98]. Metal ions in the subphase have been identified as one definite cause, but organic contamination is also a possibility. The problem can be solved either by removing the contamination, or ensuring that the monolayer is renewed regularly and completely. This may affect the monolayer or the preparation of the substrate.

4. DEPOSITION AND FILM STRUCTURE

4.1 Significance of the Isotherm

According to conventional thermodynamic theory, if compression is performed slowly enough, the resulting curve

156

of surface pressure versus average area per molecule is
independent of compression rate and represents an equation
of state or constitutive equation of the two-dimensional
layer. For many years, the isotherms of insoluble monolayers
were one of their few measurable properties. Because of
their theoretical significance and experimental accessibility,
isotherms of a vast range of materials are to be found in the
literature, and there is a long history of efforts to extract
information from them about molecular properties and
conformation.

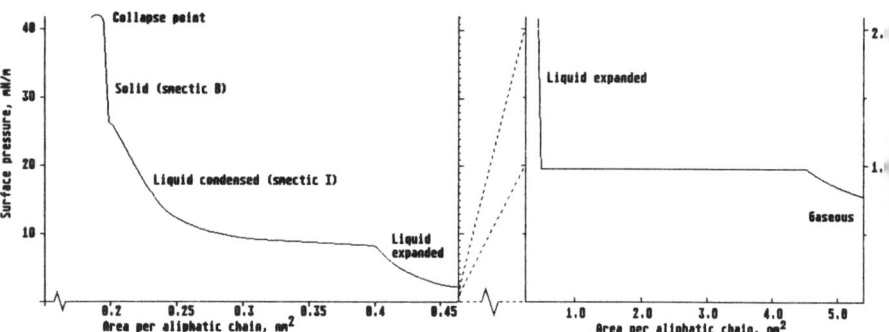

Fig 8. Typical isotherm of a lipid or C12 –C15 fatty acid.

Figure 8 shows a room-temperature isotherm which is
typical for biological lipids such as DPPC (dipalmitoyl
phosphatidyl choline) or DMPA (dimyristoyl phosphatidic
acid) [99], or for fatty acids with 12-15 carbon atoms
[100,101]. Longer-chain materials show this behaviour at
higher temperatures, and conversely, with similar isotherms
occurring when the temperature differs by roughly 10°C times
the difference in chain length. The isotherm structure is
usually interpreted in terms of transitions between four
different phases. There are several systems of nomenclature
[102]: the names gas, liquid expanded, liquid condensed and
solid shown in figure 8 are due to Harkins [103] and are the
most widely accepted. While adequate monolayer films can be

can be deposited on solid substrates from any of these phases, multilayer films of these materials can often only be deposited from the 'solid' phase.

The above list does not exhaust the repertory of aliphatic chain phases. Figure 9 shows the phase diagram for docosanoic (behenic) acid over a range of temperature and pressure. Three condensed phases in addition to those named by Harkins are apparent. Analogous phases are also found for other long-chain materials [104].

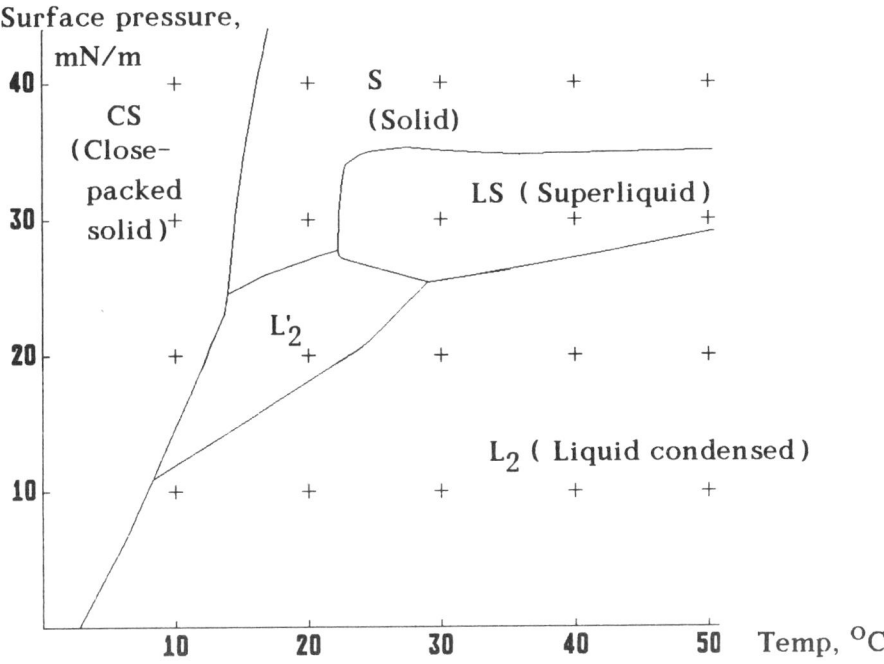

Fig 9. Surface pressure – temperature phase diagram of docosanoic acid, after [108].

The area per aliphatic chain at the collapse point of long chain derivatives is often just under 0.20 nm^2 , and this corresponds to the cross-sectional area per chain in the well-known crystalline rotator phase [105]. In many studies [6,106] this correspondence is made into a principle, and the minimum isotherm area is compared to various measured cross-sections of molecular models in order to determine the molecular conformation on the water surface. The approach is not completely illogical, and clearly if the measured isotherm area is less than the minimum possible molecular area then the layer on the subphase cannot be a monolayer. However, even in the aliphatic derivative case, there is nearly a 10% discrepancy between the area of close-packed alkyl chains in their most common crystalline packing of 0.1835 nm^2 [107] and that at a typical collapse point. Hence, without any further information relevant to molecular packing, the area at collapse can be considered no more than a guide, and tilt angles derived from it may be as much as 20° in error.

The auxiliary information necessary for a scientifically meaningful comparison of molecular packing in monolayers and bulk crystals includes diffraction patterns and thermodynamic data, in particular using differential scanning calorimetry. The latter has been done by Lundquist for a number of alkyl derivatives [104].

Conclusive X-ray diffraction studies of the floating monolayer have only become practicable recently with the development of the synchrotron as a very bright source at atomic-scale wavelengths. These results, complemented and cross-checked by fluorescence microscopy and a range of auxiliary techniques, have shown that the designations 'liquid condensed' and 'solid' are misnomers. Dervichian long ago provided evidence [109] that the liquid condensed phase is not isotropic as is normally implied by the term 'liquid', and this has now been confirmed by the observation of diffraction patterns consisting of separate spots [110], and in some cases of macroscopic anisotropy in the phase coexistence region [111]. The solid phase has a negligibly small shear-stress threshold for plastic flow [80,112], and cannot be conventionally polycrystalline because its correlation length for orientational order is 5 orders of magnitude larger than that for translational order [113]. Instead, these phases both appear to satisfy the definition of smectic liquid crystals. Hence in these cases, it would appear preferable to replace the Harkins nomenclature by the smectic category involved. In the case of DMPA [114] the solid phase appears to be identical to smectic B. This is also the case for eicosanoic acid [115], and in addition its liquid condensed phase can be identified as smectic I. Figure 10 shows the in-plane molecular environment for these two phases together with the closely-related smectic F. It is probably premature to extend this identification to other monolayer systems.

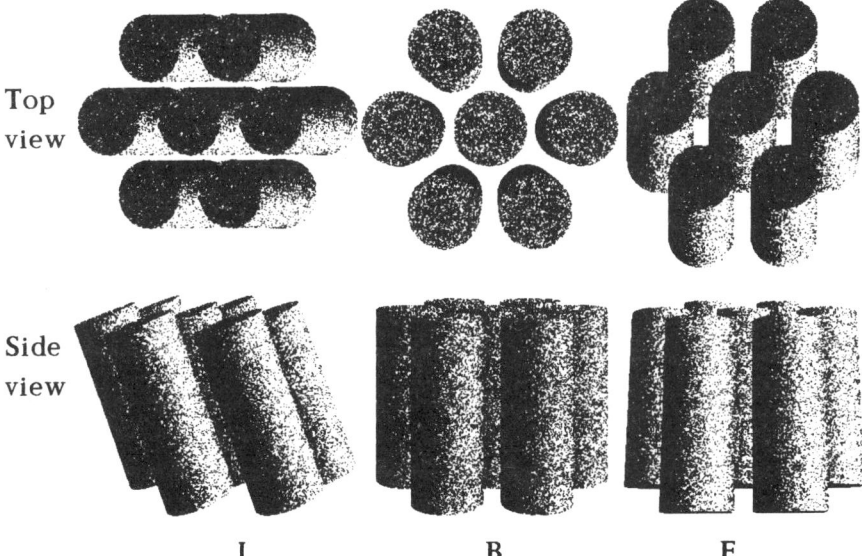

Top view

Side view

I   B   F

Figure 10. Local molecular organisation within smectic planes in the three hexatic smectic phases I,B and F.

In LB film investigations of new materials, it is common to infer both the molecular organisation of the monolayer and its readiness to form multilayers from its compressibility. If this parameter is comparable to that of a long-chain fatty acid in the Harkins' solid phase, the monolayer is said to be in a solid phase, and this is believed to correlate with the ability to deposit it as a multilayer film. There is, however, no scientific basis for this belief. Firstly, even the fatty acids in the Harkins' solid phase are much softer (in terms of the shear-stress threshold for plastic flow, assuming that the surface pressure is uniformly distributed across the monolayer thickness) than typical organic solids. Only the close-packed solid (CS) phase appears to be a true polycrystalline solid. Secondly, 22-tricosenoic and docosanoic acids can readily be deposited from the L_2' phase whose compressibility is similar to that of the non-depositing liquid-condensed phase L_2 . Finally, LB multilayers of the long-chain alcohols cannot be deposited from the Harkins' solid phase, because even at pressures near the collapse point, the first monolayer deposited on a substrate floats off again on reimmersion into the subphase.

This erroneous concept can perhaps be replaced by that of equilibrium spreading pressure (ESP). In principle, a material should be easy to deposit as a multilayer if its collapse pressure is much higher than its ESP, so that there is a comfortable pressure range over which the material is stable on the water surface, yet has no tendency to come off a substrate once deposited. To complicate matters, it is known that the ESP of collapsed film and fine bulk crystals is higher than that of large crystals of bulk material [116,117]. There have been very few studies of the ESP of LB materials, but a convenient method of measuring it has just been reported [118].

For fundamental isotherm-based investigations it is often assumed that the monolayer is homogeneous. However it has recently been shown by a number of independent groups [119-122] that monolayers of many biological lipids are inhomogeneous on a scale of tens of micrometres over most of the pressure range of the liquid condensed phase, and indirect studies have shown that the inhomogeneity extends into the 'solid' phase. Hence the measured area per molecule merely represents an average, with no guarantee that it corresponds to the area of any particular molecule. X-ray diffraction studies of floating monolayers of conventional materials [114] have revealed that the compressibility deduced from changes in the hexagonal lattice spacing is up to 60% less than that measured from the isotherm. This can be interpreted to mean that there is a significant fraction of disordered material which does not contribute to the observed diffraction peaks but which is much more compress-ible than the ordered material.

One of the fundamental assumptions of all isotherm-based investigations is that all of the material in the spreading solution ends up in the water-surface monolayer. However, in many cases the film material in the solution can partly exist as aggregates, which after the evaporation of the solvent sit on top of the monolayer as isolated clumps. This is a plausible explanation for reports in a number of recent papers [123,124] of improbably small isotherm areas per molecule, particularly for aromatic derivatives.

Isotherms of non-lollipop materials, based on polymers or lightly-substituted aromatics, hardly ever display fine structure attributable to changes of monolayer organisation. The only apparent exception to this has been structure ascribable to a transition between a monolayer and a bilayer [125]. This should not be interpreted to mean that there are no changes of lateral organisation. It is well known that small amounts of contamination – as little as 1% – are sufficient to smear out fine details in the isotherm structure of fatty acids. The contaminant does not remove the distinction between the different phases, but merely allows them to coexist over a range of surface pressures. Since polymeric materials are always a mixture of different oligomers, they may be considered to be intrinsically contaminated. Moreover it is extremely rare for the products of organic synthesis to be better than 99% pure. The fatty acids are obtainable with this purity, but are natural products. The best–studied synthetic LB material, 22–tricosenoic acid, has only recently been obtained in a state pure enough for the transitions between its phases to be observable [126].

4.2 The Carpet Model of Deposition

The illustration in Fig. 1 conforms to a widely–accepted idea of the steps in the LB process. According to this picture, during the flow of the monolayer from the horizontal subphase to the vertical substrate it flexes slightly and deforms, but undergoes no major change in organisation, in the same way as a carpet preserves its basic structure during unrolling and laying on a floor. This is believed to be true, not only for the first monolayer deposited on an incommen-surate substrate, eg glass , but also for subsequent monolayers deposited on top of an already existing film. Each monolayer is also supposed to preserve its identity, meaning that at any stage after deposition it is possible to identify *the* monolayer deposited on any particular up– or downstroke, forming a continuous, smooth sheet parallel to the substrate, and that essentially all molecules in that monolayer were in fact deposited on that stroke.

The carpet model should not be considered as a full theory

of LB deposition, covering the deposition of all possible substances under all possible conditions. Instead of being descriptive, it is prescriptive: what ought to happen during deposition, rather than what actually does happen. Under conditions of contamination or excessive vibration, a film can be completely disordered, with no analysis into distinct monolayers possible. However, for specific materials, suitably deposited, the description of the carpet model has been validated by a number of lines of evidence. One easily-tested consequence is that the transfer ratio should be equal to unity. Transfer ratios significantly outside the range 0.95 to 1.05 are in fact usually indicative of poor film homogeneity [127].

The experiments of the Göttingen group under Kuhn have already been mentioned in Section 1.3 as crucial tests of the carpet model. Information on layer identity [33,128] was provided by labelling particular monolayers on the water surface with fluorescent amphiphilic molecules. These molecules do not fluoresce immediately upon being excited by a UV photon, and the excitation, or 'exciton', can hop many times from one dye molecule to the next before a photon is finally emitted. When two different types of fluorescent dye are present, the exciton will tend to hop to the one with lower excitation energy, from which it can no longer get back to any of the higher-energy molecules. Hence the spectrum of the fluorescence provides information about the hopping kinetics, and the separation between the two dye types. In this series of experiments the measured spacings corresponded closely to the values expected from the number of intervening layers. The precisely specifiable dimensions and their observed absence of change with time are necessary requirements for any molecular electronic fabrication technique.

In a related experiment, layers of different fatty acids and soaps with characteristic individual layer thickesses were deposited in a mixed multilayer assembly [129]. This gave rise to a specific X-ray diffraction pattern which fitted well the behaviour expected from the model. The constancy of the diffraction pattern with time again confirmed that diffusion rates normal to the film in these systems were negligibly slow.

A further stipulation of the carpet model is that layer thickness should be equal at all points and should vary linearly with the number N of deposition cycles, with the statistical deviation of thickness much less than would be expected for a film deposition process, eg vacuum deposition, in which material is randomly distributed over a substrate. Early studies [130] confirmed that the effective optical thickness of the film is proportional to N. However this would also be expected to be the case for conventional film deposition processes where random amounts of material are deposited at different times and places, with merely the average thickness growing linearly with time. Optical measurements cannot distinguish this case, as they are not sensitive to film inhomogeneity on a scale much smaller than the wavelength of light. The same objection applies to more recent quartz oscillator measurements of the mass of deposited film. A more stringent test is to measure the capacitance of the film sandwiched between the substrate and an evaporated top metal contact. Since capacitance varies as the reciprocal of thickness, film inhomogeneities are expected to show up as deviations from a linear 1/C versus N plot. In many cases the expected linear plot is obtained [131,132].

Evidence that the molecular organisation in the deposited layer preserves the structure found on the water surface comes from studies of LB films of the fatty acid soaps, particularly cadmium octadecanoate and cadmium eicosanoate. At normal deposition pressures, as measured by X-ray diffraction and reflection studies on the floating monolayer [133], the molecules are perpendicular to the water surface. Very many studies have shown [91,134,135] that, within experimental error, the molecules in deposited films of the cadmium soaps make the same angle to the substrate, so that in this case, this particular aspect of monolayer structure is indeed conserved on deposition.

Although the above discussion shows that the carpet model is based on many independent lines of evidence, it has now become apparent that it does not correctly describe the deposition of many important LB materials, even though the resulting films are uniform with well-ordered molecular structure. For example, it is now well established that the tilt elevation of molecules in films of both docosanoic and 22-tricosenoic acids deposited from either the L_2' or S phase onto a hydrophilic substrate decreases progressively in the first few layers, reaching a steady value after about 6 layers [136,137]. Hence the molecular tilt in the deposited film can differ significantly from that on the water surface, and depends upon the nature of the substrate as well as the substance deposited.

The evidence that the molecules in an LB film are organised into well-defined layers corresponding to the individual deposition cycles comes mainly from shallow-angle X-ray scattering (SAXS).However this evidence does not prove the absence of occasional layer discontinuities, or that the interface planes are as smooth as shown in figure 1. RHEED studies have now shown [138,139] that the interface plane can be inclined to the substrate, and parallelism on average is achieved by discontinuous 'jogs'. No evidence is available from this source about the roughness of the interface plane, although one optical study has produced the anomalously high value of 3 monolayer thicknesses for its amplitude [140], and some SAXS studies have observed anomalies in the first and last monolayers which may be interpreted as roughness [141]. Neutron diffraction studies on codeposited layers of normal and deuterated fatty acids have shown that, at low deposition speeds, molecular exchange can occur between the monolayer being deposited and previously deposited film material [96].

The lateral organisation of molecules can also change. Again, films of the fatty acids have been most thoroughly investigated. In all the cases for which evidence exists, each monolayer has been shown to have the same local orientation of its molecular lattice as that of the underlying monolayer of the same, or chemically similar, substance. The proof is of two sorts: either the patterns of electron diffraction spots from all layers coincide [142], or the birefringent retardations of the layers combine constructively [143]. This relationship is one of the many meanings of the word 'epitaxy', although it should not be concluded that epitaxy in the strong sense, ie translational order extending from layer to layer, also occurs. Orientational correlation not associated with translational correlation has been observed in monolayers of the noble gases on graphite [144,145] and can be explained by plausible mechanisms [146,147]. Irrespective of this uncertainty about the exact nm-scale relationship between neighbouring molecules in adjacent layers, it is clear that the molecules undergo extensive rearrangement and a change of sideways neighbours as they move from the water surface to be incorporated into the growing film.

It has already been mentioned that the tilt elevation of the molecules in well-deposited films of some materials differs from the predictions of the carpet model, and since the deposition is epitaxial it is clear that the tilt azimuth will also bear little relation to its direction on the water surface prior to deposition. The actual behaviour of the tilt azimuth has been studied in connection with the occurrence of in-plane anisotropy. In this common, but not universal, phenomenon, which again has been most thoroughly studied in fatty acid films, the average size of orientational features is determined by the first deposited monolayer, while the average orientation of the film is parallel to the dipping direction of subsequent layers. It appears that while the directions from one molecule to its neighbour are highly correlated from layer to layer, the azimuth of molecular tilt is in addition influenced by hydrodynamic forces. In this respect epitaxy in LB films is quite different from epitaxy in polycrystalline materials.

The carpet model is correct in one puzzling particular. The tilt angle appears to be frozen at the value pertaining immediately after deposition and depending on deposition conditions [126]. The film is thus not in equilibrium, but displays properties depending on its history. This is the defining characteristic of a glass.

4.3 X-, Y-, and Z-Deposition
In the simple theory of LB deposition, the hydrophilic underside of the floating monolayer sticks to the hydrophilic substrate on the upstroke of the deposition cycle, thus changing its surface properties to hydrophobic. Conversely, on the downstroke the hydrophobic top surface of the monolayer adheres, and the surface properties become hydrophilic again. This type of deposition, which accounts for the majority of all known LB materials and conditions, is called Y-type, as are the resulting films with their alternating

molecular orientation.

There are however, substances known which do not comply to this picture. With specific metallic counterions, particularly Pb^{2+}, the fatty soaps display behaviour called X-type, in which the trough area decreases (and therefore presumably deposition occurs) only on the substrate downstroke. This phenomenon was first observed by Langmuir and Blodgett [148]. The opposite type of behaviour, Z-type, in which the trough area decreases only on the upstroke, has been observed recently in a large number of aromatic materials [149,150].

The phenomenon has been the subject of mild controversy. On the theory that the molecules of the deposited film retain much of the order imposed while on the water surface, X- and Z- type films should have an overall molecular polarity, and a repeat spacing normal to the substrate equal to the length of a single molecule, rather than twice that length in a Y-type film. However, a shallow-angle X-ray diffraction study by Fankuchen [24] observed only the normal, Y-type, periodicity. This latter study was actually not definitive, because the films were not deposited using any of the recipes for X-type deposition specified by Langmuir and Blodgett, but using a different arrangement which, in the present author's opinion, would result in random deposition of disordered material. Moreover, a small amount of Y-deposited material in this experiment could mask the presence of significant quantities of genuine X-film unless the diffraction line intensities were measured, which they were not.

The question has not been considered sufficiently interesting to justify further effort until recently, with the upsurge of interest in nonlinear optics, discussed further in Section 5. Organic materials are especially suited for second-order effects, which, however, require an overall molecular polarity which is not normally achieved with Y-type films. Polar films also have possible applications as pyro- and ferroelectrics. A number of groups have managed to demonstrate these effects in single-component films , all Z-type, indicating that molecular polarity can in fact be achieved [149,150]. However, a recent examination of X-deposition in lead soaps [151] concurred with Fankuchen that the resulting films consist mainly of the symmetrical Y-structure. In view of the motivation to produce these structures which now exists, it is certain that the last word has not yet been spoken.

4.4 Molecular Organisation of Deposited Films

The lipid bilayer illustrated in Figure 2 has been shown with a perfectly regular, hexagonally close-packed arrangement of the lipid molecules within each layer. This is by no means pure artistic licence: electron diffraction patterns of mono- and multilayers, of which a typical example is shown in Fig 11, often show a hexagonal pattern of spots, sometimes slightly distorted from perfect sixfold symmetry [22,90,152].

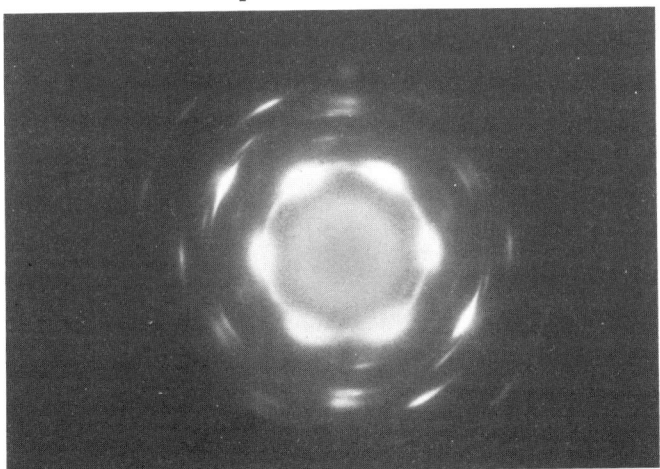

Fig. 11. A transmission electron diffraction pattern obtained from a multilayer film of 22-tricosenoic acid.

X-ray diffraction produces compatible results [153], except that the spatial resolution is much inferior so that rings rather than spots are seen. From this information it is possible to define a crystalline unit cell.

Electron diffraction cannot be used to study the molecular arrangement on the water surface, but X-ray diffraction can [98,111]. There is a very sharp radial dependence indicative of crystalline order extending over many molecules. However contrary to what would be observed for a single crystal, no variation of the scattered intensity is seen when the floating monolayer is rotated relative to the probe beam.

These results, both from the floating monolayer and from the deposited films, have often been interpreted in terms of a polycrystalline, 'domain' texture [154] where orientational variation occurs only at grain boundaries. However, direct observations of the macroscopic orientation of the crystalline lattice of fatty acid films using polarised microscopy do not conform to this view. Although there do exist discontinuities similar in some respects to grain boundaries, they do not form closed loops, and orientational variation is found everywhere [138,155]. In fact both the floating monolayer and the deposited film of fatty acids and soaps behave in all respects like thin films of smectic liquid crystals [156]. Smectic subcategories with diffraction patterns as well defined as those of LB films are now known: these include the hexatic smectics B,F and I and the crystalline smectics B, E, G, J, H and K [157]. To obtain such a diffraction pattern, the material must first be aligned to reduce the variation of orientation over the volume illuminated by the incident beam to a few degrees or less. The difference between a polycrystalline material and a smectic then lies in the irreducible radial and tangential width of the smectic reflections, corresponding to a correlation length of translational order much smaller than that of orientational order. In the case of Fig. 11, the incident beam diameter was 200 μm, while the correlation length was a few tens of lattice constants.

Liquid crystal researchers have tended to be content with phenomenological theories for the behaviour of smectics or other types of mesophase, because for their major application in displays, only average properties over large ensembles of molecules are involved. For LB films the understanding must go deeper. For example, the electrical characteristics of films produced by currently normal deposition procedures are dominated by effects associated with specific arrangements of molecules found only in a few areas of the films [158]. This is further discussed in Section 5.2. An understanding of the structure of these inhomogeneities on a molecular scale, how they are allowed by molecular interactions, and at what stage in the deposition process they arise, will be necessary to achieve significant improvements in film performance.

From the evidence which has already been presented, it is possible to reach a number of qualitative conclusions about molecular organisation in the films. Firstly, since the electron diffraction pattern consists of fairly well-defined spots, then the lattice has long-range orientational order and must almost everywhere consist of 'good crystal' in Burgers' sense [159]. This does not mean a 'perfect crystal' where there is exact translational symmetry extending over the whole material, with the separation between any two molecules an exact sum of multiples of the unit cell vectors. A perfect, strain-free crystal must necessarily give perfectly sharp diffraction spots, rather than the broad peaks seen from LB films. Instead, Burgers' sense describes crystals which have been deformed by internal stresses, but which are still topologically perfect, with the molecules maintaining the same relative positions they would have in a perfect crystal. The broadening of the diffraction spots from an LB film indicates that the regions of good crystal must be deformed significantly, and hence that large internal stresses are present.

If a sample of material consisted everywhere of good crystal, and was not subjected to any external forces, then there would be nothing to stop it relaxing rapidly to become a perfect crystal. The fact that this does not happen in LB films of fatty acids or smectics indicates the presence of regions of bad crystal, not topologically equivalent to a perfect crystal. Fig. 12 shows some simple examples of lattices with extensive crystalline organisation, but which are not everywhere good crystal. One easy method of checking this is to look at the figure edge on, along the directions of the molecular rows. In both the examples, one or more of the rows terminates.

Mathematically, a region of 'bad crystal' is best described by its Burgers vector, which is the lattice vector required to close a circuit of molecular positions equivalent at each molecule to a closed path in a perfect crystal. In each of the

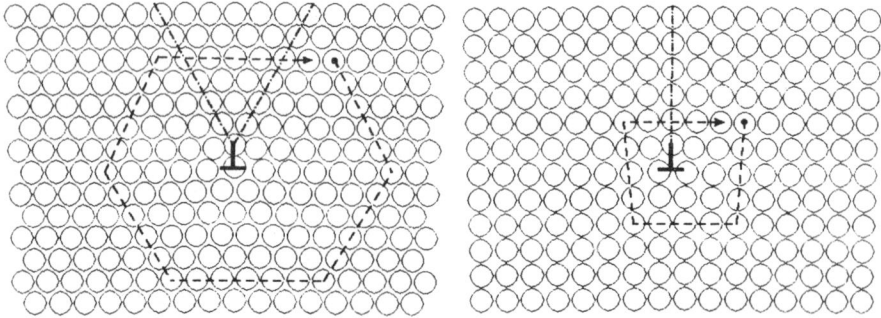

Fig. 12. Simple dislocations in a hexagonal and a square lattice showing non-closing Burgers' circuits (--) and terminating rows (-----). In each case the closure vector for the Burgers' circuit is one of the unit cell vectors.

two examples shown in Fig. 12 the closure vector is a unit cell vector. Isolated regions of bad crystal such as those shown in the figure are called dislocations. In metallurgy and semiconductor technology, the reality of dislocations has been accepted for the last thirty years [160], and the concept has proved very fruitful in explaining the strength of materials [159] and carrier lifetimes in semiconductors [161].

The spatial distribution of dislocations is very important, and information about it can be obtained using polarised microscopy, which directly provides information about the local orientation of the crystalline lattice over a continuous region [155,162]. The connection with dislocation density arises from the fact that other sorts of lattice defects cannot give rise to macroscopic variations of lattice orientation, and when there are such changes, the associated Burgers vector density is equal to the gradient of the orientational field.

Polarised microscopy can be used to determine the lattice orientation, because local field effects make the rod-shaped molecules significantly (~20%) more polarisable along their axis than perpendicular to it. When the molecular axes are not normal to the substrate, the film is birefringent, with its slow axis in the direction of the molecular tilt azimuth. Fig. 13 shows a typical image between crossed polarisers obtained from a fatty acid film. As the film is rotated relative to the polariser and analyser the transmitted intensity undergoes a cyclic variation, maximum when the preferred axes of the film make an angle of 45° to the polariser or analyser, and a null when either is parallel. The variations in brightness in Fig. 13 are due to the fact that the local film orientation varies from point to point.

So far, such images have only been reported for films of a few types of material : in the cadmium soaps the molecular

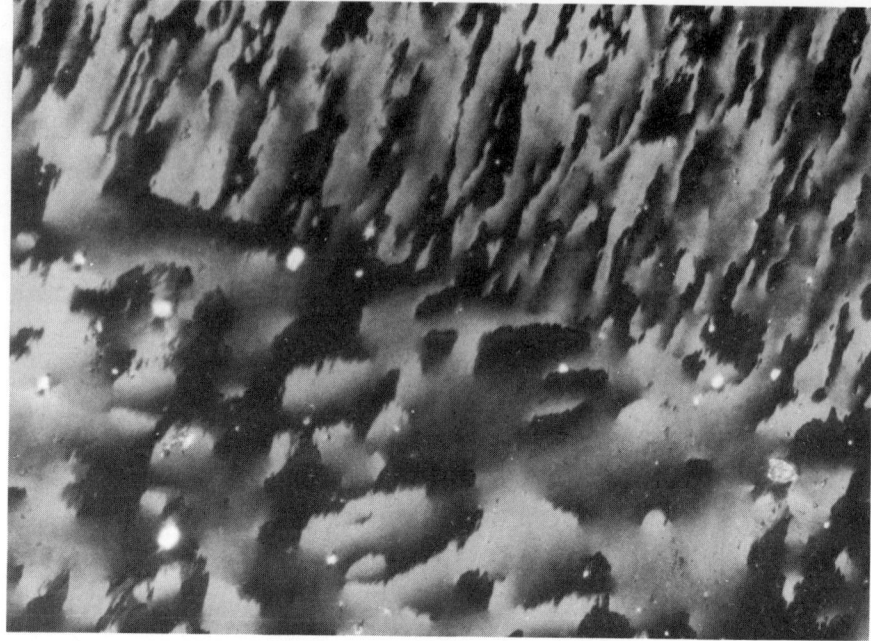

Fig. 13. Image obtained between crossed polarisers from a multilayer film of 22-tricosenoic acid.

axes are almost perpendicular to the substrate so that the films are not birefringent, while in films of many aromatic materials the correlation length of orientational order is less than the ~10 µm image resolution in the crossed-polariser mode. However there is reason to believe that if a film does not rearrange after deposition, the main features of its orientational variation resemble Fig. 13.

The continuous variation of macroscopic orientation visible in Fig. 13 means that there are dislocations almost everywhere, distributed more or less at random throughout the whole material. The pattern of stresses which results can explain most details of the observed broadening of the diffraction patterns [164].

A theoretical model explaining how dislocations can be statistically distributed, rather than highly correlated as part of grain boundaries, has been provided by Nelson and Halperin [165,166] . Essentially, this can only occur in 2-dimensional phases, where dislocations are point defects and have interaction energies comparable to or smaller than kT, or in pseudo-2D phases of highly anisotropic molecules. Fig. 14 is a computer-generated image showing the lattice positions in such a phase. The gas of weakly-interacting dislocations destroys long-range translational order, so that the diffraction spots are diffuse, and their slowly-decaying strain field gives a statistical spread of local lattice orientation. However, there are no sharp changes of orientation, so that there are still correlations of orientational order over macroscopic distances, and the spots are not smeared out into a ring.

Figure 14. A hexagonal lattice with dislocation gas.

4.5 Film Stability

It has already been mentioned that the LB films of the fatty soaps are mechanically soft and melt at fairly low temperatures. The thermal stability of the films has been extensively investigated. They melt at around $100^{\circ}C$ [167,168], and desorb from the substrate at around $150^{\circ}C$ [169].

At ambient temperature, and up to $20^{\circ}C$ below the melting point, the structure of cadmium soap films is quite stable with time. Two series of experiments demonstrating this have already been mentioned in Section 1.3. In a first series, two specific monolayers of a multilayer assembly were labelled with different fluorescent dye molecules. The fluorescence spectrum provided information about the separation between the layers [33]. In a second series, the layer labelling was achieved with fatty acids of differing chain length [129]. In other, unexpected, ways the film structure is also stable. The molecular tilt elevation of films deposited from the L_2' phase of 22-tricosenoic acid varies with deposition pressure, and it remains constant at the value fixed at deposition [126].

These findings are certainly comforting for molecular electronics applications, as they indicate that stable films are possible. However they cannot be extrapolated too widely. Even films of such closely related materials as the calcium and barium soaps have been reported in many studies to be unstable [170,171], as have films of other materials [172]. In some cases, the observed instability may have resulted from contamination of the films. In addition, thick films of materials which are fluid in the bulk state, such as the nematic liquid crystal 4-pentyl-4'-cyanodiphenyl, [173] cannot be expected to be stable.

Many researchers have attempted to find film materials with much higher temperature capability than the fatty acids and soaps. One common approach is the use of high-molecular weight materials formed by polymerisation. In the early 1980's there was a lot of enthusiasm for materials which could be polymerised after film deposition, principally molecules containing the diacetylene group [174,175], but also vinyl, acrylic and diene derivatives [41,176]. This waned when it became apparent that polymerisation did not preserve the perfection of the as-deposited film structure: an initially smooth film became quite irregular in thickness and in many cases developed cracks.

One alternative to post-deposition processing is to polymerise the material on the water surface, while the molecules are firmly anchored [177]. This approach solves the problem of film thickness variation, but not that of cracks. A possible improvement for this water-polymerisation technique, which would also be applicable to post-polymerisation, is to improve film structure by water-surface annealing [178]. Increases in the degree of polymerisation and in the domain size of the resulting film were reported, probably due to a decrease in both dislocation and disclination densities in the monomer film.

Somewhat greater interest has been shown in materials which are polymerised prior to spreading. As discussed in Section 1.4, there are now many polymeric materials known which can be deposited as multilayer LB films [179]. The uniformity of thickness of these pre-polymerised films can be much better than that of post-polymerised films, and they are completely free of cracks . However in shallow-angle X-ray scattering experiments their layer structure has been found to be much less distinct than in films of monomeric materials [180], while their electrical insulating properties are also not as good [181]. Because the molecules are significantly larger, and must form random entanglements on the water surface, the monolayers are much more viscous than those of the fatty acids, and the maximum speed at which they can be deposited is correspondingly reduced. Typically such materials are deposited at substrate withdrawal and immersion speeds of less than 10 $\mu m/s$.

Aromatic derivatives do not need to have such high molecular weights to achieve equivalent temperature capability, and in addition display strong features in their visible and UV absorption spectrum, together with significant charge carrier mobility. Because the charge transport in particular is expected to be strongly affected by the presence of long aliphatic chains, many studies have been made of lightly-substituted porphyrin [181,182] and phthalocyanine [183,184] derivatives, which are stable in air up to 400°C, and to even higher temperatures in a vacuum. However, possibly because these molecules have no strong amphiphilic character but are merely surface active, multilayer films of this type appear to have no layer structure at all [185], and there is evidence that they readily crystallise, thus destroying the painstakingly achieved uniformity of thickness.

There are lightly substituted aromatics known whose films do display good layer structure. One well-documented example is (9-butyl-10-anthryl)-3-propionic acid [186]. Films of a phthalocyanine substituted with medium length (C_8) alkyl chains have been reported to be extremely stable and in addition to show interesting orientational effects [187]. In this compound the usual metal atom in the middle replaced by silicon, by which the macrocycles are polymerised together into a thick rod resembling a shishkebab. However aromatic materials do not need to be lightly substituted to achieve temperature stability. The azostilbene chromophore hemicyanine [188] has been widely investigated in connection with its nonlinear optical properties and its C_{22}-substituted films are stable to above 200°C.

4.6 Inhomogeneities

Various types of gross inhomogeneities occurring in LB films have been discussed in Section 3.4. There are other, more subtle inhomogeneities. All LB films so far investigated display local crystalline packing of the molecules (even the preformed polymers have close-packed aliphatic chains). Since this short range order has never been observed to extend over the whole trough, some types of orientational inhomogeneity must occur. On a 2-dimensional scale, and affecting almost all points of the monolayer, is stress giving rise to curvature. This has been identified as a dominant cause of signal attenuation in LB films used as optical waveguides [189]. 1-dimensional discontinuities of orientation, usually called grain boundaries, can also occur. Of these, only the particularly low-energy variety called the twin boundary has been positively identified in fatty acid films.

Smectic-organisation films also have 0-dimensional inhomogeneities called disclinations [190,191]. These well-known and characteristic structures of liquid crystals are points where the orientational ordering interaction is frustrated. Figure 15 shows a locally-hcp lattice of tilted molecules with two disclinations of opposite sign joined by a twin boundary.

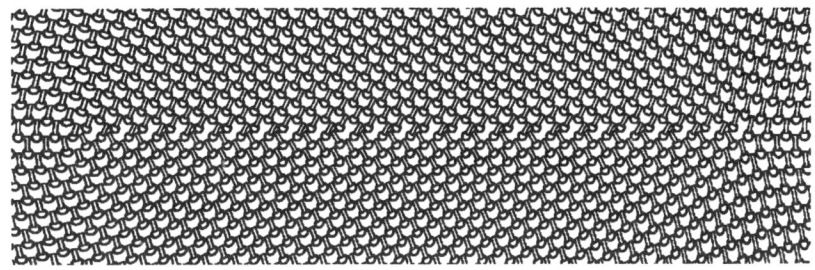

Fig. 15. A locally-hcp lattice of tilted molecules displaying two disclinations of opposite sign joined by a twin boundary.

In one respect this illustration is not true to life: in real lattices the shear stresses caused by the lattice curvature would exceed the yield stress of the material unless relieved by a density of dislocations. Twin boundaries and disclinations have been identified as dominant causes of defect conduction in mono- and multilayers of the fatty acids and soaps [155,192].

These inhomogeneities are not an inescapable fact of life. A number of studies have now shown that it is possible to process the monolayer on the water surface so as to improve the ordering of the deposited film [168,193], and in particular to achieve significant reductions in the density of disclinations [194,195].

5. APPLICABLE PROPERTIES

5.1 The Two Molecular Electronics

This chapter has so far discussed the science of LB films, and there is clearly a lot to understand about molecular interactions in low-dimensional structures, sufficient by itself to justify an interest in the field. However, a constant theme in reviews of the field over the last decade, which cannot be too far removed from the industrial interest which has been attracted, is the expectation of important applications. There is a dichotomy between the types of applications considered, which can be related to the 'broad' and 'narrow ' meanings of the term 'Molecular Electronics' discussed in Section 1.1. Much of the motivation and excitement in the field is related to the 'narrow' sense, while in the present research climate in the English-speaking world, funding has often been sought by emphasising the 'broad'.

The modern interest in LB films can be traced to the work of Kuhn, and it is clear that his inspiration was Molecular Electronics in the narrow sense of independently functional molecules [32]. His work in the field can be largely interpreted as a demonstration that LB films fulfil the technological requirements for a stable, low-defect structure with molecular-scale features and designable characteristics.

However the promise of these results has not yet come to fruition. Although 'molecular machines' may have been made, they are machines only in the sense of pulleys or levers. For example, although the transfer of excitons between different chromophore types has been fine-tuned by adjusting the distance between the layers containing them, in any particular film this distance is fixed once and for all at the time of fabrication. The transfer process cannot subsequently be influenced by any other input signal, as is necessary in an active device.

In fact molecular systems with self-sustaining, controlled behaviour will require many separate components, active as well as passive, all interconnected in complicated patterns and functioning simultaneously. A transistor in isolation is useless, and to get it to display useful behaviour, it must be connected into a circuit with other transistors and resistors, and fed by a low-impedance power supply. Ideas have been proposed for approaching some of the problems, but it is fair to say that crucial aspects still remain to be solved.

Largely because many people have realised how intractable molecular electronics in the narrow sense is, the mainstream of subsequent applied LB research has been much less visionary, and has concentrated on applying specific features of the films to solve problems in more conventional technology. For example, the layer perfection claimed by Kuhn is not only useful in assemblies of functional molecules, but makes the films suitable in principle as ultrathin insulating layers or high-resolution resists in what is conceptually only a minor modification of conventional microelectronic technology. Since organic materials have recently been demonstrated to have large, fast second-order nonlinear optical coefficients [196], another possible application is in integrated optics where the extremely good control of layer thickness will in principle allow very fine tolerances on phase matching between modes.

Because these proposed applications apparently involve coordinating a much smaller number of film properties, with simpler demands on film structure, they are often claimed to be much closer to the market place. They are not, however, soft options. In this approach LB films have to compete with other technologies which can already more or less adequately perform the desired function, and there may be important systems requirements to which LB films are not especially suited. The films must provide a significant premium of improved performance if they are to replace existing practice.

The idea of using LB films as ultrathin insulating layers as part of a more conventional semiconductor process, dealt with in more detail in Section 5.2, is a good example of the pitfalls. This proposal has attracted a lot of interest, is soundly based on the requirements of III–V microelectronic technologies. Unfortunately, in spite of this, no manufacturer has yet seen fit to incorporate LB insulators into a production device.

One widely perceived reason for this is the limited high-temperature capacity of the LB insulating materials which have been so far investigated. The response from many quarters has been to aim at improving the mechanical and temperature stability of the films, by the use of aromatic or polymeric materials. However, as already discussed in §4.5, essentially all the progress which has been achieved in this direction appears to have been at the expense of layer uniformity, which was the initial big attraction. Moreover, it is also possible that they have not been adopted because their insulating characteristics are not as good as was first believed. While inversion of semiconductor surfaces has been achieved using LB insulators, the leakage currents in these cases were much higher than those claimed for defect-free monolayer cells.

In fact the wisdom of trying to achieve modest improvements in thermal stability is questionable, when this characteristic in an insulator is available much more easily from silicon dioxide, alumina and other refractory materials. Moreover, there are a number of other aspects in which LB films do not rate highly compared to other methods of preparing thin films. These include extremely slow deposition rates for some materials, the need for uniform substrate surface properties, and mechanical softness. LB films will never be able to compete with conventional insulators in all these repects.

However, there are other accepted technologies with limited temperature range. Superconductors have much worse temperature limitations, yet because the power losses involved in current flow through them are so much lower than those of any other conductor, even the old-fashioned liquid-helium-temperature variety have carved out for themselves an applications niche where their temperature needs are catered for by refrigeration. Biotechnology is another area. On this viewpoint, the correct approach for LB films would be to try instead to achieve reproducibly excellent insulating characteristics. While processing compatibility requirements might possibly rule out soap films in LSI integrated circuits, this would not apply to extremely low-noise discrete insulated-gate field effect transistors.

5.2 Insulation

Microelectronics techology is based on ways of fabricating and manipulating thin layers, and insulating layers are one of the important categories. For insulated-gate field-effect transistors (IGFETs), the use of thin gate insulating layers is of interest to maximise transconductance, minimise offset voltage and to obtain fast switching. In the present-day technology of III-V semiconductors, a general deposition technique giving satisfactory insulating layers does not yet exist, which explains why only depletion-mode MESFETs are available, instead of the much more easily integrable enhancement-mode IGFETs. This can be traced to the poor insulating quality of III-V native oxides, and to the unacceptably-high interface state densities produced by other high-temperature deposition techniques. Since LB deposition takes place at low temperature, interface states should not be a problem.

Molecular electronics in the narrow sense also requires good insulators capable of withstanding volts of potential difference over distances of nanometres. When using an oriented LB monolayer to make electrical contact to two ends of a molecule, as discussed in Section 1.1, the leakage current must be small. Only the chemisorption technique (Section 1.4) provides an alternative method of preparing an oriented monolayer.

A factor limiting essentially all present-day insulating layer technologies is the uniformity with which the layer can be deposited. Clearly, pinhole defects result in short circuits, leading to intolerably large leakage currents. Hence the report by Kuhn and coworkers of cadmium soap metal-monolayer-metal (MMM) cells free of short circuits [29,197,198] was very exciting. Subsequently, this report was followed up by many attempts to apply LB films as insulators. Surface inversion of, and minority carrier injection into, a number of III-V semiconductors was reported [38,199,200]. Unfortunately, the leakage currents of the films in these cases were not nearly as low as those reported by Kuhn, and in the latter case the optimum film thickness was so great it could clearly not have resulted from tunnelling through bulk film.

The insulating characteristics of LB films have been the subject of many studies, and are still a matter for dispute. The films of Kuhn et al were all deposited on aluminium substrates which are well known to be covered with an insulating layer of native oxide, but this was ignored in their theoretical analysis. Their claim that metal filaments were absent was based on the agreement between the measured current density and the expected value of tunnel current through the monolayer. However Gundlach and Kadleč reported that the effect of the oxide on MMM characteristics could not be ignored [201].

Tredgold and Winter [51] found that the current in their aluminium-substrate MMM cells decreased monotonically with time, indicating that the dominant conduction mechanism could not be tunnelling through homogeneous areas of the monolayer. They interpreted this as indicating the presence of penetrating filaments of top-contact metal. In fact, all MMM cells on noble metal substrates made before 1985 were short-circuited [202], even those made by Kuhn and coworkers, and a number of workers [203,204] have since measured the density and diameter of the short-circuiting filaments. Clearly Kuhn et al were mistaken on two counts: it is not permissible to ignore the Al_2O_3 layer, and their monolayers probably did have conducting defects (certainly those on noble metal substrates did).

Nonetheless it can be readily calculated from the quoted values of filament density and diameter, taking into account the effect of the aluminium oxide in limiting the current, that the filamentary contribution to the conductance of aluminium-substrate cells must be much less than any reported measurement, even the extremely low values measured by Kuhn et al. Moreover, penetrating filaments should not be nearly as important in multilayer films. On the basis of measurements on cadmium soap mono- and multilayer cells, Sugi deduced the presence of a high density of electronic states at energies accessible to thermal charge carriers, both at the hydrophilic and hydrophobic interfaces in the films [205,206]. Chemically, this result is difficult to understand, because the saturated fatty acid molecules involved are chemically inert. Notwithstanding this expectation, it remains true that the conductivity normal to the plane of 50-layer LB films of an "electroactive" aromatic material, C4-anthracene [186], was found to be three orders of magnitude lower than those of a soap film of the same thickness [203]. Films of this thickness are found to be high impedance even on noble-metal substrates, so there is no question of penetrating metal filaments.

It has recently proved possible to produce high-impedance, ie filament-free, cells containing 2 or more monolayers of 22-tricosenoic acid on a noble-metal substrate [202,207]. The conductivities measured were perhaps an order of magnitude greater than those reported by Sugi for the layers of similar thickness, indicating that Sugi's results were not dominated by penetrating filaments. The energy barrier for charge injection into the oxide-free film was only 1 eV. Clearly, Sugi's proposal of the electroactive nature of aliphatic chains must be taken seriously, even if the underlying physics is not completely understood.

Perhaps the most encouraging of the new results are those indicating a link between film conductivity and orientational alignment. In the earliest such report, it was shown that the conductivity of a 15-layer cell of cadmium octadecanoate could be reduced by a factor of 30 by changing the conditions of preparation of the first layer only [168]. The improved preparation method was known to give much better orientational alignment of the films [208]. Subsequently it was shown that conduction through a monolayer is not homogeneous, but is concentrated at points of orientational discontinuity in the films: certainly at twin boundaries, and possibly even restricted to the cores of disclinations [155].

The molecular nature of the electronic states involved with conduction at these orientational discontinuities is still not clear, but it is almost certain to be similar to the intrinsic polymer states known to be involved in triboelectricity [209] and slow dielectric breakdown [210]. The reduced crystalline order at the discontinuities must allow the aliphatic chains to take up different conformations, one of which is responsible for the observed effects. Thermal motion of the chains leading to fairly slow creation and destruction of these conformations can explain the extremely high levels of low-frequency noise observed in the filament-free cells, which cannot be explained on any homogeneous conduction mechanism. The precise chain conformation or conformations is of great interest, because if it were possible to control its occurrence, this could be the basis for a molecular switch.

It can be concluded that the films of Kuhn et al had much better orientational alignment than those produced in other laboratories, and that even further improvements may be expected if disclination-free monolayers can be achieved. Some progress in this direction has been reported [194,195].

5.3 Dielectric Constant

If LB films displaying their theoretical insulating characteristics can be made, then they will also be of interest as dielectrics in capacitors. For integrated capacitors, the amount of capacitance per unit area depends inversely on insulator thickness. In dynamic RAMs, each memory cell capacitor must store sufficient charge to provide an adequate error margin, and this is a limiting factor in recent designs, so that thin insulating layers are of considerable interest. Since the main structural feature of many LB film molecules, the aliphatic chain, has a much lower bulk electron affinity than SiO_2, Si_3N_4 or other conventional insulating materials, LB films can potentially be made much thinner before leakage currents become a problem.

An attempt has been made to use LB films as the dielectric in a capacitor microphone [211]. In these devices, the extremely high electric field achievable should lead to high sensitivity; the high capacitance per unit area to low impedance; and the thinness of the dielectric to very high frequency response. However, the results obtained could not be understood. It is probable that Sugi's electronic states were in some way involved.

5.4 Conduction and Semiconduction

A major reason for interest in films of aromatic materials is their suitability for charge transport. Much of the work in this area has been connected with gas sensors, particularly for oxidising, reducing, acidic and basic gases, and the film molecules of choice have been the aromatic macrocycles [183,212]. The gas molecules bind to the film molecules, hopefully reversibly, creating charge carriers which are detected by measuring the film conductivity. The speed of response of LB sensors of this type is much higher than those of films of similar materials made by conventional film-forming techniques, presumably because diffusion of the gas molecules through the active layer is the limiting factor, and the LB films are typically two orders of magnitude thinner.

With the discovery a decade ago that derivatives of diacetylene could be polymerised after LB deposition, there has been considerable interest in using the films as charge transport media. Wilson [43,213] has proposed using them in a 3-dimensional memory. There have been reports of conductivity in doped films [214], but from the very low values it appears that the films are quite disordered.

Recently, many reports have emerged of 'metallic' conduction in LB films [215-218]. The molecules are variants on those used in known charge-transfer salts like tetrathiofulvalene (TTF) and tetracyanoquinodimethane (TCNQ), and are made suitable for LB deposition by substitution of a long aliphatic chain. Some of these are conductive without further processing, but in the vast majority the conductivity is induced later by treatment with an oxidising or reducing agent, typically iodine. The aim of much of this work is stated to be molecular electronics, perhaps as a way of interfacing between macroscopic wires and individual molecules or groups of molecules as shown in Fig. 16, or to achieve the ultimate in connection packing density. While this line of research is therefore of great interest, the results so

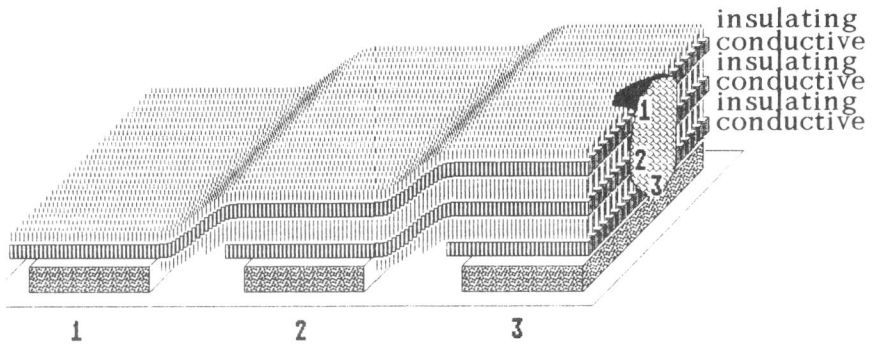

Fig. 16. Using LB films consisting of conducting aromatic planes separated by insulating layers of aliphatic chains to interface between macroscopic wires and individual molecules.

far have a long way to go: the bulk conductivity of these materials is comparable to that of a weakly-doped semiconductor and hence is not suitable for high-speed interconnections. Moreover the initial layer structure of the film is disrupted by the doping process, so that the specificity of connection illustrated in Fig. 16 cannot be achieved. In fact the reported microstructure of films of these materials is extremely inhomogeneous, and there has been no report as yet of individual monolayers being used as distinct connecting elements.

5.5 Microlithographic Resists

All microelectronic technology involves photographic methods of patterning semiconductor surfaces. Intrinsic to these methods is a sensitive layer, the resist, which converts an image into a processing mask which protects some regions while exposing others to treatment of various kinds.

The resist layer must be homogeneous in thickness and exposure characteristics on all length scales between the minimum line width and the total wafer size. At present, essentially all resists are polymeric resins. The wafer is coated by spreading a drop of polymer dissolved in a volatile organic solvent onto it while spinning at high speed. This technique produces layers which are sufficiently homogeneous at layer thicknesses down to 100 nm.

Up until a few years ago, optical methods were quite adequate to achieve the required line widths, and in many existing processes these have merely been extended by the use of shorter wavelengths in the ultraviolet. However, in recent years a number of companies have switched to manufacturing submicron-feature integrated circuits using radiation of much smaller wavelength, either electrons, X-rays or ions. Of these three types of radiation, two, namely electrons and ions, require very thin layers of resist: in the case of electrons, because the very strong low-angle scattering of electrons in the layer limits the minimum line width to roughly half the layer thickness; and in the case of ions, because the penetration depth of ions is small so that only the top of the resist layer is exposed. For uniformity of contrast and sensitivity, the layer thickness must be maintained as nearly constant as possible across the semiconductor wafer. This is an area in which LB films excel, and could conceivably provide a performance advantage over spin coating [41].

This appears a very promising area for future development, and is still an area of active research. However, since spin coating methods are also being continuously refined, LB resists may now only be called upon to achieve line widths smaller than 100 nm.

5.6 Nonlinear Optics

Especially with the advent of fibre optics as a practical communications technology, there has arisen an interest in components for processing optical signals. One such component, which will probably be introduced into telecommunications networks widely within the next few years, is the routing switch, which can permute a number of incoming optical signals between destinations. The devices now in development are based on the perovskite ferroelectrics, which show a large interaction between the slowly-varying and optical-frequency components of the electromagnetic field in the vicinity of the Curie point [196]. There is also considerable interest in using charge-transfer organic dyes which show a slightly weaker coupling but which have much improved dielectric dispersion characteristics and could be significantly cheaper to fabricate [44]. In order to achieve the required effect, the dye molecules must be oriented, either as part of a crystal or within a polymer matrix.

In the longer term, there is a prospect that all-optical nonlinear effects could be used to process the signals as well, with signal bandwidths orders of magnitude higher than the limits now imposed by conventional electronics. For these purposes, the charger-transfer dyes have much better figures of merit than the perovskite ferroelectrics. Again, in order to achieve the desired effects, the dye molecules must be oriented.

Inevitably , in view of the success of integrating semiconductor devices on a common substrate, many attempts have been made to integrate optical signal-processing elements in a similar way. For such a technology, it must be possible to define waveguides on the substrate surface. Since the Langmuir-Blodgett technique offers the possibility of oriented molecules within a thin layer of highly-precise thickness, it has attracted significant interest.

So far, LB films of some long-chain-substituted charge-transfer dyes have been demonstrated to have very large second-order all-optical coefficients, roughly an order of magnitude higher than those of more conventional materials [44]. Waveguiding over usable distances in these films has, however, not yet been demonstrated because of their very high levels of Rayleigh scattering. The dominant inhomogeneity causing this scattering in the charge-transfer films has not been positively identified, but measurements in fatty acid films were compatible with changes of local orientation as the major cause.

Electro-optical effects have also been demonstrated in LB films, with coefficients which are an order of magnitude higher than those of competing organic solutions, and comparable to those of the perovskite ferroelectrics. Again, optical scattering is the major impediment to the application of the films.

6. CONCLUSIONS AND PERSPECTIVES

The Langmuir–Blodgett technique is a way of making ultrathin organic layers with a combination of characteristics not found with any other method. Perhaps the layers do not display the absolute precision of thickness and perfection of structure depicted by some protagonists, or at least, only when prepared under certain ideal conditions which we cannot yet define. Nevertheless it is possible to produce layers with thickness in the range 1 nm to 1 µm to a precision of better than 1% over large areas. The molecules in the films are highly oriented and in many cases have very low rates of diffusion. The latest discoveries concerning the role of structure in charge transport and optical scattering indicate that there should exist ways reproducibly to achieve characteristics which at the moment are only obtained as a lucky coincidence.

The films are not the answer to the technologist's prayer. They have grave drawbacks, including mechanical softness, limited high temperature range, and often a very slow rate of deposition easily perturbed by extremely small levels of contamination. It may at some time in the future prove possible to design a film molecule so robust in these respects that the films can substitute for polymer dielectrics in high volume capacitor production, or be compatible with subsequent processing steps for integrated circuits. However at the moment these are distant prospects, and the major selling point of the films, which will make it worthwhile to attempt these further developments, is their potentially low defect density. There do exist niches for applications in which the ability to make ultrathin, low-leakage insulators can provide a competitive edge and where the disadvantages of LB films can be tolerated. Perhaps the most exciting of these is the prospect for molecular electronics in the narrow sense, but there are others not requiring such large leaps of the imagination.

The first period of significant industrial investment in LB films in Britain is now over, without any major impact on electronics as yet. However there are parallels with the history of liquid crystals to cheer the despondent. Liquid crystals were discovered in 1888, that is to say, 46 years before LB multilayers. Scientific research into their properties proceeded at a leisurely pace, involving a systematic study of their properties and a theoretical understanding of their behaviour [218]. There was a first wave of industrial interest during the sixties, in which a wide range of applications of liquid crystals was considered: although displays were certainly one of these, Fergason aroused great interest with his demonstration of a temperature indicator, and Merck thought them to be most promising as anisotropic solvents for molecular spectroscopy. The display demonstrated by RCA in 1966/67 only functioned at 80°C. Shortly thereafter doubts arose as to whether they would ever be practical, and industrial interest waned. The breakthrough came in 1973 with the discovery of the first materials with the necessary durability and operating temperature range.

Development of LB films for practical applications is a challenge, requiring an interdisciplinary outlook which neither balks at the physics involved in understanding assemblies of partially disordered and highly-anisotropic molecules nor at the cookery involved in making them. Although LB films are not and cannot be adapted to all purposes, there are signs that with sufficient understanding their behaviour can be optimised. In spite of all present difficulties, LB films have a unique potential for controlling the structure of organised matter on the ultimate scale of miniaturisation and must surely find a niche where this potential is fulfilled .

REFERENCES

1. Aristotle, 'Problemata' Book 23 No 38, Engl. transl. W.S.Hett, Heineman (New York, 1937).
2. Pliny the Elder,'Natural History', Engl. transl. H.Rackham, Vol 1, p361, William Heineman (London, 1964).
3. A.Pockels, Nature, 43, 437 (1891).
4. A.Pockels, Nature, 50, 223 (1894).
5. Lord Rayleigh, Philos. Mag., 48, 321 (1899).
6. N.K.Adam, 'The Physics and Chemistry of Interfaces', Oxford University Press, 3rd Edn (1941).
7. J.Richard, A.Barraud, M.Vandevyver and A.Ruaudel-Teixier, Thin Solid Films, 159, 207 (1988).
8. M.Matsumoto, H.Yoshimura, S.Endo and K.Nagayama, Paper GO3, 4th Intl Conf. on LB Films, Tsukuba Japan, April 1989.
9. I.Langmuir, Trans. Faraday. Soc., 15, 62 (1920).
10. K.B.Blodgett, J. Am. Chem. Soc., 56, 495 (1934).
11. K.B.Blodgett, J. Am. Chem. Soc., 57, 1007 (1935).
12. I.Langmuir and K.B.Blodgett, Kolloid-Z., 73, 257 (1935).
13. K.B.Blodgett and I.Langmuir, Phys. Rev., 51, 964 (1937).
14. I.Langmuir, V.J.Schaefer and D.M.Wrinch, Science, 85, 76 (1937).
15. I.Langmuir and V.J.Schaefer, J. A. Chem. Soc., 59, 1406 (1937).
16. I.Langmuir, V.J.Schaefer and H.Sobotka, J. Am. Chem. Soc., 59, 1751 (1937).
17. K.B.Blodgett, J. Phys. Chem., 41, 975 (1937).
18. K.B.Blodgett, Science, 89, 60 (1939).

196

19. K.B.Blodgett, Phys. Rev., 55, 391 (1939).

20. K.B.Blodgett, Rev. Sci. Instrum., 12, 10 (1941).

21. K.B.Blodgett, Low-reflectance glass, U.S. Patent 2,220,861 (1940).

22. L.H.Germer and K.H.Storks, J. Chem. Phys., 6, 280 (1938).

23. C.Holley and S.Bernstein, Phys. Rev., 49, 403 (1936).

24. I.Fankuchen, J.J.Bikerman and J.H.Schulman, Phys. Rev., 53, 909 (1938).

25. C.E.Buchwald, D.M.Gallagher, C.P.Haskins, E.M.Thatcher and P.A.Zahl, Proc. Nat. Acad. Soc., 24, 204 (1938).

26. R.M.Handy and L.C.Scala, J. Electrochem. Soc., 113, 109 (1966).

27. H.Sobotka, J. Colloid Sci., 11, 435 (1956).

28. G.L.Gaines, Jr., 'Insoluble Monolayers at Liquid-Gas Interfaces', Wiley Interscience, New York (1966).

29. B.Mann and H.Kuhn, J. Appl. Phys., 42, 4398 (1971).

30. H.Kuhn, Naturwiss., 54, 429 (1967).

31. D.Möbius, Acc. Chem. Res., 14, 63 (1981).

32. H.Kuhn, Thin Solid Films, 99, 1 (1983).

33. H.Kuhn, D.Möbius and H.Bücher, in *Physical Methods of Chemistry* Eds A.Weissberger and B.Rossiter, Vol 1 Part 3B Chap 7, Wiley NY (1972)

34. H.Kuhn, Naturwiss., 54, 429 (1967).

35. A.Aviram and M.A.Ratner, Chem. Phys. Lett., 29, 277 (1974).

36. F.L.Carter, J. Vac. Sci. Technol., B1, 959 (1983).

37. M.Conrad, Commun. A.C.M., 28, 464 (1985).

38. K.K.Kan, G.G.Roberts and M.C.Petty, Thin Solid Films, 99, 291 (1983)

39. J.R.Drabble and S.M.Al-Khowaildi, Thin Solid Films, 99, 271 (1983).

40. K.Heckmann, C.Strobl and S.Bauer, Thin Solid Films, 99, 265 (1983).

41. I.R.Peterson, IEE Proc. 130I, 252 (1983).

42. B.M.J.Kellner and G. Czornyj, Proc. Symp. on Surface and Colloid Sci. in Computer Technol., Potsdam, NY, June 24-28 (1985)

43. E.G.Wilson, Mol. Cryst. Liq. Cryst., 121, 271 (1985).

44. R.A.Hann and D.Bloor, Eds, 'Organic Materials for Nonlinear Optics', Special Publication 69, Royal Society of Chemistry, London (1989).

45. J.H.Hildebrand, Chem. Revs., 44, 37 (1949).

46. D.W.Van Krevelen and P.J.Hoftyzer, 'Properties of Polymers', Elsevier, Amsterdam (1976), pp129-148

47. A.F.M.Barton, 'CRC Handbook of Solubility Parameters and Other Cohesion Parameters', CRC Press, Boca Raton (1983).

48. F.M.Fowkes, Indust. Eng. Chem., 56, 41 (1964).

49. I.Langmuir, Proc. Roy. Soc. Lond. A, 170, 1 (1939)

50. A.Cemel, T.Fort and J.B.Lando, J. Polym. Sci., A1-10, 2061 (1972).

51. R.H.Tredgold and C.S.Winter, J. Phys. D, 15, L55 (1982).

52. Y.Nishikata, M.A.Kakimoto, A.Morikawa and Y.Imai, Thin Solid Films, 160, 15 (1988)

53. A.J.Vickers, R.H.Tredgold, P.Hodge, E.Khoshdel and I.Girling, Thin Solid Films, 134, 43 (1985).

54. J.Gun and J.Sagiv, J. Coll. Interface Sci., 112, 457 (1986).

55. H.O.Finklea, L.R.Robinson, A.Blackburn, B.Richter, D.Allara and T.Bright, Langmuir, 2, 239 (1986).

56. R.Maoz, L.Netzer, J.Gun and J.Sagiv, J. Chim. Physique, 85, 1059 (1988).

57. H.Lee, L.J.Kepley, H.G.Hong, S.Akhter and T.E.Mallouk, J. Am. Chem. Soc., 92, 2597 (1988).

58. H.Lee, L.J.Kepley, H.G.Hong, S.Akhter and T.E.Mallouk, J.Phys. Chem., 92, 2597 (1988).

59. G.G.Roberts, P.S.Vincett and W.A.Barlow, Phys. Technol., 12, 69 (1981).

198

60. B.R.Malcolm, Paper AO2, Proc. 4th Intl. Conf. on LB Films, Thin Solid Films, 178, 17 (1989).

61. G.Veale and I.R.Peterson, J. Coll. Interface Sci., 103, 178 (1983).

62. M.R.Buhaenko, J.W.Goodwin, R.M.Richardson and M.F.Daniel, Thin Solid Films, 134, 217 (1985).

63. R.C.Weast and M.J.Astle, eds, 'Handbook of Chemistry and Physics', 60th Edn., Chem. Rubber Co., Cleveland, (1980).

64. I.Langmuir, J. Am. Chem. Soc., 39, 1848 (1917).

65. W.D.Harkins and T.F.Anderson, J. Am. Chem. Soc., 59, 2189 (1937).

66. P.Fromherz, Rev. Sci. Instrum., 46, 1380 (1975).

67. G.L.Gaines, Jr., 'Insoluble Monolayers at Liquid-Gas Interfaces', Wiley Interscience, New York (1966), pp66-73.

68. I.R.Peterson, Thin Solid Films, 134, 135 (1985).

69. H.D.Göbel and H.Möhwald, Thin Solid Films, 159, 63 (1988).

70. H.Oyanagi, M.Yoneyama, K.Ikegami, M.Sugi, S.I.Kuroda, T.Ishiguro and T.Matsushita, Thin Solid Films, 159, 435 (1988).

71. O.Albrecht, Thin Solid Films, 99, 227 (1983).

72. H.E.Ries and W.A.Kimball, Proc. 2nd Intl. Congr. on Surface Activity, Vol 1, Butterworths Scientific (1957), p75.

73. H.E.Ries, Nature, 281, 287 (1979).

74. A.Barraud, J.Leloup, P.Maire and A.Ruaudel-Teixier, Thin Solid Films, 133, 133 (1985).

75. P. de Keyser and P.Joos, J. Phys. Chem., 88, 274 (1984).

76. E.P.Honig, J.H.T.Hengst and D.Den Engelsen, J. Coll. Interface Sci., 45, 92 (1973).

77. G.L.Gaines, Jr., J.Phys.Chem., 60, 1533 (1956).

78. G.Veale, I.R.Girling and I.R.Peterson, Thin Solid Films, 127, 293 (1985).

79. I.R.Peterson, Aust. J. Chem., 33, 1713 (1980).

80. M.F.Daniel and J.T.T.Hart, J. Mol. Electron., 1, 97 (1985).

81. B.R.Malcolm, J. Phys. E 21, 603 (1988).

82. F.Kajzar, I.R.Girling and I.R.Peterson, Thin Solid Films, 160, 209 (1988).

83. R.H.Tredgold, R.Jones, S.D.Evans and P.I.Williams, J. Mol. Electron., 2, 147 (1986).

84. R.Benferhat, B.Drévillon and P.Robin, Thin Solid Films, 156, 295 (1988).

85. R.M.Handy and L.C.Scala, J. Electrochem. Soc., 113, 109 (1966).

86. T.M.Ginnai, D.P.Oxley and R.G.Pritchard, Thin Solid Films, 68, 135 (1980).

87. B.Rothenhäusler and W.Knoll, Opt. Comm., 63, 301 (1987).

88. F.P.Mertens and R.C.Plumb, J. Phys. Chem., 67, 908 (1963).

89. X.Q.Yang, T.Inagaki, T.A.Skotheim, Y.Okamoto, L.Samuelsen, G.Blackburn and S.Tripathy, Mol. Cryst. Liq. Cryst., , 253 (1988).

90. I.R.Peterson and G.J.Russell, Philos. Mag., 49A, 463 (1984).

91. G.J.Russell, M.C.Petty, I.R.Peterson, G.G.Roberts, J.P.Lloyd and K.K.Kan, J. Mat. Sci. Lett., 3, 25 (1984).

92. C.Vogel, J.Corset and M.Dupeyrat, J. Chim. Phys. Phys. Chim. Biol., 76, 903 (1979).

93. O.Albrecht, T.Ginnai, A.Harrington and D.Marr-Leisy, Paper BP-3, Proc. 4th Intl Conf. on LB Films, Thin Solid Films, 178, 171 (1989).

94. O.Albrecht, Thin Solid Films, 99, 227 (1983).

95. F.C.Saunders, J.Staromlynska, G.W.Smith and M.F.Daniel, Mol. Cryst. Liq. Cryst., 122, 297 (1985).

96. R.Grundy, at the one-day meeting 'The Physics of Langmuir-Blodgett Films', Institute of Physics, Nov 1988.

97. S.E.Clark and D.C.Emmony, J. Phys. D, 22, 600 (1989)

200

98. B.Lin, J.B.Peng, J.B.Ketterson, P.Dutta, B.N.Thomas, J.Buontempo and S.A.Rice, J. Chem. Phys., 90, 2393 (1989).

99. O.Albrecht, H.Gruler and E.Sackmann, J. Coll. Interface Sci., 79, 319 (1981).

100. N.L.Gershfeld, J. Coll. Interface Sci., 85, 28 (1982)

101. N.R.Pallas and B.A.Pethica, Langmuir, 1, 509 (1985).

102. G.L.Gaines, Jr., 'Insoluble Monolayers at Liquid-Gas Interfaces', Wiley Interscience, New York (1966), p158.

103. W.D.Harkins, 'Physical Chemistry of Surface Films', Reinhold, New York (1952), pp106-117.

104. M.Lundquist, Chemica Scripta, 1, 5, 197 (1971)

105. G.Ungar, J. Phys. Chem., 87, 689 (1983).

106. G.L.Gaines, Jr., 'Insoluble Monolayers at Liquid-Gas Interfaces', Wiley Interscience, New York (1966), pp163-6,214.

107. A.I.Kitaigorodsky, 'Molecular Crystals and Molecules', Academic Press New York (1973).

108. S.Ställberg-Stenhagen and E. Stenhagen, Nature, 156, 239 (1945).

109. D.G.Dervichian, J. Chem. Phys., 7, 931 (1939).

110 . A.Fischer and E.Sackmann, J. Coll. Interface Sci., 112, 1 (1986).

111. K.Kjær, J.Als-Nielsen, C.A.Helm, P.Tippmann-Krayer and H.Möhwald, Thin Solid Films, 159, 17 (1988).

112. B.M.Abraham, K.Miyano, J.B.Ketterson and S.Q.Xu, Phys. Rev. Lett., 51, 1975 (1983).

113. S.Garoff, H.W.Deckmann, J.H.Dunsmuir, M.S.Alvarez and J.M.Bloch, J. Physique, 47, 701 (1986).

114. K.Kjær, J.Als-Nielsen, C.A.Helm, L.A.Laxhuber and H.Möhwald, Phys. Rev. Lett., 58, 2224 (1987).

115. K.Kjær, J.Als-Nielsen, C.A.Helm, P.Tippmann-Krayer and H.Möhwald, J. Phys. Chem., 93, 3200 (1989).

116. M.Iwahashi, N.Maehara, Y.Kaneko, T.Seimiya, S.R.Middleton, N.R.Pallas and B.A.Pethica, J. Chem. Soc. Faraday Trans., 81, 973 (1985).

117. J.H.Brooks and A.E.Alexander, J. Phys. Chem., 66, 1851 (1962).

118. O.Albrecht, Paper AP-6, 4th Intl Conf. on LB Films, Thin Solid Films, 178, 93 (1989).

119. M.Lösche, E.Sackmann and H.Möhwald, Ber. Bunsenges. Phys. Chem., 87, 848 (1983).

120. H.Möhwald, Thin Solid Films, 159, 1 (1988).

121. R.M.Weiss and H.M.McConnell, Nature, 310, 5972 (1984)

122. R.Peters and K.Beck, Proc. Natl. Acad. Sci. USA, 80, 7183 (1983)

123. J.W.Barton, M.Buhaenko, B.Moyle and N.M.Ratcliffe, J. Chem. Soc. Chem. Commun., 488 (1988).

124. S.Baker, M.C.Petty, G.G.Roberts and M.V.Twigg, Thin Solid Films, 99, 53 (1983).

125. B.R.Malcolm, Thin Solid Films, 134, 201 (1985).

126. I.R.Peterson, G.J.Russell, J.D.Earls and I.R.Girling, Thin Solid Films, 161, 325 (1988).

127. J.A.Spink, J. Coll. Interface Sci., 23, 9 (1967).

128. L. v. Szentpály, D.Möbius and H.Kuhn, J. Chem. Phys., 52, 4618 (1970).

129. A.Matsuda, M.Sugi, T.Fukui, S.Iizima, M.Miyahara and Y.Otsubo, J.App. Phys., 48, 771 (1977).

130. K.B.Blodgett, J. Am. Chem. Soc., 57, 1007 (1935).

131. G.G.Roberts, P.S.Vincett and W.A.Barlow, J. Phys. C, 11, 2077 (1978).

132. G.G.Roberts, K.P.Pande and W.A.Barlow, Sol. St. Electron. Dev., 2, 169 (1978).

133. K.Kjaer, J. Als-Nielsen, C.A.Helm, P.Tippman-Krayer and H.Möhwald, J.Phys. Chem., 93, 3200 (1989).

134. J.F.Rabolt, F.C.Burns, N.E.Schlotter and J.D.Swalen, J. Electron. Spectrosc. Rel. Phenom., 30, 29 (1983).

135. J.F.Rabolt, F.C.Burns, N.E.Schlotter and J.D.Swalen, J. Chem. Phys., 78, 946 (1983).

202

136. A.Bonnerot, P.A.Chollet, H.Frisby and M.Hoclet, Chem. Phys., 97, 365 (1985).

137.W.L.Barnes and J.R.Sambles, Surf. Sci., 183, 189 (1987).

138. I.R.Peterson and G.J.Russell, Brit. Polym. J., 17, 364 (1985).

139. I.Robinson, J.R.Sambles and I.R.Peterson, Thin Solid Films, 172, 149 (1989).

140. W.L.Barnes and J.R.Sambles, J. Phys. D, 20, 1125 (1987).

141. V.Skita, W.Richardson, M.Filipkowski, A.Garito and J.K.Blasie, J. Physique, 47, 1849 (1986).

142. I.R.Peterson, in *Polydiacetylenes*, NATO ASI E102 Eds D.Bloor and R.R.Chance, Martinus Nijhoff Dordrecht (1985) p377.

143. P.A.Chollet, Thin Solid Films, 52, 343 (1978).

144. M.D.Chinn and S.C.Fain, Phys. Rev. Lett., 39, 146 (1977).

145. C.G.Shaw, S.C.Fain, M.D.Chinn, Phys. Rev. Lett., 41, 955 (1978).

146. J.P.McTague and A.D.Novaco, Phys. Rev. B, 19, 5299 (1979).

147. A.D.Novaco, Phys. Rev. B, 19, 6493 (1979).

148. K.B.Blodgett, J. Am. Chem. Soc., 57, 1007 (1935).=4B

149. S.Allen, R.A.Hann, S.K.Gupta, P.F.Gordon, B.D.Bothwel, I.Ledoux, P.Vidakovic, J.Zyss, P.Robin, E.Chastaing,and J.C.Dubois, Proc. Soc. Photo-Opt. Eng., 682, 97 (1987).

150. S.T.Kowel, L.M.Hayden, R.H.Selfridge, Proc. Soc. Photo-Opt. Eng., 682, 103 (1987).

151. M.Prakash, J.B.Peng, J.B.Ketterson and P.Dutta, Chem. Phys. Lett., 128, 354 (1986).

152. J.F.Stephens and C. Tuck-Lee, J. Appl. Cryst., 2, 1 (1969).

153. M.Seul, P.Eisenberger and H.M.McConnell, Proc. Natl. Acad. Sci. USA, 80, 5795 (1983).

154. P.Haasen, *Physikalische Metallkunde*, Springer Berlin (1974), Chaps. 3 and 12.

155. I.R.Peterson, J. Mol. Electron., 2, 95 (1986).

156. I.R.Peterson, G.J.Russell, I.R.Girling and W.L.Barnes, J. Phys. D, 21, 773 (1988); I.R.Peterson, Brit. Polym. J., 19, 391 (1987).

157. G.W.Gray and J.W.Goodby, *Smectic Liquid Crystals: Textures and Structures*, Leonard Hill, Glasgow UK (1984).

158. I.R.Peterson, J. Mol. Electron., 3, 103 (1987).

159. J.Friedel, *Dislocations*, Pergamon, Oxford (1964).

160. F.R.N.Nabarro, *Theory of Crystal Dislocations*,Clarendon, Oxford (1967) 53

161. A.D.Kurtz, S.A.Kulin and B.L.Averbach, J. Appl. Phys., 27, 1287 (1956).

162. I.R.Peterson, Thin Solid Films, 116, 357 (1984).

163. F.L.Vogel, W.G.Pfann, H.E.Corey and E.E.Thomas, Phys. Rev. 90, 489 (1953).

164. I.R.Peterson, R.Steitz, H.Krug and I.Voigt-Martin, J. Physique, *in press*

165. D.R.Nelson and B.I.Halperin, Phys. Rev. B, 19, 2457 (1979).

166. D.R.Nelson and B.I.Halperin, Phys. Rev. B, 21, 5312 (1979).

167. D.R.Saperstein, J. Phys. Chem., 90, 1408 (1986).

168. R.H.Tredgold, A.J.Vickers and R.A.Allen, J. Phys. D, 17, L5 (1984).

169.L.A.Laxhuber, B.Rothenhäusler, G.Schneider and H.Möhwald, Appl. Phys. A, 39, 173 (1986).

170. F.Kopp, U.P.Fringeli, K.Mühlethaler and H.H.Günthard, Biophys. Struct. Mechanism., 1, 75 (1975).

171. E.Stenhagen, Trans. Faraday Soc., 34, 1328 (1938).

172. A.Banerjie and J.B.Lando, Thin Solid Films, 68, 67 (1980).

173. M.F.Daniel, O.C.Lettington and S.M.Small, Thin Solid Films, 99, 61 (1983).

174. G.Lieser, B.Tieke and G. Wegner, Thin Solid Films, 68, 77 (1980).

175. J.B.Lando, in *Polydiacetylenes*, Eds D.Bloor and R.R.Chance, NATO ASI E102, Martinus Nijhoff Dordrecht (1985).

176. A.Cemel, T.Fort and J.B.Lando, J. Polym. Sci., A1-10, 2061 (1972).

177. B.Hupfer, H.Ringsdorf and H.Schupp, Makromol. Chem., 182, 247 (1981).

178. K.Miyano and A.Mori, Thin Solid Films, 168, 141 (1989).

179. M.M.Carpenter, P.N.Prasad and A.C.Griffin, Thin Solid Films, 161, 315 (1988).

180. R.H.Tredgold, A.J.Vickers, A.Hoorfar, P.Hodge and E.Khoshdel, J.Phys. D, 18, 1139 (1985).

181. R.Jones, R.H.Tredgold, A.Hoorfar and P.Hodge, Thin Solid Films, 113, 115 (1984).

182. R.H.Tredgold, S.D.Evans, P.Hodge and A.Hoorfar, Thin Solid Films, 160, 99 (1988).

183. S.Baker, M.C.Petty, G.G.Roberts and M.V.Twigg, Thin Solid Films, 99, 53 (1983).

184. J.H.Kim, T.M.Cotton, R.A.Uphaus and C.C.Leznoff, Thin Solid Films, 159, 141 (1988).

185. R.H.Tredgold, A.J.Vickers, A.Hoorfar, P.Hodge and E.Khoshdel, J. Phys. D, 18, 1139 (1985).

186. G.G.Roberts, T.M.McGinnity, W.A.Barlow and P.S.Vincett, Thin Solid Films, 68, 223 (1980).

187. E.Orthmann and G.Wegner, Angew. Chem., 98, 114 (1986).

188. P.Stroeve, M.P.Srinivasan, B.G.Higgins and S.T.Kowel, Thin Solid Films, 146, 209 (1987).

189. I.R.Peterson, J.D.Earls, I.R.Girling and W.L.Barnes, J.Phys. D, 21, 773 (1988).

190. W.F.Harris, Scientific American, 237, 130 (Dec1977).

191. I.R.Peterson, Brit. Polym. J., 19, 391 (1987).

192. I.R.Peterson, J. Chim. Physique, 85, 997 (1988).

193. S.Miyata, H.Kumehara, T.Kasuga and A.Satomi, Paper BO-1, 4th Intl Conf. on LB Films, Tsukuba, Japan (1989).

194. I.R.Peterson, J.D.Earls, I.R.Girling and G.J.Russell, Mol. Cryst. Liq. Cryst., 147, 141 (1987).

195. A.M.Bibo and I.R.Peterson, Paper AP-5, 4th Intl Conf. on LB Films, Thin Solid Films, 178, 81 (1989).

196. J.Zyss and D.S.Chemla, Nonlinear Optical Properties of Organic Molecules and Crystals, Vol. 1, Academic Press Orlando (1987).

197. E.E.Polymeropoulos, J. Appl. Phys., 48, 2404 (1977).

198. E.E.Polymeropoulos, Sol. St. Comm., 28, 883 (1978).

199. G.G.Roberts, K.P.Pande and W.A.Barlow, Electron. Lett., 13, 581 (1977).

200. J.Batey, G.G.Roberts and M.C.Petty, Thin Solid Films, 99, 283 (1983).

201. K.H.Gundlach and J.Kadlec, Chem. Phys. Lett., 25, 293 (1974).

202. N.J.Geddes, J.R.Sambles, D.J.Jarvis and N.R.Couch, Proc. 3rd Workshop on Molecular Electronic Devices, Washington (1986).

203. N.R.Couch, C.M.Montgomery and R.Jones, Thin Solid Films, 135, 173 (1986).

204. S.Hao, B.H.Blott and D.Melville, Thin Solid Films, 132, 63 (1986).

205. M.Sugi, Researches of the Electrotechnical Laboratory, 794 (1978).

206. M.Sugi and S.Iizima, Phys. Rev. B, 15, 574 (1977).

207. N.J.Geddes, W.G.Parker, J.R.Sambles, N.R.Couch and D.J.Jarvis, Paper EO-5, 4th Intl. Conf. on LB Films, Thin Solid Films, 179, 143 (1989).

208. I.R.Peterson, G.J.Russell and G.G.Roberts, Thin Solid Films, 109, 71 (1983).

209. T.Tanaka, J. Appl. Phys., 44, 2430 (1973).

210. A.K.Vijh and J.P.Crine, J. Appl. Phys., 65, 398 (1989).

211. J.R.Drabble and S.M.Al-Khowaildi, Thin Solid Films, 99, 271 (1983).

212. W.R.Barger, A.W.Snow, H. Wohltjen and N.L.Jarvis, Thin Solid Films, 133, 197 (1985).

213. E.G.Wilson, Electron. Lett., 19, 237 (1983).

214. D.R.Day and J.B.Lando, J. Appl. Polymer Sci., 26, 1605 (1981).

215. A.Barraud, M.Flörsheimer, H.Möhwald, J.Richard, A.Ruaudel-Teixier and M.Vandevyver, J. Coll. Interface Sci., 121, 491 (1988).

216. A.S.Dhindsa, M.R.Bryce, J.P.Lloyd and M.C.Petty, Thin Solid Films, 163, L97 (1988).

217. T.Nakamura, M.Matsumoto, F.Takei, M.Tanaka, T.Sekiguchi, E.Manda and Y.Kawabata, Chem. Lett., 709 (1986).

218. H.Kelker, Mol. Cryst. Liq. Cryst., 21, 1 (1973).

CHAPTER 4
Nonlinear Optics
S. Allen

1. INTRODUCTION

The field of research broadly classified as optoelectronics is relatively new; it is only since the availability of laser sources in the 1960's, and the subsequent development of dielectric waveguide theories that the topics described within this chapter have been studied in any depth. The use of optics in communications and in image processing are areas that are likely to be of great commercial importance in coming years. The nonlinear optical processes discussed below form the basic enabling science for a large range of devices having applications in these fields. Although a number of inorganic materials have been developed over the past two decades, as described below these do not meet all the requirements of device manufacturers. For this reason there is much ongoing research into new, improved materials. It is clear from work over the past ten years that a number of organic (molecular and polymeric) compounds have great potential. In this chapter the relative merits of the different material technologies are discussed, and progress in the development of novel organic nonlinear optical compounds are reviewed. First though, a brief review of the physics of nonlinear optics is given.

2. NONLINEAR OPTICS
2.1 PHYSICS OF NONLINEAR OPTICS

Nonlinear optical processes describe the interaction of electromagnetic radiation (light waves) with dielectric media, under circumstances where the properties of the light cannot be simply described in terms of classical reflection, refraction and absorption. The reader is referred to any of a number of recent texts [1-3] for detailed physical descriptions and mathematical derivations of the many different nonlinear optical processes that have been investigated. In this section a brief outline of the principal ideas involved in nonlinear optical interactions is presented. When the oscillating electric field E associated with a light wave is incident on a dielectric medium it induces an oscillating polarization P within that medium. In a linear medium this polarization is directly proportional to the field E:

$$P = \varepsilon_o \chi E \tag{1}$$

The constant of proportionality (χ) is known as the polarizability of the medium, and is related to the refractive index (n) by the expression

$$n^2 = 1 + \chi \tag{2}$$

In a linear medium both χ and n are material constants for a given temperature and pressure. The oscillating induced polarization P now re-radiates an electric field, and it is the sum of the re-radiated fields from all the constituent parts of the medium that forms the propagated light beam. In this case the propagated beam clearly has the same frequency as the incident one.

In a nonlinear optical medium this strict proportionality between force (electric field) and response (induced polarization) no longer holds. It is now usual to express P in terms of powers of the field E:

$$P = \varepsilon_o \{ \chi^{(1)}E + \chi^{(2)}E^2 + \chi^{(3)}E^3 + \ldots \} \tag{3}$$

$\chi^{(1)}$ is termed the linear polarizability, $\chi^{(2)}$ and $\chi^{(3)}$ the second-order and third-order hyperpolarizabilities and so on. In general the successive terms in equation (3) are of decreasing importance. As an illustration of how these additional nonlinear terms in the expression for P can lead to interesting physical effects, consider an light wave of frequency ω ($= 2\pi c/\lambda$). The oscillating electric field can be represented as $E = E_o\sin(\omega t)$. Substituting this into equation (3) and considering just the second-order term gives

$$(P/\varepsilon_o) = \chi^{(1)}E_o\sin(\omega t) + \chi^{(2)}E_o^2\sin^2(\omega t) \tag{4}$$

On using the trigonometric identity, $\sin^2\Phi = \frac{1}{2}[1-\cos(2\Phi)]$, this becomes

$$(P/\varepsilon_o) = \frac{1}{2}\chi^{(2)}E_o^2 + \chi^{(1)}E_o\sin(\omega t) - \frac{1}{2}\chi^{(2)}(E_o)^2\cos(2\omega t) \tag{5}$$

In this case the induced polarization, and hence the re-radiated output field, has a component at twice the input frequency (second harmonic generation) and a d.c. term, as well as one of the input frequency. Second harmonic generation is an example of a "two-wave-mixing" effect, with two input photons of energy $\hbar\omega$ producing an output photon of energy $2\hbar\omega$. A similar derivation to that given above, but taking into account also the third order term would lead to additional terms, with output frequencies at ω and 3ω. These third order terms involve the mixing of three input photons.

A great variety of different nonlinear optical effects has been observed, involving the mixing of inputs of the same or different frequencies. Some of the most well-known effects are summarized in Table 1.

Although all the second-order processes can be described in terms of a second-order susceptibility $\chi^{(2)}$, the magnitude of this parameter will not in general be the same for all the different processes. This is because of the involvement of different frequencies in these processes. For example, many inorganic nonlinear optical materials are also ferroelectric. In the d.c. electro-optic effect, application of a static or low frequency electric field will modify the refractive index of the material, due both to a distortion of the electronic distribution within the crystal and also to a displacement of the ferroelectric nuclei. The magnitude of this change in refractive index is related to $\chi^{(2)}$. When the same material is used for second harmonic generation, all the applied fields are at optical frequencies. The nuclei are unable to respond at these frequencies and so their contribution to the nonlinearity is lost. Thus the magnitude of $\chi^{(2)}$ is different from that obtained for the d.c. electro-optic effect. Other factors such as the proximity of input or output frequencies to atomic or molecular resonances within the material will also contribute to the dispersion of the susceptibilities. These effects will be described in more detail later.

Table 1. SUMMARY OF NONLINEAR OPTICAL EFFECTS

Input frequencies	output frequencies	effect
$0,\omega$	ω	d.c. electro-optical effect (Pockels effect)
ω,ω	2ω	second harmonic generation
ω_1,ω_2	$\omega_1\pm\omega_2$	frequency mixing
ω,ω,ω	3ω	third harmonic generation
ω,ω,ω	ω	optical bistability
ω,ω,ω	ω	optical Kerr effect
ω,ω,ω	ω	degenerate four wave mixing
ω,ω,ω	ω	phase conjugation

In order to account for the above variations in the susceptibility it is customary to use the notation $\chi^{(2)}(\omega_3; \omega_1,\omega_2)$ and $\chi^{(3)}(\omega_4; \omega_1,\omega_2,\omega_3)$ for second and third order susceptibilities respectively, when any ambiguity would otherwise arise. In both cases the first argument refers to the output frequency, whilst the others denote the input frequencies.

Other definitions also exist for nonlinear optical coefficients specific to certain effects. Those pertaining to second harmonic generation and the electro-optic effect are most commonly encountered. Considering again equation (5), the polarization $P^{2\omega}$ at the second harmonic frequency, due to an incident electric field of the form $E = E_o\sin(\omega t)$ has the form

$$P^{2\omega} = \tfrac{1}{2}\varepsilon_o\chi^{(2)}E_o{}^2\cos(2\omega t) \tag{6}$$

By convention, the nonlinear optical coefficient for second harmonic generation, denoted 'd', is defined by the equation

$$P^{2\omega} = \varepsilon_o d E_o^2 \cos(2\omega t) \qquad (7)$$

and is therefore simply related to the nonlinear susceptibility:

$$d = \tfrac{1}{2}\chi^{(2)}(2\omega;\omega,\omega) \qquad (8)$$

We can consider the electro-optic effect in a manner similar to the derivation of equation (5). In this case the electric field within the crystal is a combination of a static field E^0 and an optical field E^ω: $E = E^0 + E^\omega \sin(\omega t)$. Now the change in induced polarization at the frequency ω is given by:

$$\delta P^\omega = 2\varepsilon_o \chi^{(2)}(\omega;0,\omega).E^0 E^\omega \sin(\omega t) \qquad (9)$$

An electro-optic coefficient 'r' is defined in terms of the change in the relative impermeability tensor induced by an applied static field:

$$\delta\kappa_{ij} = r_{ijk}E_k \qquad (10)$$

where the refractive indices n_i of the material are related to the principal components of κ by

$$n_i^2 = \varepsilon_{ii} = 1/\kappa_{ii} \qquad (11)$$

It can therefore be shown [3] that the nonlinear susceptibility and electro-optic coefficient are related by the equation

$$2\chi_{ijk}(\omega;0,\omega) = \varepsilon_{ii}\varepsilon_{jj}r_{ijk} \qquad (12)$$

In general many of the above-mentioned nonlinear optical effects can occur in any material at the same time. For example, input of an optical field at frequency ω could generate frequencies 0 and 2ω by second order processes and ω and 3ω by third order processes. The relative importance of the various processes in any given situation is largely determined by the symmetry properties of the material, and the geometry (with respect to the incident beams) in which it is used.

Equation (3) is in fact an over simplification. Nonlinear optical materials are in general anisotropic, and so the magnitude of the susceptibility will depend on the direction of propagation and polarization of the light and on the direction of any applied electric fields. In equation (3) the quantities E and P are in fact vectors, and the susceptibilities $\chi^{(i)}$ are tensors of rank $(i+1)$ [4]. Thus $\chi^{(2)}$ may have 27 different components, and $\chi^{(3)}$ could have 81! In practice the number of non zero, independent components is usually much reduced due to symmetry considerations. The physics of tensorial material properties have been comprehensively described by Nye [4]. In fact, all components of $\chi^{(2)}$ vanish in any material which possesses a centre of inversion symmetry, whereas non-zero components of $\chi^{(3)}$ may be present in any material. This is a factor of great importance in the design of organic and polymeric nonlinear optical materials, as described in later sections.

Probably the most important factor in the maximization of one nonlinear optical process with respect to the others is that of phase matching [3]. This can be understood by considering the mathematical expression for the intensity ($I^{2\omega}$) of second harmonic generated by an incident beam of intensity I^{ω} passing through a crystal of length L, having a nonlinear optical coefficient d and refractive index n:

$$I^{2\omega} = 2(\omega L)^2 (\mu_o/\varepsilon_o)^{3/2} (d^2/n^3)(I^{\omega})^2 \{\sin^2(\delta k.L/2)/(\delta k.L/2)^2\} \qquad (13)$$

This equation is derived in detail in, for example, ref [3]. The final term in this equation is a 'sinc' function (sinc$\{x\}$ = sin$\{x\}/x$), which has a strong peak at x = 0. Thus the second harmonic intensity is only large when $\delta kL = 0$, where δk ($=k^{2\omega} - 2k^{\omega}$) is the mismatch in wavevectors between harmonic and fundamental waves. When the condition $\delta k = 0$ is met, the interaction is defined as "phase matched". Since $k^{\omega} = \omega n^{\omega}/c$ and $k^{2\omega} = 2\omega n^{2\omega}/c$,

$$\delta k = (2\omega/c).(n^{2\omega} - n^{\omega}) \qquad (14)$$

and the requirement for phase matching is that the refractive index of the medium is the same at the fundamental and harmonic frequencies. However, because all materials are to a greater or lesser extent dispersive, it is usually the case that $n^{2\omega} > n^{\omega}$. In birefringent materials phase matching can be achieved by, for example, allowing the fundamental and harmonic waves to have different polarizations, as indicated in Figure 1.

Consider as an example a uniaxially birefringent material. For any direction of propagation of light through such a material other than along the optical axis there are two possible values of the refractive index, corresponding to orthogonal polarizations of the light. For light polarized perpendicular to the optic axis, the refractive index is known as the "ordinary" refractive index (n_o), and is the same for all propagation directions. For light polarized within the plane of the optic axis and propagation direction, the refractive index is the "extraordinary refractive index", $n_e(\Phi)$, and its magnitude varies with the angle Φ between the propagation direction and the optic axis. The maximum birefringence ($n_o - n_e(\Phi)$) occurs for propagation perpendicular to the optic axis ($\Phi = 90°$). Figure 1 shows schematically the dispersion curves for a uniaxial material having $n_e > n_o$. The wavelengths of the fundamental (λ_1) and second

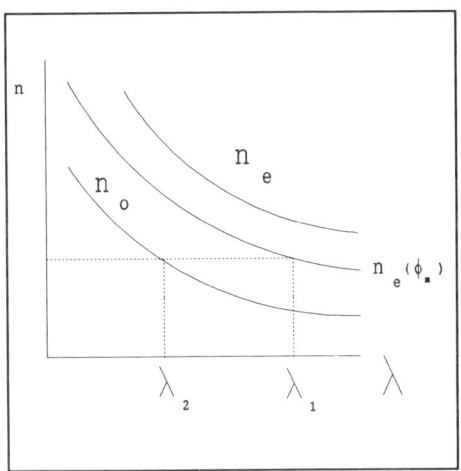

Figure 1. Typical dispersion curves for a uniaxially birefringent crystal, showing the possibility of phase matching.

harmonic (λ_2) are shown. If the birefringence is large enough, it may be possible to find some angle Φ_m at which the extraordinary refractive index at λ_1 is equal to the ordinary refractive index at λ_2, as shown. Because of the tensorial nature of the nonlinear susceptibility $\chi^{(2)}$ it is possible for a fundamental wave having the extraordinary polarization to generate a second harmonic that has the orthogonal (ordinary) polarization, so that in this case the phase matching condition, of $n_e^{\omega} = n_o^{2\omega}$ is met. Similar but more complicated arguments can be applied to other processes such as frequency mixing and parametric oscillation.

Equation (13), which is derived using the assumption that there is no significant depletion of the incident beam intensity, applied to phase matched interactions (sinc$\{\delta kL/2\} = 1$) also shows the following.

- The efficiency of second harmonic generation (defined as $I^{2\omega}/I^\omega$) is proportional to I^ω. This highlights the fact that nonlinear optical processes are only of importance for high incident intensities.
- $I^{2\omega}$ is proportional to (d^2/n^3), which is often taken as a figure of merit for second harmonic generation.
- $I^{2\omega}$ is proportional to L^2.

2.2 NONLINEAR OPTICAL APPLICATIONS

Current interest in nonlinear optical materials worldwide is intense, with significant programmes of work within all the major European, American and Japanese chemical and electronics companies. Perceived applications lie initially in the areas of optical communications and optical signal processing, as well as in data storage and other information processing markets. Some of the main applications are listed below in terms of the nonlinear optical processes described in Table 1.

Electro-optic Devices - There is currently a moderate market for bulk electro-optic devices, fabricated mainly from inorganic crystals such as lithium niobate (LiNbO$_3$) and potassium dihydrogen phosphate (KDP), for use as modulators and switches of laser beams. Q-switch devices are used within the cavities of high power lasers to control the build-up of optical power, resulting in high powers being emitted in short pulses. Modulators are used for the conversion of an electrical signal into an amplitude or phase modulation of an optical beam, for use in optical communications. Recently there has been a lot of interest in the fabrication of electro-optic waveguide structures, compatible with integrated optics, also for use as routing switches and amplitude modulators in optical communications networks.

Second Harmonic Generation - The prime interest in this area is in the generation of new laser sources. For example, KDP is frequently used with high power Nd:YAG lasers (which have a fundamental wavelength of 1.064 μm) to generate light in the green portion of the visible spectrum (532 nm). Of particular interest at present is the development of nonlinear optical materials capable of frequency-doubling low-power laser diode sources from the near infra-red into the blue region of the spectrum. This would be of use both in optical data storage, where a halving of the wavelength used results in a four-fold increase in the capacity of a disk, and also in photoreprographics (laser printers and copiers), where more sensitive photoreceptors are available at around 400 nm. As shown in the previous section, the efficiency of second harmonic generation is proportional to the power of the input beam, and so materials such as KDP provide only very low conversion of the low power diode laser beams. There is a strong drive to develop new, more efficient materials for operation in this region of the spectrum.

Frequency Mixing - Processes involving the mixing of a laser beam (usually in the infrared) carrying information, either in the form of a spatial intensity variation (*i.e.* an infrared image) or as an amplitude modulation (*i.e.* an optical signal) with a pump beam of fixed intensity can be used to convert the information carrying beam to, for example, the visible region of the spectrum. Here detection and processing may be much simpler than at the original wavelength. The inverse of frequency mixing (in which input beams of frequency ω_1 and ω_2 interact to produce one of frequency $\omega_3 = \omega_1 \pm \omega_2$) is *parametric oscillation*. In this case an input photon of frequency ω_3 results in the output of two photons at frequencies ω_1 and ω_2 (where again $\omega_3 = \omega_1 \pm \omega_2$). Thus two output beams, known as the "signal" and "idler" beams, of lower energy than the input, are produced. The frequencies obtained are determined by the phase matching conditions, which can in turn be controlled by changing the orientation of the nonlinear optical medium with respect to the input beam. In this way a tunable source of laser radiation can be produced. The process can be either spontaneous or stimulated, in which case parametric amplification can result.

Third Harmonic Generation - Applications are in principle the same as for second harmonic generation, but the efficiency of the third-order process is significantly lower.

Optical Bistability - By confinement of materials having an intensity dependent refractive index within a system that results in optical feedback such as a Fabry Perot cavity, optical bistability can be produced. This results in two output levels, or "states" being possible for a single input intensity, forming the optical analogue of the transistor. The application of such devices to optical computing and optical signal processing has been extensively described in the recent literature [5].

Degenerate Four Wave Mixing and Phase Conjugation - These are related effects, again resulting from the intensity dependence of the refractive index of a material. If two input beams of the same frequency (the "pump" beams) interfere within a third-order nonlinear optical medium, the resulting interference pattern will produce a spatial variation in refractive index. This variation in refractive index will then act as a diffraction grating, so that a further input beam (the "probe" beam) can be influenced. The diffracted (or "signal") beam has the interesting property that its phase is the complex conjugate of the phase of the probe beam. This results in many applications in signal processing, image recognition and in the reduction of beam aberrations [6].

2.3 STATUS OF INORGANIC NONLINEAR OPTICAL MATERIALS

Organic materials developed for nonlinear optical applications are in direct competition with a number of established inorganic materials. For second-order effects in particular, materials such as lithium niobate and potassium dihydrogen phosphate have been extensively studied and fabrication techniques are now well advanced. As will be discussed in following sections, the organic materials have a number of important advantages but their attractiveness to those involved in device manufacture will depend on a combination of many different properties. The status of a number of the most important inorganic nonlinear optical materials is summarized below.

Lithium niobate - This material is the current front runner for nonlinear optical waveguide applications. It has a moderately high nonlinear optical coefficient (d_{33} = 34 pm/V) [7]. Waveguides can be readily fabricated in $LiNbO_3$ by titanium ion indiffusion or by proton exchange. Waveguide second harmonic generation devices for operation with laser diodes have recently been demonstrated in $LiNbO_3$ [8], and a number of companies are producing electro-optic waveguide devices for rapid modulation and switching of light [9]. On the debit side, the material has to be grown from a non-congruent melt at high temperature which can lead to difficulties in maintaining the proper composition, and it is susceptible to optical damage at high laser powers.

Potassium dihydrogen phosphate - This material, KDP, along with its deuterated analogue KD*P and other isomorphs, has been used extensively to produce bulk crystal frequency doublers and electro-optic shutters for use in high power laser systems. Although it has a relatively low nonlinear optical coefficient (d_{36} = 0.63 pm/V) [10], large high-quality single crystals can be grown from solution. The quality of these crystals is such that they are used for frequency doubling of the extremely intense laser beams used in fusion research.

Potassium titanyl phosphate - KTP is a relative newcomer to nonlinear optics, showing great promise. Its Figure of Merit for second harmonic generation is approximately 30 times that of KDP and 3 times that of $LiNbO_3$ [11], with a phase matchable d_{23} coefficient of about 13.7 pm/V [12]. Highly efficient frequency doubling has been demonstrated in this material, and crystals have been used for intra-cavity SHG of quasi-cw Nd:YAG lasers. Recently nonlinear optical waveguides have been fabricated in KTP using ion-exchange methods [12]. The main problems at present are the limited size of crystals that can be grown and optical damage. In high power applications cumulative exposure to incident radiation results in discolouration in the bulk of the material.

ß-barium borate - BBO is another recently developed nonlinear optical crystal, having a reasonable nonlinearity (d_{11} = 2.6 pm/V, being 4.1 times that of KDP) and high optical damage thresholds (13.5 GW/cm^2 for a 1 nsec pulse at 1.06 μm)

[13]. BBO is of particular interest because of its wide transparency range, from 200 nm to 2.6 μm, which makes it the prime candidate at present for frequency doubling and frequency mixing in the ultraviolet portion of the spectrum.

Gallium arsenide - GaAs is an electro-optic material, having a fairly low coefficient (r = 1.5 pm/V). Its main attraction in this field is the ability to fully integrate the electro-optic waveguide functions onto the same wafer as other optoelectronic components such as lasers and photodetectors.

Optical bistability in GaAs and other III-V semiconductors has received much attention over recent years. Very large intensity-dependent changes in refractive index can be obtained in these materials when working with light whose frequency is very close to the absorption band edge. The main disadvantage of these materials is the large investment in equipment that is needed to produce them.

3. ORGANICS FOR SECOND-ORDER NLO APPLICATIONS

3.1 INTRODUCTION

Although, as described above, inorganic materials are currently used for nonlinear optical applications, their suitability is in many cases limited by the magnitude of their nonlinear optical coefficients, and the tendency for high optical powers to damage the materials. In recent years studies on a number of organic materials [14-16] have shown that such compounds can have significant advantages over the inorganic materials. Foremost is the fact that organic materials may have much higher nonlinear optical coefficients than compounds such as lithium niobate and KDP. A comparison of the $\chi^{(2)}$ values of several organic and inorganic materials is given in Figure 2. The full nomenclature and molecular structures for the organic compounds are given in Table 2. The origin of the high nonlinearities of the organics is described in more detail in section 3.2 below.

There is some evidence that organic crystalline compounds may also show higher resistance to optical damage than their inorganic counterparts. Urea, for example, has an "optical damage threshold" of 5GW/cm^2 [17] at 1.06 μm (for 10

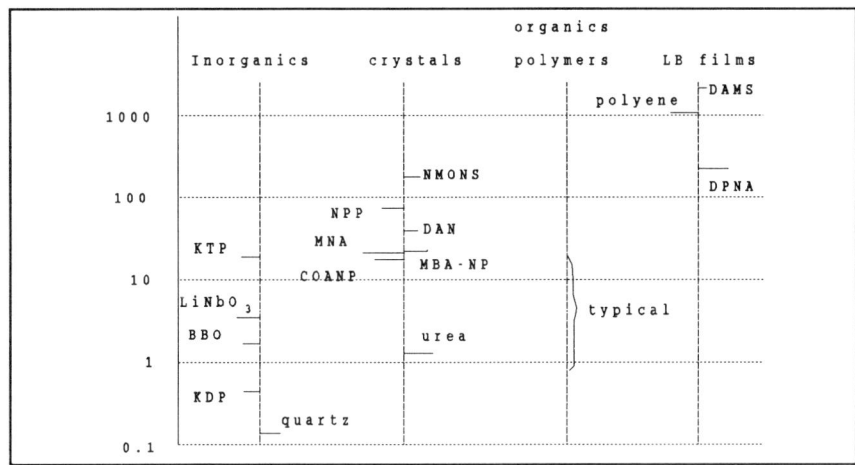

Figure 2. A comparison of the phase matchable second harmonic generation d-coefficients for a number of inorganic and organic compounds.

nsec pulses) whereas that of KDP is about 0.5 GW/cm² [18] under the same conditions.

Other advantages of organic crystals include the fact that they frequently have high intrinsic birefringence, which can facilitate phase matching for parametric processes such as second harmonic generation, and the observation that they usually have significantly lower dielectric permittivities than inorganics (which are in general also ferroelectric). This is of importance in applications such as electro-optic modulation, where high frequency electric fields have to be applied across the sample. The maximum modulation frequency that can be achieved is determined by (amongst other things) the capacitance of the sample.

Equation 3 expressed the bulk polarization induced in a material in terms of powers of the electric field of the incident light. A similar equation may be written to relate the polarization p induced in a single molecule:

$$p = \varepsilon_o \{ \alpha E + \beta E^2 + \gamma E^3 + ... \} \tag{15}$$

where now α is the "molecular polarizability", and β and γ are the second-order and third-order molecular hyperpolarizabilities and as in the case of the bulk

Table 2. MOLECULAR STRUCTURES AND NOMENCLATURE

molecule	nomenclature	acronym
benzene (structure)	benzene	
O_2N—(structure)	nitrobenzene	
H_2N—(structure)	aniline	
H_2N—(structure)—NO_2	4-nitroaniline	p-NA
(structure)—NO_2, H_2N	3-nitroaniline	m-NA
H_2N—(structure)—NO_2, H_3C	2-methyl-4-nitroaniline	MNA
CH_3, O—N(structure)—NO_2	3-methyl-4-nitro-pyridine-N-oxide	POM
O_2N—(structure)—N(prolinol), OH	N-(4-nitrophenyl)-(s)-prolinol	NPP
O_2N—(structure, N)—N(prolinol), OH	N-(5-nitro-2-pyridyl)-(s)-prolinol	PNP

Table 2 (Continued)

molecule	nomenclature	acronym
	3-(1,1-dicyanoethenyl)-1-phenyl-4,5-dihydro-1H-pyrazole	DCNP
	2,4-dinitrophenyl-(L)-alaninemethylester	MAP
	4-dimethylamino-N-methyl-4-stibazolium Iodide	SPCD
	2-N-cyclooctylamino-5-nitropyridine	COANP
	2-N-(methylbenzylamino)-5-nitropyridine	MBA-NP
	3-methyl-4-methoxy-4'nitrostilbene	MMONS

properties are in fact tensors of the second and third rank respectively. To illustrate the complex nature of the full tensorial form of equations (3) and (15), the terms for the bulk and molecular polarizabilities associated with second harmonic generation are given by

$$P_I^{2\omega} = \varepsilon_o \chi_{IJK}(2\omega;\omega;\omega).E_J^{\omega}E_K^{\omega} \tag{16a}$$

$$p_i^{2\omega} = \varepsilon_o \beta_{ijk}(2\omega;\omega,\omega).E_j^{\omega}E_k^{\omega} \tag{16b}$$

Here (IJK) refer to components of a bulk (i.e. material) coordinate system (XYZ) and (ijk) refer to coordinates within a molecular coordinate system (xyz).

It is found that in molecular compounds the forces which bind the molecules together (van der Waals forces and hydrogen bonding) are very weak compared to forces that bind the atoms together within the molecule (covalent bonds). In this case there is little interaction between individual molecules within the structure, and to a good approximation the bulk properties can be described as a sum of the individual molecular properties, taking proper account of the relative orientation of the individual molecules. This leads to a straightforward relationship between the molecular and bulk nonlinear optical coefficients ß and $\chi^{(2)}$:

$$\chi_{IJK} = (f_I^{2\omega}f_J^{\omega}f_K^{\omega}/v) \; \Sigma \; a_{Ii}a_{Jj}a_{Kk}.\beta_{ijk} \tag{17}$$

where summation over repeated indices is assumed.

The summation Σ is taken over all molecules within volume v (typically the unit cell for crystalline systems) and the factors f are "local field factors" which relate the field within the material to the incident field. All the charges and dipoles within the material will act to modify the incident field, and an exact calculation of the electric field seen by any one molecule would be complex. A number of approximate theories have been developed, which result in rather simple forms for the coefficients f [19-21]. The coefficients a_{Ii} in equation (17) are "direction cosines" relating the molecular coordinate system for any given molecule to the fixed bulk coordinate system (XYZ). If Φ_{Xx} is the angle between the x-axis of the molecular coordinate system and the X-axis of the bulk coordinate system, then

$$a_{Xx} = \cos\Phi_{Xx} \qquad (18)$$

and so on.

Thus we can consider the nonlinear optical properties of organic materials in two separate (but inter-dependent) parts. Firstly, we must design, synthesise and characterise molecules which have high hyperpolarizabilities, ß. Secondly, we must design and fabricate some bulk material in which the molecules are oriented so as to maximize the desired nonlinear optical coefficients, $\chi^{(2)}$. The relative orientation of the molecules within this system is seen from equation (17) to be critical. In the case where there is a centrosymmetric arrangement of the molecules, then the summation Σ in this equation results in a total cancellation of the individual molecular nonlinearities. The relative orientation required in fact depends on the application. For the transverse electro-optic effect it is desirable to have all the molecules aligned parallel to each other so as to maximize a single diagonal component (*e.g.* χ_{ZZZ}) of the $\chi^{(2)}$ tensor. For processes such as second harmonic generation, which require phase matching, it is necessary to maximize off-diagonal components of $\chi^{(2)}$ such as χ_{YZZ}, and this requires a more complex arrangement of the individual molecules.

A number of different techniques (namely crystal growth, Langmuir-Blodgett film deposition and polymer film fabrication) have been used to produce materials having the required degree of molecular order and are described in sections 3.3 to 3.5. Before that a description of the design, optimization and characterization of molecules with high nonlinearities (ß) is given.

3.2 DESIGN OF MOLECULAR PROPERTIES

The high intrinsic molecular hyperpolarizabilities (ß) found in many organic materials are broadly associated with bonding patterns involving π electrons. It has long been established [22] that electrons involved in π-bonds are less tightly bound to the atoms within a molecule than are those involved in σ-bonds.

Molecular structures that contain "conjugated" electronic systems, consisting of alternating single and double carbon-carbon covalent bonds contain a high density of π electrons. The electronic charge distribution associated with these loosely bound charges is readily modified on interaction with the electric fields associated with an incoming light beam. In order that second order nonlinearities are observed, the modifications of the charge distribution must be asymmetric. This is achieved by the introduction of groups that can inject charge into (donors) or withdraw charge out of (acceptors) the conjugated pathway. The largest nonlinearities are obtained when a donor group and an acceptor group are positioned such that their effects are additive.

A simple conjugated molecule is benzene. Because it is a symmetrical the β coefficient is zero. Substitution of an acceptor group such as $-NO_2$ onto the ring results in a moderate β-value of about 2×10^{-30} esu for nitrobenzene. Similarly, substitution of a donor group such as $-NH_2$ gives $\beta \approx 1 \times 10^{-30}$ esu for aniline. However if both groups are present in positions that maximize the interaction with the π-electron charge cloud, the resulting molecule (4-nitroaniline) has a β-value of around 20×10^{-30} esu. Thus the effect on β of combining the donor and acceptor groups is much more than additive. Much work has been carried out experimentally and theoretically, using methods described below, to investigate the effect on β of changing the size and nature of the conjugated pathway and the number and strength of the donor/acceptor groups. It is found that, in general, increasing the length of the conjugation path will increase the magnitude of the nonlinearity β, and that the donor and acceptor groups can be ranked in effectiveness, for example, by their *substituent constants* (σ) as shown in Table 2. The effectiveness of a donor group (negative constants) or acceptor group (positive constants) is proportional to the difference between the ground state and excited state substituent constants.

A "two-level" model has been used [23] to indicate the important factors in the determination of β. In the case when the molecular absorption spectrum is dominated by a transition to a single excited state (which is a generally a "charge transfer" state), the theoretical expression for β can be approximated as:

$$\beta \approx Wf\delta\mu/\{[W^2 -(2\hbar\omega)^2][W^2 -(\hbar\omega)^2]\} \tag{19}$$

where W is the energy, f is the oscillator strength of the molecular transition, and $\delta\mu$ the change in molecular dipole moment that occurs between the ground and excited state. Thus, to maximize β it is necessary to design a molecule that has a strong transition (large f) to an excited state having a very different charge distribution to the ground state (large $\delta\mu$). Many traditional dye molecules fit this bill [24].

Table 3. SUBSTITUENT CONSTANTS

substituent	ground state constant	excited state constant
CH_3	-0.08	-0.12
CF_3	-0.02	-0.04
NH_2	-0.47	-1.55
$N(CH_3)_2$	-0.60	-2.33
OH	-0.43	-1.00
OCH_3	-.48	-1.20
COOH	+0.17	+0.47
CHO	+0.22	+0.81
CN	+0.13	+0.19
NO_2	+0.11	+1.06
NO	+0.43	+2.29

Equation (19) shows that there can be strong dispersion of the nonlinearity. The β value increases rapidly as the energy of either the fundamental or harmonic waves approaches that of the molecular transition. In practice, of course, the molecular absorption also increases and an optimum wavelength for operation will be found, which is close enough to the wavelength of molecular absorption to provide some "resonant enhancement" of the ß-value, but far enough away that absorption losses in the material are acceptable.

Because of this dispersion it is easier to make a comparison of the nonlinearities of different molecules using a "zero-field" value ß(0), given by equation (19) in the limit $\omega \to 0$:

$$\beta(0) \approx f \delta \mu / W^3 \tag{20}$$

This shows that the molecular nonlinearity can be increased by designing molecules with lower transition energies (longer wavelengths of maximum absorption). Unfortunately, moving the absorption to longer wavelength limits the useful operating range for the material. This "transparency-efficiency trade off" [25] is a fundamental problem associated with these types of conjugated donor-acceptor molecules. The target of producing molecular species with high nonlinearities, but being transparent over the complete wavelength range 400-900 nm (for frequency doubling of diode lasers) is the basis of much chemical research at present.

A number of nonlinear optical molecules, along with their ß values, are shown in Figure 3, and these illustrate the trends discussed above.

Direct measurement of the components of the molecular ß tensor is rather difficult. In practice it is only possible to measure the bulk nonlinearity (i.e. $\chi^{(2)}$) of some system, and derive the molecular values using a known or assumed distribution of molecular orientations, and assuming a form for the local field factors that relate internal and external electric fields. The method most commonly used is *Electric Field Induced Second Harmonic Generation* (EFISH) [26]. A dilute solution of the molecule of interest is used, so as to minimize effects due to intermolecular interactions. Such a solution contains an isotropic distribution of molecular orientations, and so would not show any second-order nonlinear optical effects. In the EFISH experiment a strong d.c. electric field is applied to the solution synchronously with a pulse of incident laser light. The electric field induces a degree of orientation in the nonlinear optical molecules,

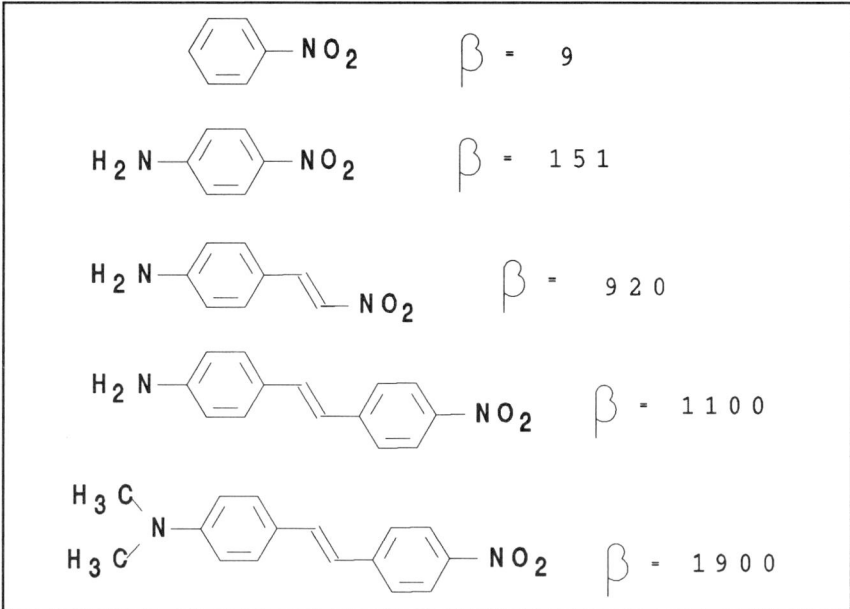

Figure 3. Representative molecular structures and their experimental molecular β values (in units of 10^{-40} m^4V^{-1}).

and some second harmonic is generated. The magnitude of the second harmonic can be related to the nonlinearity of the molecule.

The main drawbacks of this technique are the dependence of the results on accurate measurements of other properties of the molecules and solvents, such as dipole moment and refractive index, and the fact that only a "vector component", β_z, is determined, where

$$\beta_z = \beta_{zzz} + \tfrac{1}{3}(\beta_{zxx} + \beta_{zyy} + 2\beta_{xzx} + 2\beta_{yzy}) \tag{21}$$

Nevertheless, the method has been very useful in providing data for the development of the understanding of the mechanisms for high molecular nonlinearities. A comprehensive listing of reported measurements using this technique has recently been compiled [27].

As an alternative to the somewhat time and resource-consuming process of synthesis followed by experimental determination of β-values, there has been a great deal of work on the development of theoretical models to predict molecular nonlinearities. A number of different techniques have been used, but perhaps the most successful has been the use of semi-empirical methods [28-30]. These are based on a perturbation theory derivation for ß for second harmonic generation of the form

$$\beta_{ijk} + \beta_{ikj} = (-e^3/4h^2)\Sigma_{n'}\Sigma_n[(r^j_{gn'}r^i_{n'n}r^k_{ng} + r^k_{gn'}r^i_{n'n}r^j_{ng})B1$$
$$+ (r^i_{gn'}r^j_{n'n}r^k_{ng} + r^i_{gn'}r^k_{n'n}r^j_{ng})B2 + (r^j_{gn'}r^k_{n'n}r^i_{ng} + r^k_{gn'}r^j_{n'n}r^i_{ng})B3] \quad (22)$$

with

$$B1 = \{1/[(\omega_{n'g}-\omega)(\omega_{ng}+\omega)] + 1/[(\omega_{n'g}+\omega)(\omega_{ng}-\omega)]\}$$
$$B2 = \{1/[(\omega_{n'g}+2\omega)(\omega_{ng}+\omega)] + 1/[(\omega_{n'g}-2\omega)(\omega_{ng}-\omega)]\}$$
$$B3 = \{1/[(\omega_{n'g}-\omega)(\omega_{ng}-2\omega)] + 1/[(\omega_{n'g}+\omega)(\omega_{ng}+2\omega)]\}$$

Here the r's are transition moments between the various excited states of the molecules, and are related to the oscillator strengths and dipole moments of the states. The $\omega_{n'g}$ are the frequencies of the transitions between the states. It is the simplification of this equation that leads to the two-level model already described. Fortunately all the terms in this equation are readily calculated using standard techniques of quantum chemistry, and the components of the ß tensor may then be derived. CNDO methods have been reported, in which the parameterization of the program utilizes experimental data on dipole moments and transition energies for a range of molecules of differing size and complexity. The correlation then achieved between calculated and experimental values for the ß coefficients is excellent. In contrast to the EFISH experiment described above, all components of the ß-tensor are obtained, and it is possible to calculate, for example, the wavelength dependence of the coefficients, or by a simple extension, the coefficients for related processes such as the electro-optic effect.

The method has been extensively used, for example, to compare the effect of adding different substituents to a given backbone structure and to produce a "ranking" of the efficiency of different donor and acceptor groups [31], or to study the results of systematic variation of the structure or size of the conjugated pathway [32]. A combination of this theoretical approach with experimental molecular characterization has led to the building up of a large database of molecules known to have high molecular nonlinearities ß. The problem that remains is finding the most suitable solid-state form for each of these molecules, so that the macroscopic $\chi^{(2)}$ values are maximized. The approaches to this problem are described in the following sections.

3.3 SINGLE CRYSTALS

Many of the applications described in section 2 require bulk samples in the form of single crystals, fabricated as, for example, electro-optic modulators, Pockels cells and frequency doublers for lasers. As shown in section 3.1 (equation 17), the size of the $\chi^{(2)}$ coefficients will depend both on the molecular ß-tensor and on the relative orientation of the molecules.

For non zero coefficients, the molecules must crystallize in a noncentrosymmetric structure, and to maximize the size of these coefficients the molecules must align in such a way that the major ß coefficients are additive. For applications such as second harmonic generation the situation is further complicated by the need for phase matching. The optimal orientation of low dimensional molecules (those that can crudely be described by a single vector-like charge transfer axis) for phase matching has been determined for all noncentrosymmetric crystal symmetries [33]. The general

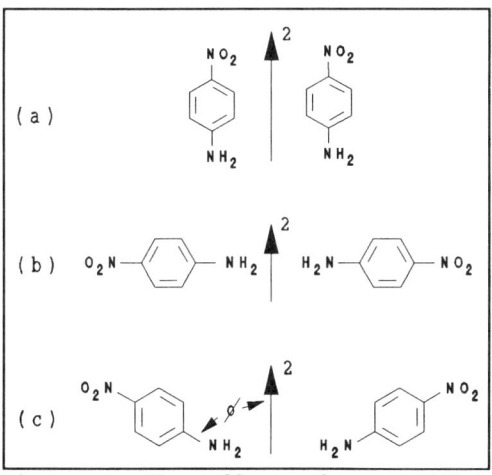

Figure 4. The effect of crystalline symmetry on the observed nonlinearity. See text for explanation.

principle is illustrated in Figure 4. The molecule shown has a large ß component along its length (β_{333} in its own frame of reference) associated with charge transfer from the $-NH_2$ donor group to the $-NO_2$ acceptor. If the molecule crystallizes in the point group 2 there will be two molecules in the unit cell of the crystal structure, related by a 180 degree rotation about a 2-fold axis parallel to the crystal's Y axis. If, as in Figure 4a, the molecules lie parallel to the 2-fold axis, then the large ß components are additive, resulting in a large χ_{YYY} coefficient but with all other components being much smaller. However, for the case of second harmonic generation utilising this coefficient, both the fundamental and harmonic waves will be polarized in the same direction (parallel to the crystal Y axis) and therefore phase matching by dispersion of refractive indices is not possible. If, as in Figure 4b, the molecules are arranged perpendicular to the 2-fold axis the large ß components cancel each other out resulting in little nonlinear optical activity. The ideal structure has the molecule inclined at an angle ϕ to the 2-fold axis as in Figure 4c. In this case the χ_{YYY} coefficient is reduced from its maximum value (being proportional to $\beta_{333}\cos^3\phi$), but the off-diagonal component χ_{YXX} is increased (proportional to $\beta_{333}\cos\phi\sin^2\phi$). This off-diagonal component of χ will allow Type-I phase matching, with the fundamental beam polarized along X generating a harmonic polarized along Y. Zyss and Oudar [33] have calculated that the optimum value of for a one-dimensional molecule in this point group is 54°.

It should be noted that in processes such as the electro-optic (Pockels) effect which do not require phase matching a parallel arrangement of molecules such as in Figure 4a may be preferred, as the largest possible $\chi^{(2)}$ coefficient is then obtained.

Unfortunately it is not possible to accurately predict from a molecular structure how molecules will crystallize. There are often gross changes in crystal structure associated with relatively minor changes in molecular structure. For example, 4−nitroaniline (p-NA) crystallizes centrosymmetrically, whereas the related molecule, 2-methyl-4-nitroaniline (MNA), crystallizes with the molecules all pointing in essentially the same direction and has a very high electro-optic coefficient. Indeed, it is often found that the same molecule will crystallize with a different structure depending on the crystallization conditions or even on the particular solvent that is used. Studies on a number of different crystal systems [34] suggest that in many cases a polar crystal structure suitable for second-order nonlinear optical processes may be achieved under non-equilibrium growth conditions (for example by fast recrystallization), but when equilibrium growth conditions are used, for the preparation of larger samples, a different polymorph is obtained, having a centrosymmetric or pseudo-centrosymmetric structure!

Although crystal structures cannot be predicted, there is much that the organic chemist can do to influence the way in which molecules will arrange. Because of the highly polar nature of these molecules there is a tendency for them to align in antiparallel pairs (possibly in solution prior to crystallization) to minimize the electrical energies. Substitution by bulky side groups which do not in themselves detract from the β value (as in the case of MNA described above) may hinder this dimerization process. Alternatively, molecules having lower dipole moments can be designed. Inclusion of chiral centres within these substituent groups will ensure that the crystal has a noncentrosymmetric structure, but this does not necessarily result in high $\chi^{(2)}$ coefficients. Finally, incorporation into the molecule of groups that encourage hydrogen bonding is advantageous, as the relatively strong forces involved in hydrogen bonding may be sufficient to overcome the dipolar forces. One efficient nonlinear optical material where hydrogen bonding is thought to play an important role is N-(4-nitrophenyl)-s-prolinol, NPP [35].

The first stage in the assessment of crystalline nonlinear optical properties is the use of the powder SHG technique of Kurtz and Perry [36]. Although only qualitative in nature, and strictly speaking not applicable to a search for electro-optic materials, this experiment allows rapid screening of potential compounds. Many hundreds of organic materials have now been screened in this way and a recent tabulation of published results provides a useful reference source [37].

For further assessment of nonlinear optical properties large crystals are required. Growth of organic crystals is usually from solution, using methods of controlled temperature lowering of a bath containing a saturated solution of the compound, in which a seed crystal is suspended [38]. Large crystals of high perfection can be grown by this method, although it can take many weeks to produce crystals large enough to allow orientation and cutting of device-sized samples. Many organic crystals can also be grown from the melt using the Bridgman technique. Clearly this method is limited to those materials that are stable at their melting temperature, and the crystals are often not as perfect as those grown from solution due to stresses that occur on cooling from the melt temperature. However, as compared to solution growth this technique is fast. Extensive characterization of a number of organic single crystals has been reported in the literature, both in terms of their linear optical properties (refractive indices, orientation of optical axes, optical transmission spectra) and also their nonlinear optical properties. It is beyond the scope of this chapter to detail all this work, but the highlights are given below, and the results are summarized in Table 4. There are three main processes that have been described, namely second harmonic generation, the electro-optic effect and parametric amplification.

Second harmonic generation - SHG studies on 2-methyl-4-nitroaniline, MNA [39] (described above) showed early on the potential of organic compounds in this field. Although in this compound the molecules are aligned close to parallel and, therefore, as discussed above are not ideally suited to phase matched SHG, the efficiency of this compound (defined by the coefficient d^2/n^3) is 45 times that of LiNbO$_3$. Clearly the use of more complicated molecules should result in still higher efficiencies. There is, however, one drawback to the use of organic materials for SHG. In general there is a trade-off between the magnitude of the nonlinearity and the transmission wavelength range of the compound. MNA, for example, is yellow with an absorption edge at about 500 nm. Larger molecules such as 3-(1,1-dicyanoethenyl)-1-phenyl-4,5-dihydro-1H-pyrazole, DCNP [40] have increased nonlinearities but the absorption edge is shifted towards the red, occurring in the case of DCNP at about 630 nm. This limits the usefulness of

Table 4. SUMMARY OF SINGLE CRYSTAL PROPERTIES

Compound	E.O. coef. r (pm/V)	SHG coef. d (pm/V)	refractive indices (at nm)	point group	d^2/n^3	Refs.
MNA	67	250 * 32	2.0, 1.6 (633)	m	7812 128	39,42
POM	5	10	1.6,1.7,1.9	222		48,49
DCNP	87	–	1.9,2.7 (633)	m		40,41
MAP	–	17	1.51,1.60, 1.84 (1064)	2		50
NPP	–	84		2	880	35
urea	0.5	1.4	1.48	42m		17
SPCD	430 **	–	1.5,1.3 (633)	mm2		51
m-NA	16.7	–	–			42
PNP	28	68	2.18,1.88, 1.49 (532)	2	580	52,109
DAN		50	1.66,1.71 1.78 (546)	2	420	53,54, 110
MBA-NP		34	1.75 (532)	2		55
COANP		15	1.70,1.85 1.68 (550)	mm2		56
MMONS	39.9	184 * 71	1.57,1.69, 2.13 (633)	mm2	850	116
LiNbO$_3$	31	5.9 34 *	2.272 (700)	3m	3.0 99	45

*: non-phase matchable coefficient

**: based on their reported refractive indices of 1.5 and 1.3, which are abnormally low, particularly close to the absorption edge these materials in the desirable goal of frequency doubling laser diode sources from the near infra-red into the blue portion of the spectrum. A number of other nitro-aniline derivatives, having molecular orientations within the crystal structure closer to the optimum for phase matched SHG, have recently been studied and show SHG efficiencies two orders of magnitude greater than LiNbO$_3$. The compound NPP [35] probably exhibits the maximum nonlinearity that can be achieved using this simple nitroaniline system, as in this case the molecules are oriented very close to the ideal angle of 54° to the crystal's 2-fold axis. Urea [17] also deserves mention as, although its nonlinear coefficient is relatively small (less than that of LiNbO$_3$) it has a very large transparency range (from 200 to 1400 nm), making it particularly suitable for frequency conversion of infra-red and visible lasers into the uv region of the spectrum.

Electro-optic effect - The linear electro-optic coefficients of m-nitroaniline (m−NA) were measured as long ago as 1972 by a number of groups [42,43], giving a coefficient r_{33} about one third that of LiNbO$_3$. It was found that the linear electro-optic susceptibility $\chi^{(2)}(\omega;\omega,0)$ is approximately equal to the susceptibility derived from SHG experiments $\chi^{(2)}(2\omega;\omega,\omega)$. This appears to be generally true for organic compounds and is in contrast to the case of inorganic compounds described in section 2. This is a consequence of the fact that in the organic compounds the origin of the electro-optic effect is purely electronic.

The compound, 2-methyl-4-nitroaniline (MNA), has a very high electro-optic effect [44]. The highly polar arrangement of molecules within this structure, with the molecular charge transfer axis lying at only 20° to the crystal's polar axis, is close to the ideal case of having the charge transfer axes parallel to the polar axis. Single crystals of MNA are rather difficult to grow, and variations in crystal quality have given poor reproducibility of measured electro-optic coefficients. An important parameter in the definition of the efficiency of an electro-optic material at any given wavelength is the half-wave voltage (V_π), this being the applied voltage required to change the transmission of an optical shutter (formed by placing an appropriately aligned sample of the electro-optic material between crossed polarizers) from maximum to minimum. It is the voltage required to introduce an additional phase shift of π radians between two perpendicularly polarized components of the electric field. This half wave voltage will in general depend on the dimensions of the crystal used, as well as the wavelength at which it is defined, and so a comparison of the merits of different materials is obtained using a reduced half wave voltage (V_π^*) which is the value of V_π applicable to a unit cube of the material. The value of V_π^* determined for MNA was 1300 volts, compared to a value of 3000 volts for $LiNbO_3$ [45].

More recently measurements on 3-(1,1-dicyanoethenyl)-1-phenyl-4,5-dihydro-1H-pyrazole (DCNP) [40,41] have shown even greater electro-optic efficiencies. In this case the molecules are indeed aligned parallel to each other, maximising the electro-optic effect. This arrangement of molecules also simplifies interpretation of the observed effects, as one coefficient (r_{33}) dominates the electro-optic tensor. At 633 nm the value of V_π^* obtained was 370 volts, nearly ten times lower than that of $LiNbO_3$. This value is somewhat enhanced by the fact that the absorption edge extends beyond 600 nm, especially for light polarized parallel to the length of the molecules. The half-wave voltage increases with wavelength, but remains less than half that of lithium niobate even at 1.2 μm. Crystals of DCNP can be readily grown from solution and there appears to be no significant deterioration in the quality of these crystals over periods of many months exposure to normal laboratory atmospheres.

Optical parametric processes - Two organic materials, 3-methyl-4-nitropyridine-N-oxide (POM) [46] and urea [47], have been extensively characterized in terms of their parametric properties, with in both cases the phase matching curves being determined over a wide wavelength range. Urea, for example, if pumped with the fourth harmonic of the Nd:YAG output, at 266 nm, is capable of optical parametric oscillation (OPO) giving an output tunable continuously from 330 nm to 1.4 μm. The large transparency window and high optical damage thresholds of urea make it particularly suited to this application. To date experiments have been carried out using the Nd:YAG third harmonic as the pump beam, and this limits the tuning range to 498 nm - 1.24 μm. However, combination of this OPO with an urea frequency doubler extends the range down to 249 nm. Conversion efficiencies of about 8 % were achieved.

In summary, although great advances have been made over recent years in the development of new organic crystalline materials, it can be seen from the above section that there are many problems associated both with the selection of the best materials for crystal growth, and also with obtaining samples of sufficient quality and size for fabrication into devices. It is only recently that large crystals of organic nonlinear optical materials have become commercially available. The process of crystal growth will always be time consuming, and is not in general well adapted to the formation of thin films suitable for use in waveguide devices. This requires control of the orientation of the molecules within the film and with respect to the substrate. In order to meet the requirements of ordered efficient nonlinear films, preferably being easily fabricated, alternative techniques to single crystal growth are being investigated, as described in the following sections.

3.4 LANGMUIR-BLODGETT FILMS

From the discussion in the previous section it can be seen that there is a need for alternative techniques for the formation of ordered thin films capable of producing nonlinear optical effects in geometries suited to waveguide applications. One approach that has been studied in some depth is the use of the Langmuir-Blodgett (LB) films. A detailed description of the physics and chemistry of LB films is given in Chapter 3 whereas in this section the application of LB films in the field of nonlinear optics is outlined. The main attraction of the technique lies in the ability to build up on a substrate a film of accurately controlled thickness and composition, in which the molecules are highly ordered. The molecular composition of the film can be determined a monolayer at a time allowing, in principle, complex structures to be engineered. By careful chemical design it is possible to fabricate multilayers having very high nonlinear optical coefficients. In particular, as described in more detail below, interleaved LB film structures can be designed so that the molecular nonlinearities from adjacent monolayers are all additive. The geometry of the LB film is naturally compatible with waveguide devices, and the precise control of thickness is ideal for processes such as second harmonic generation and parametric amplification which require phase matching. This can only be achieved in waveguides if the film thickness is tightly controlled.

There are however some drawbacks to the method. Molecules have to be specifically designed so as to be suitable for LB deposition. This, in general, involves the incorporation of a fatty hydrophobic chain onto the molecules which does not contribute to the nonlinearity, β, of the molecule and so serves to dilute the overall macroscopic nonlinearity, $\chi^{(2)}$, of the film. Also the process of fabricating films thick enough to guide light (*i.e.* > 1 μm) in monolayer steps is tedious, and with present technology would be an expensive manufacturing step. Studies on the use of such "thick" films as waveguides have shown them to have high optical losses, caused by scattering of light within the film. This appears to be largely due to boundaries between crystalline domains within the film. Nevertheless, the technique holds great promise, as the results described below will indicate.

Most experimental studies of nonlinear optical effects in LB films have been carried out on monolayers or "thin" multilayers (<30 layers thick). Second harmonic generation from LB films has been extensively reported [57-62], and is in fact a useful alternative to the EFISH experiment described in section 3.2 for the determination of the molecular nonlinearity, β. The basic experimental arrangement is

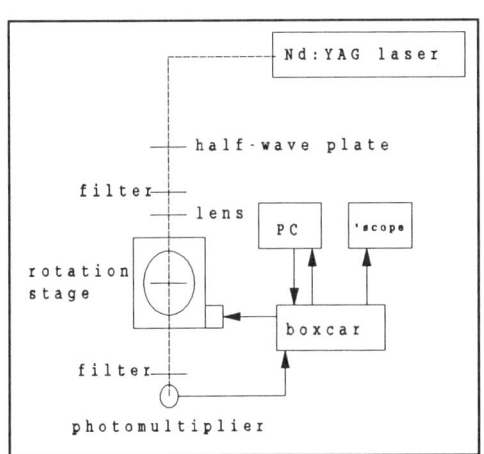

Figure 5. Experimental arrangement for the measurement of SHG from LB films.

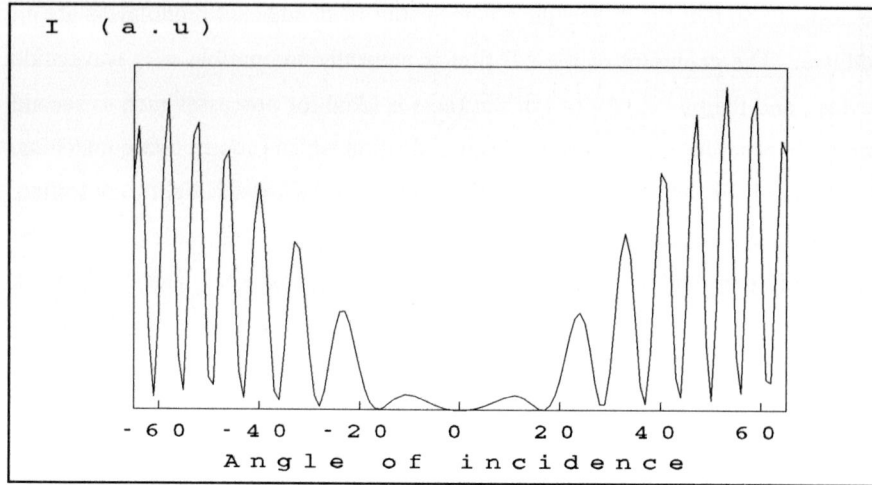

Figure 6. SHG intensity from an LB film, as a function of angle of incidence of the fundamental laser beam.

shown in Figure 5. The LB film is deposited on one or both sides of a substrate (usually glass or fused silica) and placed in the path of the fundamental laser beam. Measurements are made at either a fixed angle of incidence (using two perpendicular polarization directions for the laser beam), or as a function of the angle of incidence. In the latter case, the output second harmonic intensity has the form shown in Figure 6. Here the fringing effect is due to interference between the harmonic signals produced from the films on the front and back faces of the substrate. It is the height and shape of the envelope of these fringes which is of interest. Comparison of the signal obtained with that from a standard sample (such as a quartz crystal) allows the nonlinearity $\chi^{(2)}$ of the film to be determined [60]. Also, the shape of the envelope (as defined by the position and angular half-width of the peak) leads to the determination of the tilt angle of the nonlinear optical molecules with respect to the film normal. From these data, and knowing (by measurement or molecular modelling) the thickness of the film it is possible to calculate the molecular nonlinearity (ß).

The highest nonlinearities have been reported for "alternating Y-type films" in which the nonlinear layers are interleaved (see Table 5b). If the two component

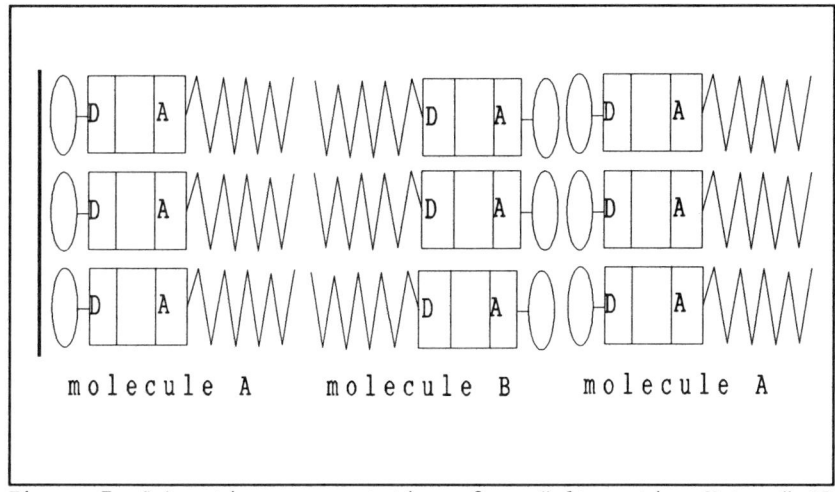

molecule A   molecule B   molecule A

Figure 7. Schematic representation of an "alternating Y-type" LB film.

molecules are designed so as to have the nonlinear optical units directed in opposite directions with respect to the hydrophobic substituents (Figure 7) then their nonlinearities will add in the resultant bilayer. There is indeed evidence that the nonlinearity is enhanced by the interaction between adjacent molecules. The second harmonic coefficients of several LB molecules are shown in Table 5. The trends described in section 3.2 can be clearly seen in this list, with the highest nonlinearities being found in long-chain conjugated polyenic molecules.

The theory of second harmonic generation in thin films (having thicknesses much smaller than the coherence length) leads us to expect a quadratic relationship between the second harmonic intensity and film thickness. In fact a number of studies [59,63] have indicated deviations from this expected behaviour, with the effective nonlinearity $\chi^{(2)}$ of the film falling off with increasing number of layers in the film. However, recently quadratic SHG enhancement has been obtained from 300 layer film structures of a hemicyanine dye interleaved with a compatible two-legged spacer. This is discussed in Chapter 1.

Measurements on active monolayers spaced from the substrate by increasing numbers of inert layers (such as arachidic acid) suggest that there is interaction

Table 5a. LANGMUIR BLODGETT MONOLAYERS

molecule	$x^{(2)}$ (pm/V)	r (pm/V)	ref.
(DPNA)	670		57
(DMAS)	2615		58
	550		59
		69	112
		0.7	111

Table 5b. ALTERNATING Y-TYPE LANGMUIR BLODGETT FILMS

molecule A	molecule B	$\chi^{(2)}$ (pm/V)	ref.
HOOC–⟨⟩–N=N–⟨⟩(NO$_2$)–N(R)(CH$_3$)	ROOC–⟨⟩(NO$_2$)–N=N–⟨⟩–N(CH$_3$)(C$_2$H$_4$COOH)	340	114
(O=⟨⟩=CH–CH=⟨⟩–N–R)	CH$_2$=CH(CH$_2$)$_{20}$COOH	550	115
HOOC(CH$_2$)$_2$–N(CH$_3$)... (polyene) ...H$_3$C, CN, COOH	HOOC(CH$_2$)$_2$N(CH$_3$)... (polyene) ...H$_3$C, CN, COOR (polyene)	2000	60
H$_3$C–(H$_3$C)N–⟨⟩–CH=CH–⟨⟩–N$^+$(R)(Br)	O$_2$N–⟨⟩–CH=CH–⟨⟩–N(H)–C(=O)–R	400	62

between the polar layer and the substrate which leads to an enhancement of the nonlinearity of the layer adjacent to the substrate. It has also been suggested that there is often a decrease in the degree of order achieved as the number of layers is increased, with an associated decrease in the nonlinearity. Nevertheless, films of moderate thickness have been produced that show nonlinearities significantly greater than those achieved in the polymeric thin films described in the next section.

The factor that has most hindered the application of LB films in waveguide devices has been the optical quality of films produced. It is a big step-up from the assessment of thin films by transverse illumination, where optical pathlengths within the film are of the order of 100 nm or less, to the propagation of light through several centimetres of a thick (e.g. 2 μm) waveguide. The few measurements made on such thick films have generally reported very high optical losses due to scattering within the bulk of the film [64]. Investigations of fatty-acid films suggest that this might be due to regions of local order (crystalline domains) within the film [65]. Recent work using preformed polymer LB films [66-68] has shown much better optical quality; nonlinear structures were made by alternating polymer layers with layers containing a highly nonlinear merocyanine dye. Thick (160 bilayer) films of these materials showed strong nonlinear optical effects. Günter and coworkers [69], utilising a more conventional nonlinear optical LB molecule, have recently demonstrated waveguiding over 1 to 2 cm in a film comprising 270 bilayers. The intensity of SHG in this system was found to have the expected quadratic dependence on number of layers, showing that a high degree of ordering was maintained through the thickness of the film. Thus it would appear that the well-publicised problems of optical quality in LB films should be solvable, given sufficient chemical effort.

There have also been a number of studies of the electro-optic effect in LB films, utilising a technique based on the observation of surface plasmon modes in the attenuated total reflection (ATR) spectrum from a composite structure consisting of the LB film deposited on top of a thin (50 nm) silver film [70,71]. The excitation of the surface plasmon mode (at the silver/LB film interface) gives rise to a characteristic dip in the reflection spectrum, and the shape and position of this feature are very sensitive to the refractive index of the LB film. Modification of this refractive index via the electro-optic effect, by the application of an electric field, gives rise to variations in the reflected intensity. Analysis of the field dependence of the reflected signal leads to the determination of the electro-optic coefficient, and thence to the molecular β coefficient. Values of r determined using this method are also given in Table 5. Equivalent $\chi^{(2)}$ values are close in magnitude to the values determined by second harmonic variation. The similarity of the results indicates that the nonlinearities in these compounds are largely electronic in origin, as discussed in section 3.2.

From the above, it can be seen that thin LB films having very high nonlinearities can be produced. Work is currently under way extending this to the thick films required. The problems remaining to be solved before waveguide devices using such films can be routinely produced are mostly concerned with the quality (optical and mechanical) of these thick films.

3.5 POLYMER FILMS

Another approach to the formation of thin ordered films having high nonlinear optical coefficients that is currently being followed is the use of polymeric compounds. In such materials molecules having high nonlinearities (β) are dispersed within a polymeric matrix, either by simply mixing the two components together ("guest-host" systems) [72,73] or by attaching the nonlinear optical unit onto the polymer chain, for example, as a pendant side group [74-77] as shown in Figure 8. This results in an isotropic polymer which does not possess any second-order nonlinear optical properties. A degree of orientational order can be obtained by poling the material using a strong d.c. electric field. Poling is carried out at an elevated temperature, above the polymer's glass transition temperature, T_g. The polymer is in a rubbery state at this temperature, with the polymer

244

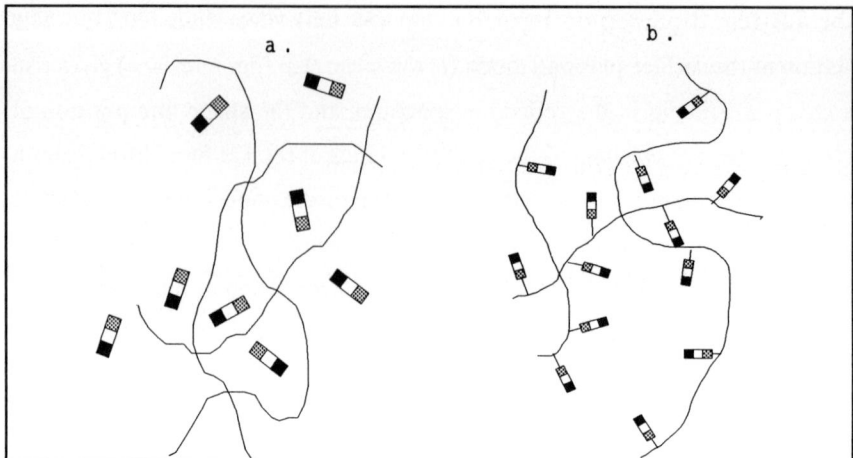

Figure 8. Schematic representation of (a) a guest-host polymer and (b) a side-chain polymer.

backbone having a certain amount of freedom to move about. The nonlinear optical units (free or attached) are also able, to a certain extent, to rotate under the influence of the applied field so as to tend to align their permanent dipole moments with the field direction. If the polymer is now cooled well below its glass transition temperature, with the electric field still applied, then motion of the backbone is inhibited and the orientation of the nonlinear optical units is frozen in. On removal of the electric field the active molecules remain locked within the cage of the polymer chains and the material is now noncentrosymmetric. The advantages of this technique can be summarized as follows.

- In principle any nonlinear optical molecule can be used; we are not limited to that small subset of molecules that form noncentrosymmetric crystal structures.
- Standard techniques of polymer processing such as spin coating and lithography are well suited to the formation of thin film waveguides in these materials and may easily be adapted to mass production.
- The polymeric matrices can have very good linear optical properties, with low scattering and absorption losses. The low dielectric constants of the polymers (compared to lithium niobate for example) reduce velocity mismatches

between electric and optical signals in the travelling wave modulator, allowing higher operating frequencies to be reached.

On the debit side, however, there are a number of points relating to the maximum nonlinearity that can be achieved in such as system.

♦ The nonlinearity is diluted by the inactive polymer matrix and the concentration of nonlinear optical units is limited by their solubility in the matrix, usually in the region of 10 to 20 %. Higher concentrations can be achieved in side-chain polymers, but this may be at the cost of, for example, reducing the glass transition temperature of the compound.

♦ The degree of orientation that can be achieved is limited by thermal effects. Thermodynamic analysis [78] has shown that, in many cases, the molecules have an average angle of about 80° to the field direction (as compared to 90° for a totally isotropic material). The degree of orientation can be increased somewhat by using polymers which are liquid crystalline, but this introduces other problems. In particular, the texture of liquid crystal polymers tends to lead to high levels of scattering.

♦ It is clearly important that the orientation induced in the material is retained after the electric field has been removed. This requires proper chemical design of the polymer so as to minimize the possibility of chain motion over all temperatures that may be encountered by a device. In general the long term stability of side chain polymers, where the nonlinear optical unit is chemically attached to the backbone, is far superior to that of guest host compounds.

From the above it is clear that the most promising technology is that of side-chain polymers, and it is here that most work has been concentrated. The most advanced systems reported to date have been from workers at Lockheed, using materials developed by Hoechst Celanese Research Corporation [74,79]. These materials are side-chain polymers, which are deposited onto substrates by spin coating. It is necessary to fabricate a multilayer structure, as shown in Figure 9,

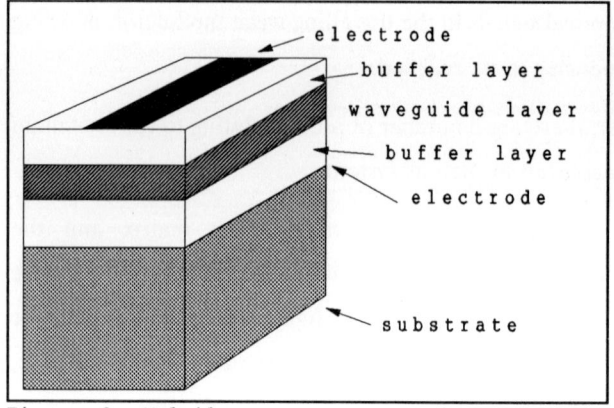

Figure 9. Multilayer structure of a polymeric planar waveguide.

so that there is no overlap of the electric field of the guided light with the lossy metallic electrodes. The buffer layers must be chosen so as to have a refractive index lower than the waveguide layer, and to be compatible with the processing requirements. (For example, the lower buffer layer must be insoluble in the solvent used for spin-coating of the waveguide layer.)

In the Lockheed devices, the poling process required to produce the noncentrosymmetric structure is also used to define the channel waveguides [80]. As shown in Figure 10, a waveguide pattern may be deposited directly onto the substrate, before application of the buffer layer. Poling takes place by the application of an electric field between this patterned electrode and a planar electrode deposited on top of the structure. As explained above, the nonlinear optical molecules are oriented to some extent by this field. Because the molecules are highly anisotropic with, in general, a much higher polarizability (and hence refractive index) associated with the axis along the length of the molecule than with an axis perpendicular to its length, the poling process also induces some local birefringence in the material. The refractive index for light polarized parallel to the poling direction will be slightly raised, whilst that for light polarized perpendicular to the poling direction will be lowered. Thus, for the geometry shown in Figure 10, the poling process will define a channel that will guide light having TM polarization, but not guide TE modes.

A number of prototype electro-optic waveguide devices have been fabricated using this principle [79], including a Mach-Zender interferometer, which has shown

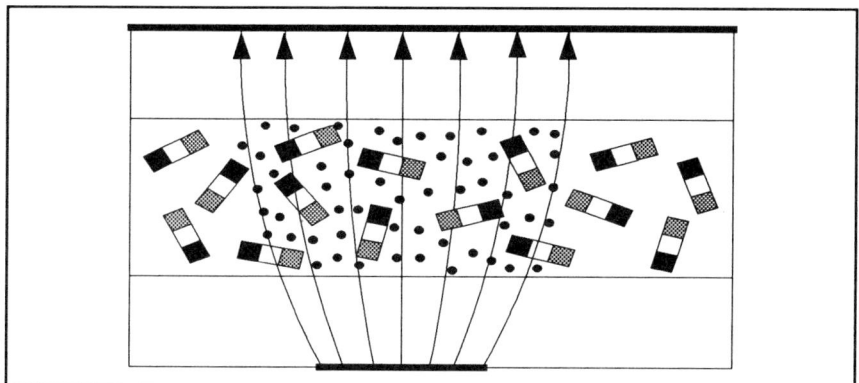

Figure 10. Waveguide formation by electrode poling. The dotted region will have a raised refractive index for TM modes.

kHz modulation of the light beam, and a directional coupler that demonstrated significant switching of light from one channel to the other. Although not yet optimized, these devices form the basic building blocks for the construction of complex integrated optic networks capable of the encoding and routing optical signals.

Many chemical and electronics companies throughout the world have significant research programmes in this area. A variety of side chain polymers are under development, of which a representative selection is shown in Figure 11. Electro-optic coefficients comparable with those of $LiNbO_3$ are now being reported, along with stability data indicating 90 % storage of the electro-optic coefficients over periods of greater than a year [81]. It seems likely that organic electro-optic waveguide devices will become commercially available in the near future. These devices will have certain properties, such as high switching or modulation speeds, that are an improvement on the inorganic, $LiNbO_3$, and as such a niche in the market seems assured.

4. ORGANICS FOR THIRD-ORDER NLO APPLICATIONS
4.1 INTRODUCTION

In recent years there have been extensive studies of third-order nonlinear optical

248

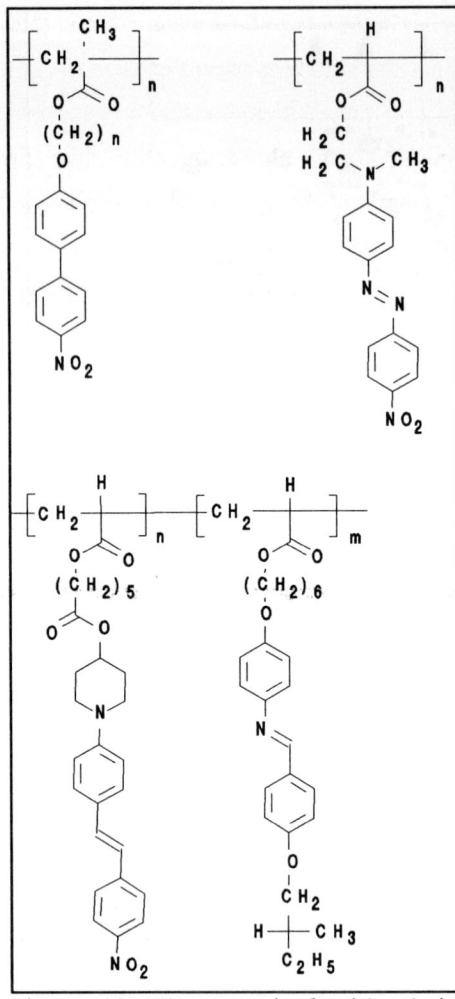

Figure 11. Three typical side-chain polymer structures.

effects in semiconductor compounds such as GaAs, GaAlAs and InSb. Optical bistability has been obtained using these compounds, usually in a Fabry-Perot resonator configuration, operating at room temperature. The effective nonlinearities, as indicated by the magnitude of the nonlinear (intensity dependent) refractive index n_2, are very high. This index is related to the effective $\chi^{(3)}$ coefficient by the equation

$$n_2 = (4\pi^2/c).(\chi^{(3)}/n^2) \qquad (23)$$

where $\chi^{(3)}$ is in units of cm^4/esu^2 (commonly abbreviated as esu). Values of $\chi^{(3)}$ of about 6×10^{-2} esu in GaAlAs multiple quantum well (MQW) devices have been reported, resulting in optically bistable operation even at mW input power levels [82,83]. These huge nonlinearities are not, however, intrinsic, suggesting that they are not due to a simple anharmonic oscillatory response of the constituent charges to an input electromagnetic field. They are instead due to complicated excitation and relaxation processes involving absorption of light within the tail of the semiconductor band edge. This leads to a change in the profile of this absorption edge, which in turn results in the modification of the refractive index. Such an origin of the nonlinear effects means that devices utilising them must always operate at a specific wavelength and thus have limited versatility. It also means that the switching speeds of such devices (in particular the switching off speed) may be limited by the relaxation processes involved within the materials' conduction bands. In contrast to this, the nonlinearities studied to date in organic compounds have been intrinsic in nature (as described below). The maximum nonlinearities observed (10^{-9} to 10^{-8} esu) are many orders of magnitude lower than those of the semiconductor systems and this of course must raise questions as to whether they can ever be useful. However, organic compounds have some points in their favour.

Firstly, because the processes responsible for the nonlinearities in the organic compounds are not dependent on a real excitation of the molecule, they are not specific to a particular wavelength, and so the materials can be used at any point of the spectrum where they do not significantly absorb the incident or emitted light. Secondly, again because there is no direct involvement of excitation and relaxation processes, the response times of these materials can be very fast and are in principle only limited by the electronic response times. Sub-picosecond responses are predicted [84].

Finally, the processability of polymeric compounds, as mentioned for the case of second-order materials earlier, means that the ease and cost of fabrication of devices out of these compounds would be a small fraction of the cost of manufacturing semiconductor devices. It seems likely therefore that semiconductor and organic materials may both have a niche within this area, and that the properties of devices from the two types of compound may to a certain extent be complementary.

4.2 MOLECULAR PROPERTIES

Many fewer organic compounds have been characterized in terms of third order molecular coefficients, γ, than have been in terms of β. For this reason there is probably a less complete understanding of the precise molecular properties (in terms of charge distribution, structure and size) required for high nonlinearities,

γ, than in the second-order case. Because γ is a fourth rank tensor whereas β is third rank, the symmetry constraints on the various components are different. In particular all molecules, whether or not they have inversion symmetry, may have non zero γ coefficients. The magnitude of the coefficients depend on the nature of the anharmonic potential seen by the electrons in the system, under the influence of an incident electromagnetic field. It appears, as in the case of second-order effects, that the highest nonlinearities are obtained from molecules having large conjugated π-electron systems. In this case it is not necessary to have donor and acceptor substituents attached to the conjugated pathway, although there is evidence that such substituents enhance the nonlinearity still further.

There have been relatively few reports of direct measurement of γ values for single molecules; work has concentrated on the characterization of polymers, often in the form of thin films suitable for waveguide fabrication. As the size of the polymer chain is often not accurately known, the concept of a molecular coefficient must be modified, and a sensible definition is to quote the nonlinearity associated with a single repeat unit of the polymer chain. In practice it is often just the bulk nonlinearity $\chi^{(3)}$ that is quoted.

Measurements that have been made have concentrated on simple molecules (conjugated and unconjugated) and have generally used the technique of Electric Field Induced Second Harmonic Generation, either in solutions or in the gaseous state. This is the same experiment described earlier for the determination of β. For symmetrical molecules, where $\beta = 0$, γ is determined *via* the nonlinear optical coefficient $\chi^{(3)}(2\omega;\omega,\omega,0)$. For molecules with finite β coefficients the analysis is somewhat more complicated. The available results show that γ increases with increasing conjugation [85,86]; saturated molecules such as methane (CH_4) and propane (C_3H_8) have values of 0.5 x 10^{-36} and 1.5 x 10^{-36} esu respectively whereas for benzene (C_6H_6), nitrobenzene (C_6H_5–NO_2) and nitroaniline (H_2N–C_6H_4–NO_2) γ increases from = 2.3 x 10^{-36} esu to 43.3 x 10^{-36} esu to 596 x 10^{-36} esu. Thus, at least for these small molecules, the rules governing γ appear to be similar to those governing β. Measurements have also been made on a few "long chain molecules"

[87], the γ values for non-conjugated molecules of the form C_nH_{2n} and C_nH_{2n+2} being less than 5×10^{-36} esu up to n = 15. However in polyene systems, having from 3 to 19 conjugated double bonds in the chain, higher values up to 1.7×10^{-32} esu have been found. The role of extended conjugated systems has been explored in greater detail during studies of polymeric systems (particularly polydiacetylenes) described below.

Calculations of the molecular γ values have also been rather scarce; for small molecules, such as benzene and other ring systems, again the degree of conjugation is the most important factor in obtaining a high γ [88,89]. The calculation of γ is considerably more difficult than that of β, as many more terms (similar in form to those of equation 22) are involved and it also appears that it necessary to involve many more of the possible excited states of the molecule. For β, the calculations can in general be limited to summations over the lowest energy $\pi-\pi^*$ transitions, but this is not true for γ.

Calculations on long chain polyenes (of the type described above) and related cyanine and aza-cyanine chains (which have additional delocalized charges) [90] suggest that γ-values two orders of magnitude higher than those reported for the polyene chains can be obtained using the aza-cyanine system.

More recently a number of different theoretical approaches (mainly semi-empirical) have been developed to initiate more systematic investigations of the role of, for example, the nature and length of the conjugated pathway in the determination of γ-values [91-93].

4.3 POLYMER FILMS

The relationship between the macroscopic coefficient $\chi^{(3)}$ and the molecular γ coefficients is analogous to equation (17). The main difference, which is a consequence of the fact that the coefficients are in this case fourth-rank tensors, is that the summation over the appropriate direction cosines does not in this case necessarily lead to a cancellation of the molecular coefficients in an isotropic (centrosymmetric) material. Thus polymer films, for example, will exhibit third-order nonlinearities without the need for d.c. electric field poling.

A number of polymer films have been characterized recently. Polymers are of interest from the point of view of their potentially good mechanical properties and ease of deposition, the prime target of these studies being to find a material which combines the above attributes with a high optical nonlinearity and good optical quality. The technique used has generally been that of third harmonic generation (THG). Electromagnetic wave propagation theory has been used to analyse the third harmonic generation from a nonlinear optical composite consisting of a substrate (usually chosen to be silica, for calibration purposes) and polymer film surrounded by air [94], and comparison of the observed angular profiles (of third harmonic intensity as a function of the angle of incidence of the fundamental laser beam to the film) with those determined theoretically enables the film nonlinearity to be measured relative to that of the substrate. An alternative technique is that of degenerate four wave mixing (DFWM). This technique has been used to determine the nonlinear optical response on very fast (subpicosecond) timescales [95].

The coefficients determined by THG and DFWM are different (being $\chi^{(3)}(-3\omega;\omega,\omega,\omega)$ and $\chi^{(3)}(-\omega;\omega,\omega,\omega)$ respectively) and are only comparable if there are no absorption bands between ω and 3ω for the material of interest.

The most frequently studied polymer system is polydiacetylene, largely because it displays the highest $\chi^{(3)}$ values! This material will be described in more detail below. The $\chi^{(3)}$ values of a number of other polymer films are listed in Table 6, along with that determined for a Langmuir-Blodgett film of a merocyanine dye. It can be seen that the polymers all have higher nonlinearities than the monomeric merocyanine, due to their greater conjugation lengths.

Polyacetylene $(CH)_x$ has been studied for many years as an electrically conducting material. Recent measurements of its nonlinear optical coefficients [96,97] are very promising, with the highest values (about 1×10^{-9} esu) being comparable with those obtained from polydiacetylenes. Measurements on oriented samples [98] show even higher nonlinearities. These high values are partly due to resonant enhancement by two and three-photon absorption of the fundamental beam. When off-resonance, the $\chi^{(3)}$ value is reduced by an order of magnitude (and is still therefore high compared to other polymers). The origin of these very high values of $\chi^{(3)}$ again lies in the high degree of conjugation that occurs along the polyacetylene polymer chains. Polyacetylene, however, has the disadvantage that its mechanical and optical properties are rather poor.

The other polymers listed in Table 6 all have comparable nonlinearities of around 5 x 10^{-12} esu. Their relative merit for fabrication into waveguide devices utilising the third-order nonlinearities will therefore depend on other properties already mentioned, and in particular on the optical quality (in terms of scattering and absorption losses, uniformity *etc*) that can be obtained. Poly-*p*-phenylene benzobisthiazole (PBT) is reported [99] to have good mechanical strength, stability and optical damage thresholds, due to the "rigid rod" nature of the polymer chains.

Polydiacetylenes (PDA s) have been extensively studied, in terms of their many different properties, since the discovery [100] that certain diacetylene monomers undergo solid state polymerization, in the form of single crystals, resulting in macroscopic crystals of the polymer. The polymerization reaction is controlled by

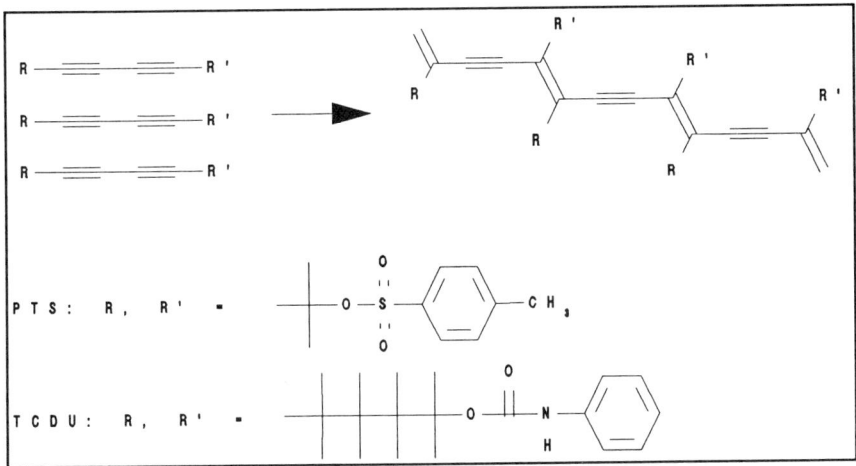

Figure 12. Generic diacetylene and polydiacetylene structures.

Table 6: Third-order nonlinearities of polymers

polymer	name	$X^{(3)}$ (10^{-12} esu)	λ (μm)	ref
	polyacetylene	1300	1.92	96
		100	0.8	97
	poly-p-phenylene-benzobisthiazole (PBT)	9	0.604 (DFWM)	99
	polysilane	1.5	1.9	117
		1.3	2.05	118
	poly-p-phenylenevinylene (PPV)	7.8	1.9	119
	merocyanine	0.5	1.06	120

the packing of the monomer and the mechanism is now well understood. The polymer crystals consist of a series of conjugated backbones (consisting of double, single and triple carbon-carbon bonds), as shown in Figure 12. A variety of sidegroups R can be substituted onto the backbone, giving the opportunity for "molecular engineering" of the properties of the resultant polymer.

Extensive reviews of the physical, chemical and linear optical properties of PDAs have appeared in the literature [101,102], and so these will not be covered in any detail here. The absorption spectra of the polydiacetylenes are, however, of some relevance to the nonlinear optical properties. Most of the PDAs have a strong absorption peak in the visible region, with the band edge (i.e. the long wavelength limit of absorption) at around 500 to 630 nm. As described below, the nonlinear optical properties vary considerably as fundamental or harmonic wavelengths are tuned in and out of these absorption bands.

Varying the sidegroups of the diacetylene monomer can result in polymers with very different properties and this can be used, for example, to produce soluble or insoluble polymers, polymers functionalized for Langmuir-Blodgett deposition, and polymers having noncentrosymmetric side-groups for the study of second order nonlinear optical effects. Third order nonlinearities have been studied by a variety of different methods, including third harmonic generation from solutions and sol-gels, from thin Langmuir-Blodgett, solvent cast and vacuum deposited films and from single crystals, and by degenerate four wave mixing, as described below.

Early measurements on PTS and TCDU (see Figure 12) single crystals [103], using THG with fundamental laser wavelengths of 1.89 and 2.62 μm, showed the effect of resonant enhancement. At 2.62 μm the third harmonic wavelength (873 nm) lies above the band edge of both PTS (633 nm) and TCDU (559 nm). In this case $\chi^{(3)}$ values of 3.7 x 10^{-11} esu and 1.6 x 10^{-10} esu were found for TCDU and PTS respectively when the laser light was polarized parallel to the polymer chains. The degree of orientation within PDA crystals is such that for perpendicular polarization the $\chi^{(3)}$ coefficient is two orders of magnitude smaller. At 1.89 μm the third harmonic (630 nm) lies much closer to the absorption peak, particularly of PTS, and the $\chi^{(3)}$ values increase to 7 x 10^{-11} and 8.5 x 10^{-10} esu respectively. More recently DFWM experiments [104] have been carried out on single crystal PTS at wavelengths ranging from 651.5 nm (where the absorption in a 1 μm thick crystal is about 50%) out to 701.5 nm, where the absorption is less than 2%. The resonant $\chi^{(3)}$ value at 651.5 nm is 9 x 10^{-9} esu whereas at the long wavelengths a value of about 5 x 10^{-10} esu is obtained. These measurements suggest that the non-resonant value for PTS is reached at about 700 nm. Measurements of the time response of the nonlinearity, again in single crystal PTS [105], show that off resonance it is faster than the 300 fsec resolution of the laser used, whilst the resonant response time is about 1.8 psec.

Kajzar and Messier have also studied the frequency dependency of the nonlinearity [106], this time in a Langmuir-Blodgett polydiacetylene (R = $CH_3(CH_2)_{15}-$, R' = $-(CH_2)_8COOH$) by THG using laser wavelengths from 0.85 to 1.907 μm. Two and three-photon resonances were observed at 1.35 μm (0.92 eV) and 1.907 μm (0.65 eV), with a maximum $\chi^{(3)}$ of 2.2 x 10^{-10} esu at 1.907 μm. The non-resonant $\chi^{(3)}$ was in this case about 2 x 10^{-11} esu. The long sidegroups on this molecule, required for Langmuir-Blodgett deposition, decrease the concentration of conjugated backbone chains, which may be responsible for the reduced $\chi^{(3)}$ as compared to PTS.

Measurements on other polydiacetylene systems are in general agreement with the above results, showing non-resonant nonlinearities in the range 10^{-11} to 10^{-10} esu and significant enhancement on-resonance. Thus, they are of interest for fabrication of nonlinear optical devices. Studies on Langmuir-Blodgett films of PDAs have shown that they can support waveguide propagation of light, and more recently waveguides have been fabricated by spin-coating of soluble PDA films [107,108]. In general PDA films are not as mechanically robust as, for example films of the polymers described in the previous section. They do, however, seem to be much more stable than the polyacetylene films, these being the only other materials having such high nonlinearities. Clearly much work is still to be done before these compounds can be routinely incorporated into devices.

5. CONCLUSIONS

In this chapter the potential advantages of organic materials for second and third order nonlinear optical processes have been described and the origin of their high nonlinear optical coefficients explained in terms of molecular structure. An attempt has been made to review the current state of progress in the development of these organic materials from the status of an "exciting new discovery" to that of well characterized components of nonlinear optical devices, capable of being used in efficient production processes. This field of research can no longer be described as being in its infancy. Indeed, almost every chemical and electronics company in Europe, the USA and Japan now has a research programme in this area! Many of these programmes are now producing "prototype devices" which, on a laboratory scale, have performances superior to the equivalent devices currently available, fabricated from inorganic materials such as $LiNbO_3$. However, further increases in efficiency and performance are required in order to persuade manufacturers that it is worthwhile abandoning technologies that they now know so well (despite the limitations of such technologies) and investing in the new processes required for exploitation of the organic materials. In fact, there is still much research and development needed before techniques such as Langmuir-Blodgett deposition [111-115] can even be considered on a production scale.

Nevertheless, advances have been made over the last five years and given the effort being expended on this topic world-wide, it can only be a matter of time before nonlinear optical devices utilising organic materials are commonplace.

REFERENCES

[1]   Y.R.Shen, *"The Principles of Nonlinear Optics"*, John Wiley, New York (1984).

[2]   F.A.Hopf and G.I.Stegeman, *"Applied Classical Electrodynamics"*, Vol. 2: *Nonlinear Optics*, John Wiley, New York (1986).

[3]   A.Yariv and P.Yeh, *"Optical Waves in Crystals"*, John Wiley, New York (1984).

258

[4] J.F.Nye, *"Physical Properties of Crystals"*, Oxford University Press, Oxford (1957).

[5] E.M.Garmire (ed.), *"Special Issue on Optical Bistability"*, IEEE J. Quant. Electron., **QE-21(9)** (1985); H.M.Gibbs, P.Mandel, N.Peyghambarian and S.D.Smith (eds.), *"Optical Bistability III"*, Springer Proceedings in Physics Vol. 8, Springer-Verlag, New York (1986).

[6] R.A.Fisher (ed.), *"Optical Phase Conjugation"*, Academic Press, New York (1983); *"Special Issue on Dynamic Gratings and Four-wave Mixing"*, IEEE J. Quant. Electron., **QE-22(8)** (1986).

[7] R.S.Weis and T.K.Gaylord. *Appl. Phys. A*, **37**, 191 (1985); G.D.Boyd, R.C.Miller, K.Nassau, W.L.Bond and A.Savage, *Appl. Phys. Lett.*, **5**, 234 (1964).

[8] *Laser Focus*, August 1987, p.30; *Laser Focus World*, September 1989, p.36.

[9] R.A.Becker and W.J.Silva, *Proc. SPIE Int. Soc. Opt. Eng.*, **704**, 214 (1987).

[10] D.Eimerl, *Ferroelectrics*, **72**, 95 (1987).

[11] R.F.Belt, G.Gashurov and Y.S.Liu, *Laser Focus*, 110 (October 1985).

[12] J.D.Bierlein, A.Ferretti, L.H.Brixner and W.Y.Hsu, *Appl. Phys. Lett.*, **50**, 1216 (1987); J.D.Bierlein and H.Vanherzeele, *J. Opt. Soc. Am. B*, **6**, 622 (1989).

[13] C.Chen and G.Z.Liu, *Ann. Rev. Mater. Sci.*, **16**, 203 (1986).

[14] D.J.Williams, *Angew. Chem. Int. Ed. Engl.*, **23**, 690 (1984).

[15] D.S.Chemla and J.Zyss (eds), *"Nonlinear Optical Properties of Organic Molecules and Crystals"*, Vols 1 & 2", Academic Press, New York (1986).

[16] P.N.Prasad and D.R.Ulrich (eds), *"Nonlinear Optical and Electroactive Polymers"*, Plenum, New York (1987).

[17] C.Cassidy, J.-M.Halbout, W.Donaldson and C.L.Tang, *Opt. Commun.*, **29**, 243 (1979); J.Zyss, D.S.Chemla and J.F.Nicoud, *J. Chem. Phys.*, **74**, 4800 (1981).

[18] S.I.Wax, M.Chodrow and H.E.Puthoff, *Appl. Phys. Lett.*, **16**, 157 (1970).

[19] L.Onsager, *J. Am. Chem. Soc.*, **58**, 1486 (1936).

[20] B.F.Levine and C.G.Bethea, *J. Chem. Phys.*, **63**, 2666 (1975).

[21] M.Hurst and R.W.Munn, *J. Mol. Electron.*, **2**, 35 (1986).

[22] M.J.S.Dewar, *"The Molecular Orbital Theory of Organic Chemistry"*, McGraw Hill, New York (1969).

[23] J.L.Oudar and D.S.Chemla, *J. Chem. Phys.*, **66**, 2664 (1977).

[24] P.F.Gordon and P.Gregory, *"Organic Chemistry in Colour"*, Springer Verlag, New York (1983).

[25] J.F.Nicoud and R.J.Twieg, in *"Nonlinear Optical Properties of Organic Molecules and Crystals"*, Vol. 1 (D.S.Chemla and J.Zyss, eds.), p. 227, Academic Press, New York (1986).

[26] J.L.Oudar, *J. Chem. Phys.*, **67**, 446 (1977).

[27] J.F.Nicoud and R.J.Twieg, in *"Nonlinear Optical Properties of Organic Molecules and Crystals"* (D.S.Chemla and J.Zyss, eds.), Vol. 2, p. 255, Academic Press, New York (1986).

[28] S.J.Lalama and A.F.Garito, *Phys. Rev. A*, **20**, 1179 (1979).

[29] V.J.Docherty, D.Pugh and J.O.Morley, *J. Chem. Soc., Faraday Trans. II*, **81**, 1179 (1985).

[30] J.A.Morrell and A.C.Albrecht, *Chem. Phys. Lett.*, **64**, 46 (1979).

[31] J.O.Morley, V.J.Docherty and D.Pugh, *J. Chem. Soc. Perkin Trans. II*, 1351 (1987).

[32] J.O.Morley, V.J.Docherty and D.Pugh, *J. Chem. Soc. Perkin Trans. II*, 1357 (1987).

[33] J.Zyss and J.L.Oudar, *Phys. Rev. A*, **26**, 2028 (1982).

[34] S.R.Hall, P.V.Kolinsky, R.Jones, S.Allen, P.Gordon, B.Bothwell, D.Bloor, P.A.Norman, M.Hursthouse, A.Karaulov, J.Baldwin, M.Goodyear and D.Bishop, *J. Cryst. Growth.*, **79**, 745 (1986).

[35] J.Zyss, J.F.Nicoud and M.Coquillay, *J. Chem. Phys.*, **81**, 4160 (1984).

[36] S.K.Kurtz and T.T.Perry, *J. Appl. Phys.*, **39**, 3798 (1968).

[37] J.F.Nicoud and R.J.Twieg, in *"Nonlinear Optical Properties of Organic Molecules and Crystals"* (D.S.Chemla and J.Zyss, eds.), Vol. 2, p. 221, Academic Press, Orlando (1986).

[38] B.J.McArdle and J.N.Sherwood, in *"Advanced Crystal Growth"* (P.M.Dryburgh, B.Cockayne and K.C.Barraclough, eds.), Prentice Hall, New York (1987).

[39] B.F.Levine, C.G.Bethea, C.D.Thurmond, R.T.Lynch and J.L.Bernstein, *J. Appl. Phys.*, **50**, 2523 (1979).

[40] S.Allen, T.D.McLean, P.F.Gordon, B.D.Bothwell, M.B.Hursthouse and S.A.Karaulov, *J. Appl. Phys.*, **64**, 2583 (1988).

[41] S.Allen, in *"Organic Materials for Non-linear Optics"* (R.A.Hann and D.Bloor, eds.), p.137, Royal Society of Chemistry, London (1989).

[42] D.Kalymnios, *J. Phys. D*, **5**, 667 (1972).

[43] J.L.Stevenson, *J. Phys. D*, **6**, L13 (1973).

[44] G.F.Lipscombe, A.F.Garito and R.S.Narang, *J. Chem. Phys.*, **75**, 1509 (1981).

[45] E.R.Tirner, F.R.Nash and P.M.Bridenbaugh. *J. Appl. Phys.*, **41**, 5278 (1970).

[46] J.Zyss, I.Ledoux, R.Hierle, R.Raj and J.L.Oudar, *IEEE J. Quantum Electron.*, **QE21**, 1286 (1985).

[47] M.J.Rosker and C.L.Tang, *J. Opt. Soc Am. B*, **2**, 691 (1985).

[48] I.Ledoux, D.Josse, R.Hierle and J.Zyss, Paper WW1, CLEO (1988). (Advance Program, p 87).

[49] M.Sigelle and R.Hierle, *J. Appl. Phys.*, **52**, 4199 (1981).

[50] J.L.Oudar and R.Hierle, *J. Appl. Phys.*, **48**, 2699 (1977).

[51] T.Yoshimura, *J. Appl. Phys.*, **62**, 2028 (1987).

[52] R.J.Twieg and C.W.Dirk, *J. Chem. Phys.*, **85**, 3537 (1986).

[53] P.A.Norman, D.Bloor, J.S.Ohbi, S.A.Karaulov, M.B.Hursthouse, P.V.Kolinsky, R.J.Jones and S.R.Hall, *J. Opt. Soc. Am. B*, **4**, 1013 (1987).

[54] J.-C.Baumert, R.J.Twieg, G.C.Borklund, J.A.Logan and C.W.Dirk, *Appl. Phys. Lett.*, **51**, 1484 (1987).

[55] R.T.Bailey, F.R.Cruikshank, S.M.G.Guthrie, B.J.McArdle, H.Morrison, D.Pugh, E.A.Shepherd, J.N.Sherwood, C.S.Yoon, R.Kashyap, B.K.Nayar and K.I.White, *Opt. Commun.*, **65**, 229 (1988).

[56] P.Günter, C.Bosshard, K.Sutter, H.Arend, G.Chapuis, R.J.Twieg and D.Dobrowski, *Appl. Phys. Lett.*, **50**, 486 (1987).

[57] I.Ledoux, D.Josse, P.Vidakovic, J.Zyss, R.A.Hann, P.F.Gordon, B.D.Bothwell, S.K.Gupta, S.Allen, P.Robin, E.Chastaing and J.-C.Dubois, *Europhys. Lett.*, **3**, 803 (1987).

[58] D.Lupo, W.Prass, U.Scheunemann, A.Laschewasky, H.Ringsdorf and I.Ledoux. *J. Opt. Soc. Am. B*, **5**, 300 (1988).

[59] I.R.Peterson, in *"Organic Materials for Non-linear Optics"* (R.A.Hann and D.Bloor eds.), pp. 317-333, Royal Society of Chemistry, London (1989).

[60] S.Allen, T.D.McLean, P.F.Gordon, B.D.Bothwell, P.Robin and I.Ledoux, *Proc. SPIE Int. Soc. Opt. Eng.*, **971**, 206 (1988).

[61] J.Zyss, *J. Molecular Electronics*, **1**, 25 (1985).

[62] D.B.Neal, M.C.Petty, G.G.Roberts, M.M.Ahmad, W.J.Feast, I.R.Girling, N.A.Cade, P.V.Kolinsky and I.R.Peterson, *Elect. Lett.*, **22**, 461 (1986).

[63] I.Ledoux, D.Josse, J.Zyss, T.McLean, R.A.Hann, P.F.Gordon, S.Allen, D.Lupo, W.Prass, U.Scheunemann, A.Laschewsky and H.Ringsdorf, in *"Nonlinear Optical Effects in Organic Polymers"* (J Messier et al., eds.), pp 79-91, Kluwer Academic Publishers (1989).

[64] W.L.Barnes and J.R.Sambles, *J. Phys. D*, **20**, 1125 (1987).

[65] I.R.Peterson, J.D.Earls, I.R.Girling and W.L.Barnes, *J. Phys. D*, **21**, 773 (1988).

[66] R.H.Tredgold, M.C.J.Young, P.Hodge and E.Khoshdel, *Thin Solid Films*, **151**, 441 (1987).

[67] R.H.Tredgold, M.C.J.Young, R.Jones, P.Hodge, P.Kolinsky and R.J.Jones, *Elect. Lett.*, **24**, 308 (1988).

[68] M.C.J.Young, R.H.Tredgold and P.Hodge, in *"Organic Materials for Nonlinear Optics"* (R.A.Hann and D.Bloor eds.), pp 354-360, Royal Society of Chemistry, London (1989).

[69] C.Bossard, G.Drecher, B.Tieke and P.Günter, *Proc. SPIE Int. Soc. Opt. Eng.*, **1017**, 141 (1989).

[70] G.H.Cross, I.R.Girling, I.R.Peterson and N.A.Cade, *Elect. Lett.*, **22**, 1111 (1986).

262

[71] G.H.Cross, I.R.Peterson and I.R.Girling, *Proc. SPIE Int. Soc. Opt. Eng.*, **824**, 79 (1987).

[72] G.R.Meredith, J.G. van Dusen and D.J.Williams, *Macromolecules*, **15**, 1385 (1985).

[73] K.D.Singer, J.E.Sohn and S.J.Lalama, *Appl. Phys. Lett.*, **49**, 248 (1986).

[74] R.DeMartino, D.Haas, G.Khanarian, T.Leslie, H.T.Man, J.Riggs, M.Sansone, J.Stamatoff, C.Teng and H.Yoon, in *"Nonlinear Optical Properties of Polymers"* (A.J.Heeger, J.Orenstein and D.R.Ulrich eds.), *Mat. Res. Soc. Symp. Proc.*, **109**, 65 (1988).

[75] S.Esselin, P.LeBarny, P.Robin, D.Broussoux, J.C.Dubois, J.Raffy and J.P.Pocholle in *"Nonlinear Optical Properties of Organic Materials"* (G.Khanarian, ed.), *Proc. SPIE Int. Soc. Opt. Eng.*, **971**, 120 (1988).

[76] J.R.Hill, P.Pantelis, F.Abbasi and P.Hodge, in *"Organic Materials for Nonlinear Optics"* (R.A.Hann and D.Bloor, eds.), pp. 404-411, Royal Society of Chemistry, London (1989).

[77] A.C.Griffin, A.M.Bhatti and R.S.L.Hung, in *"Nonlinear Optical and Electroactive Polymers"* (P.N.Prasad and D.R.Ulrich, eds.), pp. 375-392, Plenum, New York (1988).

[78] K.D.Singer, M.G.Kuzyk and J.E.Sohn, in *"Nonlinear Optical and Electroactive Polymers"* (P.N.Prasad and D.R.Ulrich, eds.), pp. 189-204, Plenum, New York (1988).

[79] R.Lytel, G.F.Lipscomb, M.Stiller, J.I.Thackara and A.J.Ticknor, in *"Organic Materials for Nonlinear Optics"* (R.A Hann and D.Bloor, eds.), pp. 382-389, Royal Society of Chemistry, London (1989).

[80] J.Thackara, M.Stiller, A.J.Ticknor, G.F.Lipscomb and R.Lytel, *Appl. Phys. Lett.*, **52**, 1031 (1988).

[81] D.R.Ulrich, in *"Organic Materials for Nonlinear Optics"* (R.A.Hann and D.Bloor, eds.), pp. 241-263, Royal Society of Chemistry, London (1989).

[82] D.A.B.Miller, D.S.Chemla, P.W Smith, A.C.Gossard and W.Wiegmann, *Opt. Lett.*, **8**, 477 (1983).

[83] B.S.Wherrett and S.D.Smith (eds.), *"Optical Bistability, Dynamical Nonlinearity and Photonic Logic"*, Royal Society, London (1985).

[84] P.W.Smith, *Bell Syst. Tech. J.*, **61**, 1975 (1982).

[85] J.L.Oudar, D.S.Chemla and E.Batifol, *J. Chem. Phys.*, **67**, 1626 (1977).

[86] J.L.Oudar and D.S.Chemla, *J. Chem. Phys.*, **66**, 2664 (1977).

[87] J.P.Hermann and J.Ducuing, *J. Appl. Phys.*, **45**, 5100 (1974).

[88] E.N.Svendsen, T.Stroyer-Hansen and H.F.Hameka, *Chem. Phys. Lett.*, **54**, 217 (1978).

[89] E.F.McIntyre and H.F.Hameka, *J. Chem. Phys.*, **68**, 5534 (1978).

[90] S.C.Mehendale and K.C.Rustagi, *Opt. Comm.*, **28**, 359 (1978).

[91] J.-M.Andre, C.Barbier, V.Bodart and J.Delhalle, in *"Nonlinear Optical Properties of Organic Molecules and Crystals"*, (D.S.Chemla and J.Zyss, eds.), pp. 137-158, Academic Press, Orlando (1987).

[92] B.M.Pierce, in *"Nonlinear Optical Properties of Organic Materials"* (G.Khanarian, ed.), *Proc. SPIE Int. Soc. Opt. Eng.*, **971**, 25 (1988).

[93] I.J.Goldfarb and J.Medrano, in *"Nonlinear Optical Effects in Organic Polymers"* (J.Messier, F.Kajzar, P.Prasad and D.Ulrich, eds.), *NATO ASI Series E*, Vol. 162, pp. 93-99, Kluwer Academic Press, Dordrecht (1989).

[94] F.Kajzar and J.Messier, *Phys. Rev. A*, **32**, 2352 (1985).

[95] D.N.Rao, R.Burzynski, X.Mi and P.N.Prasad, *Appl. Phys. Lett.*, **48**, 387 (1986).

[96] M.Sinclair, D.Moses, A.J.Heeger, K.Vilhelmsson, B.Valk and M.Salour, *Solid State Commun.*, **61**, 221 (1987).

[97] F.Kajzar, S.Etemad, G.L.Baker and J.Messier, *Solid State Commun.*, **63**, 1113 (1987).

[98] M.R.Drury, *Solid State Commun.*, **68**, 417 (1988).

[99] D.N.Rao, J.Swiatkiewicz, P.Chopra, K.Ghoshal and P.N.Prasad, *Appl. Phys. Lett.*, **48**, 1187 (1986).

[100] G.Wegner, *Z. Naturforsch.*, **24b**, 824 (1969).

[101] D.Bloor and R.R.Chance (eds.), *"Polydiacetylenes: Synthesis, Structure and Electronic Properties"*, Martinus Nijhoff, Dordrecht (1985).

264

[102] D.S.Chemla and J.Zyss (eds.), *"Nonlinear Optical Properties of Organic Molecules and Crystals"*, Vol. 2, Academic Press, Orlando (1986).

[103] C.Sauteret, J.-P.Hermann, R.Frey, F.Pradere, J.Ducuing, R.H.Baughman and R.R.Chance, *Phys. Rev. Lett.*, **36**, 956 (1976).

[104] G.M.Carter, M.K.Thakur, Y.J.Chen and J.V.Hryniewicz, *Appl. Phys. Lett.*, **47**, 457 (1985).

[105] G.M.Carter, J.V.Hryniewicz, M.K.Thakur, Y.J.Chen and S.E.Meyler, *Appl. Phys. Lett.*, **49**, 998 (1986).

[106] F.Kajzar and J.Messier, *Thin Solid Films*, **132**, 11 (1985).

[107] J.L.Jackel, N.E.Schlotter, P.D.Townsend, G.L.Baker and S.Etemad, in *"Nonlinear Optical Properties of Organic Materials"* (G.Khanarian, ed.), *Proc. SPIE Int. Soc. Opt. Eng.*, **971**, 239 (1988).

[108] S.Mann, A.R.Oldroyd, D.Bloor, D.J.Ando and P.J.Wells, in *"Nonlinear Optical Properties of Organic Materials"* (G.Khanarian, ed.), *Proc. SPIE Int. Soc. Opt. Eng.*, **971**, 245 (1988).

[109] K.Sutter, Ch.Bossard, L.Baraldi and P.Günter, in *"Materials for Non-linear and Electro-optics 1989"* (M.H.Lyons, ed.), *Inst. Phys. Conf. Ser. No. 103*, pp. 127-132, IOP, Bristol (1989).

[110] P.Kerkoc, M.Zgonik, K.Sutter, Ch.Bossard and P.Günter, in *"Materials for Non-linear and Electro-optics 1989"* (M.H.Lyons, ed.), *Inst. Phys. Conf. Ser. No. 103*, pp 133-138, IOP, Bristol (1989).

[111] J.Tsibouklis, J.P.Cresswell, N.Kalita, C.Pearson, P.J.Maddaford, H.Ancelin, J.Yarwood, M.J.Goodwin, N.Carr, W.J.Feast and M.C.Petty, *J. Phys. D: Appl. Phys.*, **22**, 1608 (1989).

[112] G.H.Cross, I.R.Girling, I.R.Peterson, N.A.Cade and J.D.Earls, *J. Opt. Soc. Am. B*, **4**, 962 (1987).

[113] Ch.Bossard, B.Tieke, M.Seifert and P.Günter, in *"Materials for Non-linear and Electro-optics 1989"* (M.H.Lyons, ed.), *Inst. Phys. Conf. Ser. No. 103*, pp. 181-186, IOP, Bristol (1989).

[114] I.Ledoux, D.Josse, P.Fremaux, J.P.Piel, G.Post, J.Zyss, T.D.McLean, R.A.Hann, P.F.Gordon and S.Allen, *Thin Solid Films*, **160**, 217 (1988).

[115] I.R.Girling, P.V.Kolinsky, N.A.Cade, J.D.Earls and I.R.Peterson, *Opt. Comm.*, **55**, 289 (1985).

[116] J.D.Bierlein, L.K.Cheng, Y.Wang and W.Tam, *App. Phys. Lett.*, **56**, 423 (1990).

[117] F.Kajzar, J.Messier and C.Rosilio, *J. Appl. Phys.*, **60**, 3040 (1986).

[118] S.Matsumoto, K.Kubodera, T.Kurihara and T.Kaino, *Appl. Phys. Lett.*, **51**, 1 (1987).

[119] T.Kaino, K.-I.Kubodera, S.Tomaru, T.Kurihara, S.Saito, T.Tsutsui and S.Tokito, *Electron. Lett.*, **23**, 1095 (1987).

[120] F.Kajzar, J.Messier, I.R.Girling and I.R.Peterson, *Electron. Lett.*, **22**, 1230 (1986).

[121] Although β, as defined in equation (15), is in units of m^4/V, values in the literature are generally (for historical reasons) quoted in the cgs units of cm^5/esu, commonly abbreviated to "esu". The conversion factor from "esu" to m^4/V is 4.18 x 10^{-10}. Similarly $\chi^{(2)}$ is either quoted in the SI unit of m/V (or pm/V), or the cgs unit of cm^2/esu, again usually abbreviated to just "esu". The conversion factor from "esu" to m/V is 4.18 x 10^{-4}.

CHAPTER 5
Piezoelectricity, Pyroelectricity and Ferroelectricity
J. Sworakowski

1. INTRODUCTION

Piezo, pyro and ferroelectricity have been known for many years: the piezoelectric effect in quartz and Rochelle salt ($KNaC_4H_4O_6 . 4H_2O$) was first described by J. & P. Curie in 1880 [1]; the ferroelectricity of Rochelle salt was discovered by Valasek in 1921 [2], and there is evidence that pyroelectricity was known to the ancient Greeks [3]. The piezo, pyro and ferroelectric properties of dielectrics have been reviewed in several books (*e.g.*, [3-14]), and articles (here, only three literature digests will be cited, published regularly in *"Ferroelectrics"* and giving a complete reference to work published on these properties [15-17]). Interest in these classes of substances has been enhanced by their application as transducers, sensors and memory elements.

In this chapter, a description of the effects related to piezo, pyro and ferroelectricity in dielectric materials is given with particular attention being paid to molecular solids.

Among the piezo, pyro and ferroelectric substances, molecular materials have initially gained limited interest only, in spite of there being a large number of potentially interesting substances belonging to this class of solids. To date, several millions of organic compounds have been synthesized. Most of them, when solidified, are able (at least in principle) to form molecular crystals, *i.e.* solids built of neutral molecules, interacting with one another *via* weak van der Waals-type forces. In the crystals of interest, from the point of view of this chapter, higher-term electrostatic (*e.g.*, dipole-dipole) interactions as well as hydrogen bonds often play an important role. Most molecular crystals belong to low-symmetry classes: triclinic, monoclinic and orthorhombic. It is beyond the scope of this chapter to discuss details of the interactions and arrangement of molecules in molecular crystals; the reader is referred to suitable monographs (e.g., [18]).

Piezo and pyroelectricity of molecular materials (mainly polymers) were systematically studied by Fukada and his collaborators as early as the sixties (see, *e.g.*, [19] for the review), but interest in molecular materials has increased significantly since the discovery of piezoelectricity [20], and pyro and ferroelectricity [21] of poly(vinylidene fluoride), commonly referred to as PVDF, which is now one of the most extensively studied molecular polar materials (see, *e.g.*, [22-30] for the reviews and collections of work on this topic). The research on polar molecular materials has been additionally stimulated by their interesting nonlinear optical properties. Several monographs concerning this topic have been published recently (e.g., [31-33]) and it is discussed, in depth, in Chapter 4.

2. MACROSCOPIC DESCRIPTION

Consider an electric field acting on a dielectric material. According to the macroscopic (phenomenological) description, a polarization builds up in the dielectric, the relationship between the vectors of electric field (*E*), polarization (*P*) and electric displacement (*D*) being given by the equation

$$D_i = \epsilon_o E + P_i \quad (i = 1, 2, 3) \tag{1}$$

where $\epsilon_o = 8.854 \times 10^{-12}$ F/m, the electric permittivity of free space. Note that the direction of the polarization vector does not necessarily coincide with that of the applied field, thus, in general, the vectors *D*, *E*, and *P* are not parallel.

The polarization itself is a function of external fields. In general, the dependence is superlinear ([31,33], see also chapter 4 of this book), but in many dielectrics (the so-called linear dielectrics) a linear dependence may be adopted to a good approximation in sufficiently low fields. Employing Einstein's convention, which implies the summation over repeated indices, its dependence on the electric field may be written as

$$P_i = \chi_{ij}\epsilon_o E_j \tag{2}$$

where χ is the electric susceptibility (second rank tensor). Combining equations (1) and (2), one obtains

$$D_i = (\delta_{ij} + \chi_{ij})\epsilon_o E_j = \epsilon_{ij} E_j \tag{3}$$

δ standing for the Kronecker's delta and ϵ for the electric permittivity.

All dielectrics, when exposed to an electric field, suffer deformation. This effect, referred to as *electrostriction*, is proportional to the square of the applied field and usually small. It is represented by

$$\varepsilon_{ij} = Q_{klij} E_k E_l \tag{4}$$

where ε is the strain and Q the electrostriction coefficient. However, in some crystals one may observe a *linear effect*, usually much larger: the application of an electric field gives rise to a strain (ε) or stress (σ), both proportional to the applied field.

$$\sigma_{ij} = -e_{ijk} E_k \tag{5a}$$

$$\varepsilon_{ij} = d_{kij} E_k \tag{5b}$$

When exposed to a mechanical stress or to a deformation, these crystals produce an additional polarization,

$$\Delta P_i = d_{ijk}\sigma_{jk} \tag{5c}$$

$$\Delta P_i = e_{ijk}\varepsilon_{jk} \tag{5d}$$

Equations (5a) to (5d) define the converse and direct *piezoelectric effects*, the third rank tensors d and e standing for the piezoelectric coefficients. As may be shown [34], the absence of the inversion centre among the symmetry elements characterising the unit cell is the necessary condition for a crystal to be piezoelectric. Among 32 crystal classes, only 20 are non-centrosymmetric, thus allowing piezoelectricity (*cf.* Table 2).

In the dielectrics considered so far, the polarization appears only after application of external (electrical or mechanical) fields. In some classes of dielectric crystals, however, the symmetry allows for a non-zero polarization even in the absence of the external fields; these crystals may possess a *spontaneous polarization*. When such samples are stored over a sufficiently long time, the spontaneous polarization (which should manifest itself as a charge appearing on the sample surface) is compensated by adsorbed species (ions, dipolar molecules *etc.*), and a change of spontaneous polarization may be observed only upon changing temperature.

$$\Delta P_i = p_i \Delta T \tag{6}$$

This is called the *pyroelectric effect*, and the parameter p is referred to as the pyroelectric coefficient. It is evident that only crystals with a non-zero dipole moment of the unit cell may be pyroelectric. In ionic crystals this means that the centroids of the positive and negative charges in the unit cell do not coincide. In molecular crystals, appearance of spontaneous polarization is due to a lack of compensation of molecular dipoles, as is schematically shown in Fig. 1.

Symmetry allows for only 10 of the 20 non-centrosymmetric crystal classes to have a non-zero spontaneous polarization. These polar classes are listed in Table 2. It should be noticed that the symmetry determines the direction of spontaneous polarization with respect to the crystal axes in several of the classes.

In some polar crystals, there exist a few (at least two) equivalent directions of spontaneous polarization, and under the action of a sufficiently high electric field the polarization of the crystal may be switched between different stable positions. These polar crystals are called *ferroelectrics*.

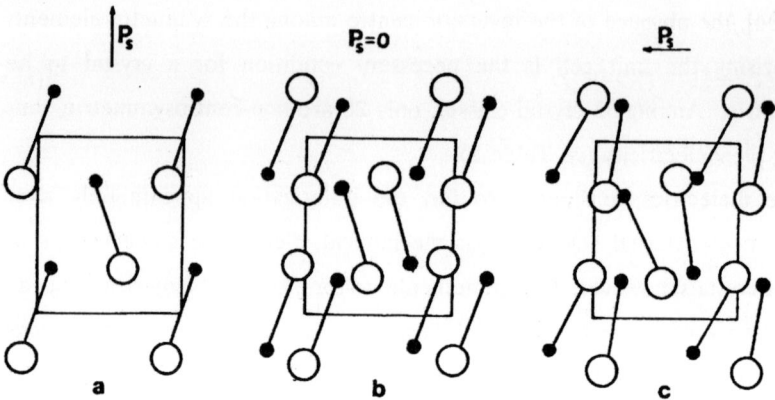

Fig. 1. Arrangements of molecules in hypothetical crystals built of simple dipolar molecules. The arrangements (a) and (c) result in uncompensated dipole moments. The directions of spontaneous polarization are indicated by arrows.

2.1. THERMODYNAMICS OF LINEAR DIELECTRICS

In general, a dielectric sample may be adequately described if its mechanical, electrical and thermal states are known. In other words, the internal energy of the sample (U) may be written in terms of three parameters: (σ or ε), (E, P or D) and (T or S where S is entropy). From the first and second laws of thermodynamics, one may demonstrate that $U = U(\varepsilon, D, S)$, *i.e.*

$$dU = \sigma_{ij}d\varepsilon_{ij} + E_k dD_k + TdS \tag{7}$$

For practical purposes, this is not the most convenient set of parameters, and one usually defines other functions of state, *e.g.*:

• *free energy (Helmholtz free energy)*

$$F = U - TS \tag{8a}$$

• *free enthalpy (Gibbs free energy)*

$$G = U - TS - \sigma_{ij}\varepsilon_{ij} - E_k D_k \tag{8b}$$

• *electrical free energy*

$$G' = U - TS - E_k D_k \tag{8c}$$

• *elastic free energy*

$$G'' = U - TS - \sigma_{ij}\varepsilon_{ij} \tag{8d}$$

Combining equations (7) and (8b) one easily obtains

$$dG = -\varepsilon_{ij}d\sigma_{ij} - D_k dE_k - SdT \tag{9}$$

i.e. $G = G(\sigma, E, T)$. In a similar way one may demonstrate that $F = F(\varepsilon, D, T)$, $G' = G'(\varepsilon, E, T)$ and $G'' = G''(\sigma, D, T)$. Comparing equation (9) with the total differential of free enthalpy

$$dG = \left(\frac{\partial G}{\partial \sigma_{ij}}\right)_{E,T} d\sigma_{ij} + \left(\frac{\partial G}{\partial E_k}\right)_{T,\sigma} dE_k + \left(\frac{\partial G}{\partial T}\right)_{\sigma,E} dT \tag{10}$$

and taking the second derivatives, one arrives at the equations:

$$-\frac{\partial^2 G}{\partial \sigma_{ij}\partial E_k} = \left(\frac{\partial D_k}{\partial \sigma_{ij}}\right)_{E,T} = \left(\frac{\partial \varepsilon_{ij}}{\partial E_k}\right)_{T,\sigma} = d_{ijk} \tag{11}$$

$$-\frac{\partial^2 G}{\partial T\partial E_i} = \left(\frac{\partial D_i}{\partial T}\right)_{\sigma,E} = p_i^\sigma \tag{12}$$

Equations (11) and (12) provide definitions of the piezoelectric coefficient and pyroelectric coefficient of a mechanically relaxed sample (total pyroelectric coefficient) equivalent to those defined by equations (5) and (6). In a similar way, from equations (7) and (8c) one obtains:

$$-\frac{\partial^2 G'}{\partial T\partial E_i} = \left(\frac{\partial D_i}{\partial T}\right)_{\varepsilon,E} = p_i^\varepsilon \tag{13}$$

$$-\frac{\partial^2 G'}{\partial \varepsilon_{ij}\partial E_k} = \left(\frac{\partial D_k}{\partial \varepsilon_{ij}}\right)_{E,T} = -\left(\frac{\partial \sigma_{ij}}{\partial E_k}\right)_{T,\varepsilon} = e_{ijk} \tag{14}$$

Equation (13) defines the pyroelectric coefficient of a mechanically clamped sample (*i.e.* the so called primary pyroelectric coefficient) and equation (14), the piezoelectric coefficient, e. It is straightforward to demonstrate (*e.g.*, [34]) that the piezoelectric coefficients d and e are related by the equations

$$e_{ijk} = d_{ilm}c_{lmjk} \tag{15a}$$

$$d_{ijk} = e_{ilm}s_{lmjk} \tag{15b}$$

where the fourth rank tensors c and s are defined below. Similarly, the total pyroelectric coefficient, p^{σ}, is related to the primary coefficient by the equation

$$p_i^{\sigma} = p_i^{\varepsilon} + d_{imn}c_{mnkl}\alpha_{kl} = p_i^{\varepsilon} + p_i^{sec} \tag{16}$$

where α is the thermal expansion coefficient. The second term in the above equation, referred to as the *secondary* pyroelectric coefficient, gives the contribution of the piezoelectric effect to the total pyroelectricity.

The primary and secondary pyroelectricity are effects related to properties of individual unit cells, *i.e.* they appear in polar crystals upon homogeneous changes of temperature. Often, the concept of tertiary pyroelectricity is introduced [3]. This effect, resulting from a temperature gradient acting on a sample, is not related directly to properties of the unit cell, and will not be considered in this chapter.

It has already been shown that the parameters E, P and D are coupled (cf. equations 1 to 3); similarly, the mechanical variables are related *via* a generalized form of Hooke's law

$$\varepsilon_{ij} = s_{ijkl}\sigma_{kl} \tag{17a}$$

$$\sigma_{ij} = c_{ijkl}\varepsilon_{kl} \tag{17b}$$

where s and c are fourth rank tensors describing the elastic compliance and stiffness, respectively. A complete set of equations relating the mechanical, electrical and thermal parameters (equations of state) may now be written (*cf.*, *e.g.*, [35]). Taking σ, E and T as independent variables, one obtains:

$$d\varepsilon_{ij} = s^E_{ijkl}d\sigma_{kl} + d_{kij}dE_k + \alpha_{ij}dT \tag{18a}$$

$$dD_i = d_{ijk}d\sigma_{jk} + \epsilon^\sigma_{ij}dE_j + p^\sigma_i dT \tag{18b}$$

The choice of ε, E and T as independent variables results in the following set of equations

$$d\sigma_{ij} = c^E_{ijkl}d\varepsilon_{kl} - e_{kij}dE_k \tag{19a}$$

$$dD_i = e_{ijk}d\varepsilon_{jk} + \epsilon^\varepsilon_{ij}dE_j + p^\varepsilon_i dT \tag{19b}$$

where, as previously, the superscripts E, σ and ε denote parameters taken at constant electric field, stress and strain respectively. Note that the electric permittivities, ϵ^σ and ϵ^ε, in equations (18b) and (19b), are in general different and are related by the equation [5,7]

$$\epsilon^\varepsilon/\epsilon^\sigma = 1 - k^2 \tag{20}$$

where the parameter k (k < 1) is the electromechanical coupling coefficient, defined by the equation

$$k = e_{i\alpha}/(\epsilon^\varepsilon_{ii}c^D_{\alpha\alpha})^{\frac{1}{2}} \tag{21a}$$

or, equivalently,

$$k = d_{i\alpha}/(\epsilon^\varepsilon_{ii}s^E_{\alpha\alpha})^{\frac{1}{2}} \tag{21b}$$

The electromechanical coupling coefficient is a measure of the efficiency of conversion of the mechanical energy into electrical energy. The electric permittivity of a mechanically clamped sample is smaller than that of a mechanically relaxed one due to its piezoelectricity. Similarly, one may show that an open-circuited sample is less compliant than a short-circuited one.

It should be stressed once again that all equations in this section were derived assuming a linear dependence between parameters. This assumption holds for linear dielectrics in moderate fields. As will be discussed later, it breaks down in the case of non-linear dielectrics.

2.2. REDUCTION OF THE NUMBER OF COMPONENTS

The number of independent coefficients of a tensor of rth rank is, in general, 3^r. However, in the case of tensors considered in this chapter, this number may be substantially reduced (and, consequently, the notation of many dependences greatly simplified) if one follows the commonly adopted assumption (but not fully justified from the theoretical point of view [36]) that the second rank tensors introduced above (α, ε, σ) are symmetric, i.e. $\alpha_{ij} = \alpha_{ji}$, etc. Consequently, the number of independent components of higher rank tensors is also reduced due to the fact that, for example, $d_{ijk} = d_{ikj}$, $s_{ijkl} = s_{jikl} = s_{ijlk} = s_{jilk}$, etc. Therefore, each symmetric second rank tensor contains 6 independent parameters at most, and each third rank and fourth rank tensor, no more than 18 and 36 independent components, respectively.

One can simplify the notation, using the so called contracted (matrix) notation which allows one to represent tensors up to the fourth rank as two-dimensional matrices. The rules of equivalence between the full notation of tensors and the contracted one are shown in Table 1. For example, the electric permittivity (second rank tensor), piezoelectric coefficient (third rank tensor), and elastic compliance (fourth tank tensor) now assume the forms:

$$\varepsilon_{ij} \;\Rightarrow\; \epsilon_\lambda \;=\; (\varepsilon_1 \; \varepsilon_2 \; \varepsilon_3 \; \varepsilon_4 \; \varepsilon_5 \; \varepsilon_6) \tag{22a}$$

$$d_{ijk} \;\Rightarrow\; d_{i\lambda} \;=\; \begin{vmatrix} d_{11} & d_{12} & d_{13} & d_{14} & d_{15} & d_{16} \\ d_{21} & d_{22} & d_{23} & d_{24} & d_{25} & d_{26} \\ d_{31} & d_{32} & d_{33} & d_{34} & d_{35} & d_{36} \end{vmatrix} \tag{22b}$$

$$s_{ijkl} \;\Rightarrow\; s_{\lambda\mu} \;=\; \begin{vmatrix} s_{11} & s_{12} & s_{13} & s_{14} & s_{15} & s_{16} \\ s_{21} & s_{22} & s_{23} & s_{24} & s_{25} & s_{26} \\ s_{31} & s_{32} & s_{33} & s_{34} & s_{35} & s_{36} \\ s_{41} & s_{42} & s_{43} & s_{44} & s_{45} & s_{46} \\ s_{51} & s_{52} & s_{53} & s_{54} & s_{55} & s_{56} \\ s_{61} & s_{62} & s_{63} & s_{64} & s_{65} & s_{66} \end{vmatrix} \tag{22c}$$

The equations of state, (18) and (19), may now be rewritten in their contracted forms:

$$d\varepsilon_\lambda = s^E_{\mu\lambda} d\sigma_\lambda + d_{k\lambda} dE_k + \alpha_\lambda dT \tag{23a}$$

$$dD_i = d_{i\lambda} d\sigma_\lambda + \epsilon^\sigma_{ij} dE_j + p^\sigma_i dT \tag{23b}$$

$$d\sigma_\lambda = c^E_{\lambda\mu}d\varepsilon_\mu - e_{k\lambda}dE_k \qquad (24a)$$

$$dD_i = e_{i\mu}d\varepsilon_\mu + \epsilon^\varepsilon_{ij}dE_j + p^\varepsilon_i dT \qquad (24b)$$

The number of independent components may be further reduced due to constraints imposed by the crystal symmetry; Table 2 lists all crystallographic classes in which piezo, pyro and ferroelectricity are allowed by symmetry, giving the number of independent components of tensors (*the explicit forms of these tensors may be found in most textbooks on crystal physics, e.g., [34,13]*).

Table 1. EQUIVALENCE BETWEEN THE FULL NOTATION AND CONTRACTED NOTATION OF TENSORS

Rank of tensor	Full notation	Contracted notation	Equivalence Rules indices	Equivalence Rules component values
1 (vector)	y_i	y_i		
2	y_{ij}	y_λ	$\lambda=i$ for $i=j$ $\lambda=9-i-j$ for $i\neq j$	$y_\lambda=y_{ij}$ for $\lambda \leq 3$ $y_\lambda=2y_{ij}$ for $\lambda >3$
3	y_{ijk}	$y_{i\lambda}$	$\lambda=j$ for $j=k$ $\lambda=9-j-k$ for $j\neq k$	$y_{i\lambda}=y_{ijk}$ for $\lambda \leq 3$ $y_{i\lambda}=2y_{ijk}$ for $\lambda >3$ ($e_{i\lambda}=e_{ijk}$ for all λ)
4	y_{ijkl}	$y_{\lambda\mu}$	$\lambda=i$ for $i=j$ $\lambda=9-i-j$ for $i\neq j$ $\mu=k$ for $k=1$ $\mu=9-k-1$ for $k\neq1$	$y_{\lambda\mu}=y_{ijkl}$ for λ and $\mu \leq 3$ $y_{\lambda\mu}=2y_{ijkl}$ for λ or $\mu >3$ $y_{\lambda\mu}=4y_{ijkl}$ for λ and $\mu >3$

$i,j,k,1 = 1,2,3$; $\lambda,\mu = 1....6$

Table 2. CRYSTALLOGRAPHIC CLASSES ALLOWING FOR
PIEZOELECTRICITY AND PYROELECTRICITY

Crystallographic system	Point group	Number of independent coefficients		
		d,e	α, ϵ	p^σ, p^ϵ
Triclinic	1	18	6	3
Monoclinic	2	8	4	1
	m	10	4	2
Orthorhombic	222	3	3	–
	mm2	5	3	1
Tetragonal	4	4	2	1
	$\bar{4}$	4	2	–
	422	1	2	–
	4mm	3	2	1
	$\bar{4}$2m	2	2	–
Trigonal	3	6	2	1
	32	2	2	–
	3m	4	2	1
Hexagonal	6	4	2	1
	$\bar{6}$	2	2	–
	622	1	2	–
	6mm	3	2	1
	$\bar{6}$m2	1	2	–
Regular	23	1	1	–
	$\bar{4}$3m	1	1	–

2.3. NON-LINEAR RELATIONS

It was mentioned in section 2.1 that all equations above were derived assuming
linear dependences between parameters characterising the materials. In other
words, expressions for, for example, free energy and free enthalpy (written below
in their integrated forms) read

$$G = G_o + s_{\mu\lambda}\sigma_\mu\sigma_\lambda + \epsilon_{ij}E_iE_j - d_{i\lambda}E_i\sigma_\lambda \tag{25}$$

$$F = F_o + c_{\mu\lambda}\varepsilon_\mu\varepsilon_\lambda + \varkappa_{ij}D_iD_j - h_{i\lambda}D_i\varepsilon_\lambda \tag{26}$$

where h is a piezoelectric coefficient, related to already introduced coefficients d and e, and $\varkappa$ standing for the inverse electric permittivity ($\varkappa_{ij}\epsilon_{jk} = \delta_{ik}$). This is justified if the values of the mechanical and electrical parameters are small. If, however, this is not the case, higher order terms should be added in the equations. In the case of the free energy F, these terms are of the form $D_iD_j\varepsilon_\mu$, $\varepsilon_\mu\varepsilon_\lambda\varepsilon_\delta$, $D_iD_jD_kD_l$, etc. In general, it is quite difficult to write a complete expansion relating the free energy to all relevant variables, thus several simplifications are usually introduced.

A good (although in some cases only qualitatively correct) description of nonlinear dielectrics, and in particular of ferroelectrics may be obtained within the framework of the Landau-Devonshire phenomenological theory [37]. We shall follow this approach (cf. [10-14,38]) to derive basic equations characterising the behaviour of ferroelectrics. The discussion will be limited to a few simple cases and we shall consider only uniaxial ferroelectrics.

As is mentioned above (see Table 2), in several classes of polar crystals the direction of the vector of spontaneous polarization is determined by symmetry. For example, let $P_{s,1} = P_{s,3} = 0$, $P_{s,2} = P_s$; furthermore, let the field be applied parallel to P, thus $E_1 = E_3 = 0$, $E_2 = E$. Let us further suppose that the free energy depends mainly on the value of the polarization, and dependences on other parameters can be neglected. A simplified expression may now be written, expanding the electric free energy in powers of the electric displacement. However, in many cases it is more convenient to choose the polarization as an independent parameter, and this convention, employed by most authors, will be adopted in this chapter. Thus we may write

$$G' = F - EP = F_o + AP^2 + BP^4 + CP^6 - EP \qquad (27)$$

where F_o is a free energy term, independent of the polarization, and A, B and C are expansion parameters which may depend on temperature. The free energy was expanded up to the sixth power terms; note that only even terms appear in equation (27) (except for the last term describing interaction with the external field) as F should be invariant with respect to the polarization reversal.

We shall first examine properties of the ferroelectric at E = 0 (thus G' = F, *cf.* equation 8c). Obviously, the thermodynamically stable crystal structures will be those which allow for a minimum (or equivalent minima) of F, *i.e.* $(\partial F/\partial P)_{P_s} = 0$, $(\partial^2 F/\partial P^2)_{P_s} > 0$. If all the coefficients in equation (24) are positive, the only possible solution is $P_s = 0$, *i.e.* the resulting crystal structure is non-polar.

Let us now consider the case A < 0, B > 0 (in this case, the term with P^6 does not add any new feature to the qualitative description of the ferroelectric, thus C may be set equal to zero). After simple calculations, one arrives at the condition for a minimum of the free energy

$$P_{E=0} = P_s = \pm(-A/2B)^{\frac{1}{2}} \tag{28}$$

thus the resulting structure is polar with two equivalent positions of spontaneous polarization.

Within the simple model, we have shown that the position of the minimum of the free energy on the P *vs.* F plane (and hence the magnitude of the spontaneous polarization) depends on the sign of coefficient A. Since A may vary with temperature, one may expect that in some crystals there occurs an inversion of its sign, and, consequently, a transition from a polar (*ferroelectric*) to a non-polar (*paraelectric*) phase. In the vicinity of an inversion temperature (T_o), A may be taken to vary linearly with temperature

$$A = a(T - T_o) \tag{29}$$

($a > 0$); the higher term parameters (usually weakly temperature dependent) will be assumed constant. The spontaneous polarization thus varies as $(T_o - T)^{\frac{1}{2}}$ in the vicinity of the phase transition

$$P_s = \pm\{a(T_o - T)/2B\}^{\frac{1}{2}} \text{ for } T < T_o; \; 0 \text{ for } T \geq T_o \tag{30}$$

The polar phase vanishes at $T = T_o$, thus in this case the phase transition temperature (T_c) corresponds to T_o.

In a similar way, other thermodynamical parameters may be expressed in terms of the expansion coefficients, and relation between them in the ferroelectric and paraelectric phases may be found (*cf.* Table 3). In particular, the temperature dependence of the inverse electric susceptibility amounts to

$$\varkappa = 4a(T_o - T) \text{ for } T < T_o; \; 2a(T - T_o) \text{ for } T \geq T_o \tag{31}$$

The latter dependence is often referred to as the Curie-Weiss law.

It should be noted that the first derivatives of F (*e.g, S, P_s etc.*) are continuous at the phase transition temperature, the discontinuities appearing only in the second derivatives of the free energy (c_v, p *etc.*); the transition described above is therefore of the second order. The function F(P) and the expected temperature dependences $P_s(T)$ and $\varkappa(T)$ are shown in Fig. 2.

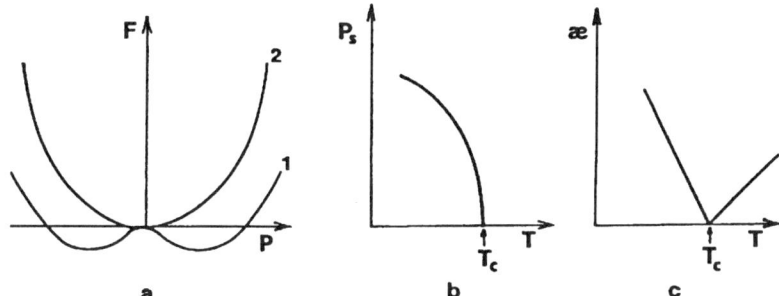

Fig. 2. (a) Dependence of free energy on the polarization as predicted by the Landau-Devonshire theory for second-order transitions (1, T < T_o ; 2, T > T_o). (b) and (c) Temperature dependences of the spontaneous polarization and reciprocal electric permittivity in the vicinity of the second-order phase transition temperature.

A dependence of P on E may be obtained from equations (27) and (29). Thus,

$$E = 2a(T - T_o)P + 4BP^3 \tag{32}$$

The P(E) dependences for T < T_c and T > T_c are shown in Fig. 3. The dependence for T < T_c needs a comment. Suppose a ferroelectric sample is characterized by a spontaneous polarization P_s. Upon applying a field (E < 0), the polarization in the sample changes towards P_1. Here, the system becomes unstable and switches to a new value P_2. If one now removes the field, the crystal will have the spontaneous polarization $-P_s$, which again may be switched to P_s if a sufficiently large positive field is applied. Thus at $T < T_c$ one should observe a hysteresis of the $P(E)$ dependence. The lowest field necessary to give rise to the polarization reversal at a given temperature is called the *coercive field* (E_c). It is straightforward to demonstrate that $E_c \propto (T_o - T)^{3/2}$.

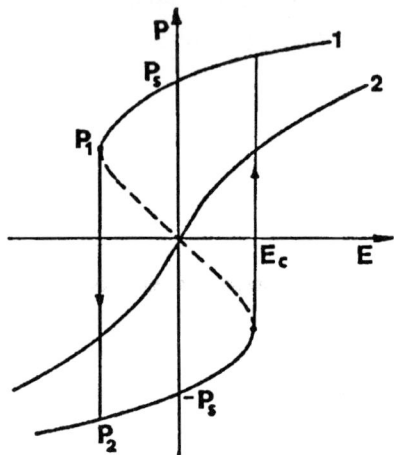

Fig. 3. *P(E)* dependence in a ferroelectric in the vicinity of a second-order phase transition (1, T $< T_c$; 2, T $> T_c$). See text for further discussion.

Let us now briefly consider the case when $B < 0$. The term with P^6 must now be added, and we set $C > 0$. As before, A is assumed to fulfil equation (29). Solution of this case leads to the following results (*cf.* Fig. 4). At sufficiently low temperatures, there exist two minima of the free energy with $P_s \neq 0$ (*i.e.* corresponding to a polar phase), the temperature dependence of the spontaneous polarization being given by the equation

$$P_s = \pm[-\tfrac{1}{3}BC^{-1}(1 + \{3aB^{-2}C(T_o - T)\}^{\frac{1}{2}})]^{\frac{1}{2}} \tag{33}$$

At $T = T_o$, a third minimum appears at $P = 0$, being initially shallower than those corresponding to the polar phase; just above T_o the paraelectric phase is metastable. The free energies of the ferro and paraelectric phases become equal at a temperature T_c given by the equation

$$T_c = T_o + B^2/4aC \tag{34}$$

At this temperature, both phases co-exist in equilibrium; T_c is thus the phase transition temperature (note that in this case $T_c \neq T_o$). The ferroelectric phase is metastable between T_c and a temperature, T_u, where

$$T_u = T_o + B^2/3aC = T_c + B^2/12aC \tag{35}$$

One should thus expect hysteresis of the properties of the dielectric: the ferroelectric phase can be overheated above T_c, the paraelectric undercooled below T_c. The temperature dependences of P_s and $\varkappa$, predicted by the Landau-Devonshire theory, are shown in Fig. 4.

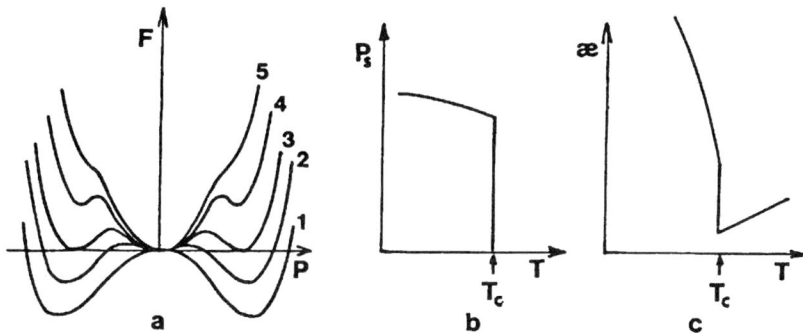

Fig. 4. (a) Dependence of free energy on the polarization as predicted by the Landau-Devonshire theory for the first-order transitions (1, $T < T_o$; 2, $T_o < T < T_c$; 3, $T = T_c$; 4, $T_c < T < T_u$; 5, $T > T_u$. (b) and (c) Temperature dependences of the spontaneous polarization and reciprocal electric permittivity in the vicinity of the first-order phase transition temperature.

The calculated $P(E)$ dependences are schematically shown in Fig. 5. Below T_c, one observes a single hysteresis loop, similarly as in the case of the second order transition. However, in the temperature range $T_c < T < T_u$ the theory predicts a double loop; indeed, the double loops have been observed in several ferroelectrics undergoing first order phase transitions.

Table 3. TEMPERATURE DEPENDENCES OF PARAMETERS IN THE VICINITY OF PHASE TRANSITIONS IN FERROELECTRICS DERIVED FROM LANDAU-DEVONSHIRE THEORY.

Parameter	Temperature dependence		Difference at $T = T_c$
	Ferroelectric phase $(T < T_c)$	Paraelectric phase $(T \geq T_c)$	
Second order transitions			
P_s	$\pm[\tfrac{1}{2}aB^{-1}(T_o-T)]^{\frac{1}{2}}$	0	0
p^ε	$\pm[8a^{-1}B(T_o-T)]^{-\frac{1}{2}}$	0	∞
$\varkappa$	$4a(T_o-T)$	$2a(T-T_o)$	0
S	$S_o - \tfrac{1}{2}a^2B^{-1}(T_o-T)$	S_o	0
c_v	$c_{v,o} + \tfrac{1}{2}aB^{-1}T$	$c_{v,o}$	$\tfrac{1}{2}a^2B^{-1}T_o$
First order transitions[*)]			
P_s	$\pm[-\tfrac{1}{3}BC^{-1}\{1+\psi(T)\}]^{\frac{1}{2}}$	0	$\pm(-\tfrac{1}{2}BC^{-1})^{\frac{1}{2}}$
p^ε	$\pm\tfrac{1}{4}[\tfrac{1}{3}a^{-2}B^3C^{-1}\psi^2(T)\{1+\psi(T)\}]^{-\frac{1}{2}}$	0	$\pm(-\tfrac{1}{2}a^2B^{-3}C)^{\frac{1}{2}}$
$\varkappa$	$8a[(T_o-T) + \tfrac{1}{3}a^{-1}B^2C^{-1}\{1+\psi(T)\}]$	$2a(T-T_o)$	$3B^2/2C$
S	$S_o + \tfrac{1}{3}aBC^{-1}\{1+\psi(T)\}$	S_o	$\tfrac{1}{2}aBC^{-1}$
c_v	$c_{v,o} - \tfrac{1}{2}a^2B^{-1}T/\psi(T)$	$c_{v,o}$	$-a^2B^{-1}T_c$

[*)] $\psi(T) = [1 + 3aB^{-2}C(T_o-T)]^{\frac{1}{2}}$

The temperature dependences of some parameters characterising both phases have been collected in Table 3. Note that both P_s and S are discontinuous at T_c, the transition is thus of the first order.

The Landau-Devonshire mean-field approach provides an adequate description of the behaviour of many ferroelectrics. As an example, results are shown for the most thoroughly investigated ferroelectric, triglycine sulphate {TGS, $(NH_2CH_2COOH)_3.H_2SO_4$} which undergoes a second order transition at 322 K.

The results of measurements of P_s, $\varkappa$ and c_v, shown on Fig. 6, demonstrate that the theory satisfactorily describes these dependences. Similarly, the behaviour of several ferroelectrics undergoing first order transitions is well fitted by the Landau-Devonshire theory in its simple form.

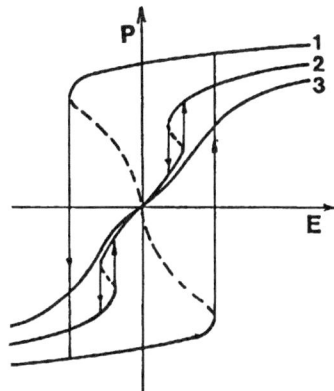

Fig. 5. P(E) dependence in a ferroelectric in the vicinity of a first order phase transition (1, $T < T_c$; 2, $T_c < T < T_u$; 3, $T > T_u$.

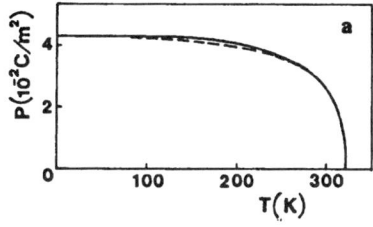

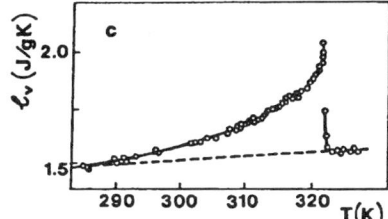

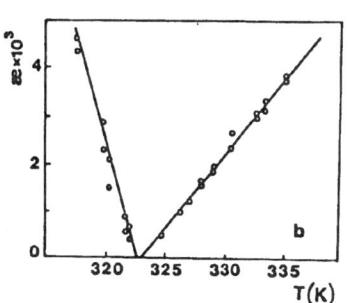

Fig. 6. Experimental data obtained on TGS single crystals: (a) $P_s(T)$ [39], broken line - experiment, full line - theory; (b) $\varkappa(T)$ [40] (copyright 1958 by IBM Corp., reprinted with permission); (c) $c_v(T)$ [41].

However, in several cases this simple approach fails to properly describe the temperature dependences of the electrical and thermal variables, and their mutual relations. Sometimes, an influence of other parameters should be taken into account. This is, for example, the case of crystals which in their paraelectric phases are piezoelectric. A term coupling the polarization to an external stress or deformation (*e.g.*, of the type $P\varepsilon$) should now be added to the polynomial. In this case, one or more components of the piezoelectric coefficient may exhibit anomalous behaviour in the vicinity of the phase transition. In the paraelectric phase, the temperature dependences of these components should assume a Curie-Weiss type dependence

$$d_{i\lambda} = d_{i\lambda}^{o} + B/(T - T_o) \tag{36}$$

where B is a constant. Another example, which will be briefly discussed below, is the so-called improper ferroelectric.

We have so far considered the behaviour of ferroelectrics in which any influence of parameters other than polarization could be neglected. The polarization is thus taken as a unique order parameter. As was indicated in the preceding sections, other variables may be equally important, and the behaviour of some dielectrics is controlled by another order parameter (it will be denoted by η without specifying its nature), the polarization being only coupled to it. If the coupling leads to ferroelectricity, this class is referred to as improper ferroelectrics. Their properties depend, in part, on symmetry relations between the order parameter and polarization. When η and P have different symmetries, one of the possible solutions gives $P_s \propto (T_o - T)$. In the simplest case, when the symmetries of both parameters are identical and $B > 0$ (*i.e.*, the Landau polynomial can be terminated at the fourth power of the order parameter), the free energy may be written as [12]

$$G' = F_o + \alpha(T - T_o)\eta^2 + B\eta^4 + CP^2 + LP\eta - EP \tag{37}$$

where L is a coupling parameter. In this case the transition is of the second order, the spontaneous polarization being given by the equation

$$P_s = \pm\{\tfrac{1}{8}aL^2B^{-1}C^{-2}\,(T_c - T)\}^{\frac{1}{2}} \text{ for } T < T_c; \; 0 \text{ for } T \geq T_c \qquad (38)$$

and the transition temperature being shifted with respect to T_o by an increment which is dependent upon the coupling between η and P

$$T_c = T_o + L^2/4aC \qquad (39)$$

3. STRUCTURES OF MOLECULAR CRYSTALS AND POLYMERS

In the preceding section, the symmetry requirements were given for pyroelectric and/or piezoelectric crystals. It is instructive to present structures of typical molecular materials, and to discuss features of molecular arrangement in their crystal unit cells.

As already mentioned, the crystals must possess a non-zero spontaneous polarization, *i.e.* the molecules should be packed in such a way so as to prevent complete compensation of their electric moments. Obviously, the molecules themselves must have non-zero dipole moments. However, it should be realized that the dipole-dipole interactions favour a "head-to-tail" (centrosymmetric) configuration of dipoles, thus additional interactions are necessary to stabilize a polar arrangement of molecules. The most obvious one is the repulsion associated with molecular shapes. In such a way polar structures of, for example, meta-disubstituted benzenes are realized. The structure of *m*-dinitrobenzene (*m*DNB) [42] is shown as an example (see Fig. 7). The crystal is orthorhombic (space group Pbn2₁). It is clearly seen from the figure that the vectorial sum of molecular dipole moments is non-zero in the c direction. Structures of several other *m*-disubstituted benzenes are similar to the one shown above: *e.g.*, resorcinol (*m*-dihydroxybenzene), *m*-dichlorobenzene and *m*-dibromobenzene crystallize in the space group Pbn2₁, identical with that of *m*DNB, the space group of *m*-nitroaniline (*m*NA) being Pbc2₁ [43-45].

Structures of several other polar crystals have been reported: most belong to the three lowest crystallographic classes: triclinic, monoclinic and orthorhombic, although there exist polar molecular crystals of higher symmetries, for example, some complexes of iodoform (CHI_3) and its analogues having a trifold axis (*e.g.*, [46]). It is interesting to compare the structures of these compounds with that of pure iodoform. The structure of the latter is formally $P6_3$ [47], *i.e.* the crystal should be polar; however, due to statistical disorder of the orientations of the CHI_3 molecules with respect to the c axis, the crystal behaves as a centrosymmetric one [48]. In crystals of complexes, charge-transfer interactions, albeit weak, stabilise the ordered (polar) phase. For example, the 1:3 complex of CHI_3 with quinoline crystallizes in the space group R3; the complexes $CHI_3:3S_8$ and $SbI_3:3S_8$ are isomorphous crystallizing in the space group R3m, *etc.* The polar arrangement of molecules can also be stabilised by the hydrogen bonding as is often the case in molecular-ionic polar crystals.

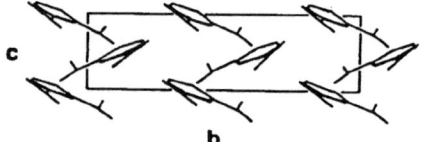

c

b

Fig. 7. Projection of the crystal structure of *m*DNB on the (bc) plane [42].

One of the few molecular ferroelectric materials identified to date is thiourea, $SC(NH_2)_2$, which is ferroelectric below 169 K and paraelectric above 202 K, and has 3 intermediate phases between these temperatures [49,50]. Crystal structures of the hydrogenated and deuterated thiourea were determined several times, the first being as early as 1928 [51]. The unit cell of the room temperature phase (phase V) is orthorhombic (space group Pnma) whereas in the low-temperature phase (phase I) the crystal remains orthorhombic but the space group is now $P2_1ma$ [52]. Fig. 8 shows the projections of both phases on the (ac) plane. The crystal may be viewed as consisting of two sublattices. Below 169 K, these are inequivalent and an uncompensated electric moment appears along the a direction of the crystal. Above 202 K, the dipole moments of all molecules cancel each other resulting in a non-polar lattice. The ferroelectric properties of molecular crystals are reviewed in section 5 of this chapter.

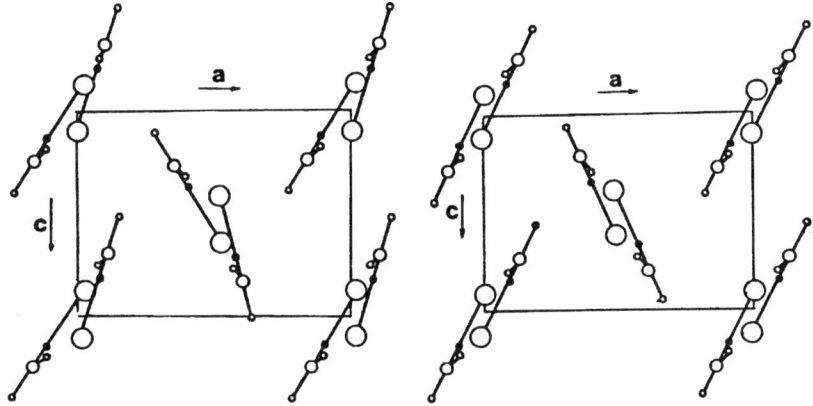

Fig. 8. Projection of the unit cell of the low-temperature phase (left) and room-temperature phase (right) of thiourea on the (ac) plane (after [52]).

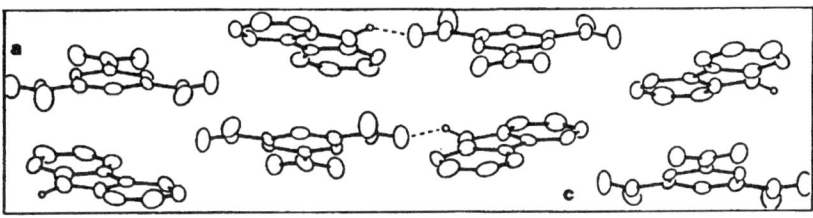

Fig. 9. Projection of the structure of the piezoelectric complex carbazole-trinitrobenzene (T = 323 K) [54].

The polar structures shown above also allow the crystals to be piezoelectric, but the piezoelectricity may appear also in non-polar crystals (see Table 2). These crystals may be built of molecules possessing the dipole moments or even of centrosymmetric ones. An example of the former class of crystals is the charge-transfer complex of carbazole with 1,3,5-trinitrobenzene, crystallizing in the orthorhombic system (space group $P2_12_12_1$) [53,54]. A projection of this structure is shown in Fig. 9. The complex units have dipole moments (are polar) but in the crystals they are arranged in centrosymmetric pairs. However, when an external stress is applied, the crystal symmetry allows for a deformation of these pairs, thus the crystal can exhibit piezoelectricity.

In a similar way, one may explain the appearance of a piezoelectric signal upon deformation of crystals built of centrosymmetric molecules. This is, for example, the case of hexamethylenetetramine (HMTA, $(CH_2)_6N_4$). The spherical molecules of HMTA crystallise in the regular system, space group $I\bar{4}3m$ [55] and, obviously, neither individual molecules nor the crystal lattice exhibit a non-zero polarization in a relaxed crystal. On applying an external stress, one distorts the molecules in the crystal, and since the distortion remains uncompensated, the external action may result in a piezoelectric charge appearing on the crystal surfaces.

Let us now review the structure of poly(vinylidene fluoride), $-(CH_2CF_2)_n-$, referred to as PVDF, which is at present one of the most extensively studied molecular materials. PVDF is a semicrystalline polymer consisting of highly polar molecular segments ($\mu \approx 2.1$ D). When solidified from the melt, it forms spherulitic structures consisting of about 50 % of lamellar crystals, typically 1 μm thick and 10 μm long, embedded in an amorphous phase. A molecule, typically 10^3 to 10^4 segments long, usually extends over several crystalline and amorphous regions as is schematically shown in Fig. 10 [24].

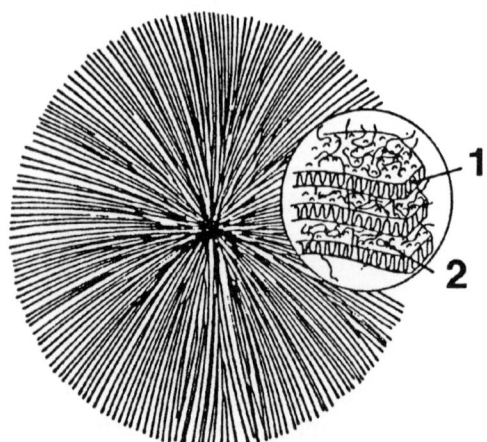

Fig. 10. Scheme illustrating the structure of a semicrystalline polymer and detail showing the lamellar structure of radiating branches (1, crystalline lamella; 2, non-crystalline components [24]).

The crystal structure of the PVDF has been studied by several groups [56-58]. Four phases have been identified, differing in conformations assumed by the PVDF macromolecules and their arrangements in the unit cells. In phase I (sometimes also labelled ß), the chain assumes an all-trans configuration (a nearly planar zig-zag with slight deviations from planarity to allow for a better packing). Phase II (α) has a *trans-gauche$^+$-trans-gauche$^-$* conformation whereas phase III (γ) has a *trans^3-gauche$^+$-trans^3-gauche$^-$* conformation. Phase IV (II_p, δ or α_p) has been identified as being similar to phase II but characterised by a polar arrangement of chains. The structures of the most widely studied phases I and II are shown in Fig.11. The unit cell of phase I is orthorhombic (space group Cm2m). There are two polymer segments per unit cell, and the uncompensated electrical moment is directed along the b axis. Phase II is monoclinic and centrosymmetric (space group $P2_1/c$). These phases can be obtained from the melt or from various solvents. They can also be interconverted as will be discussed later in this chapter.

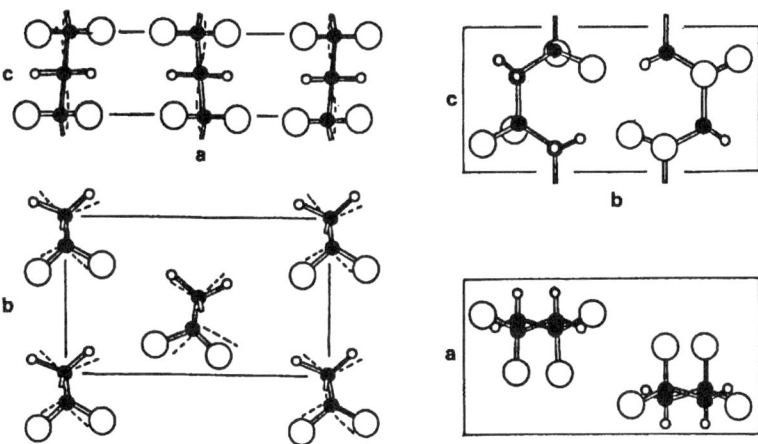

Fig. 11. Projections of the crystal structures of two phases of PVDF: polar phase I (left) and non-polar phase II (right) (after [58]).

Another group of polymers which may be a possible candidate for high piezoelectric and pyroelectric activity are polyamides (nylons). The formula of a simple polyamide may be written as $(-NH-(CH_2)_x-C(O)-)_n$, the dipole moment of the amide group amounting to *ca.* 3.7 D. The polyamide chains usually assume *all-trans* configurations, the adjacent polymer chains being linked with N—H···O—C hydrogen bonds, forming planar sheets. In polymers with x odd (*e.g.*, nylon 7, nylon 11 *etc.*) such an arrangement results in a non-zero spontaneous polarization perpendicular to the chain direction, as is schematically shown in Fig. 12.

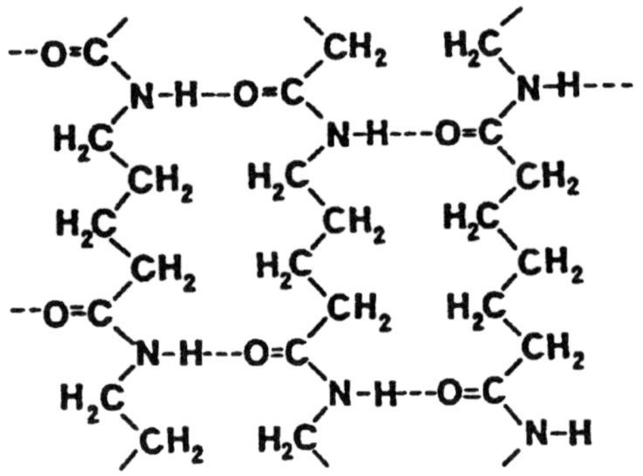

Fig. 12. Scheme of a planar hydrogen-bonded sheet of nylon 5. Dipole moments are parallel to the NH····CO bonds.

The best known representative of this group is nylon 11. Its room-temperature structure is triclinic; at 268 K it undergoes a transition to a pseudohexagonal phase with a different arrangement of hydrogen bonds [59].

Finally, we shall briefly review the structures of poly(diacetylenes), a unique class of polymers in which macroscopic-size single crystals are relatively easy to obtain. Monomeric diacetylenes have the general formula $R_1-C\equiv C-C\equiv C-R_2$, R_1 and R_2 being usually bulky and flexible side groups. Many of them can be easily grown from common solvents and then polymerized thermally, by exposure to UV light, γ-rays, electron beam *etc.*, yielding in many cases large single crystals of polymer of a reasonable quality (see, *e.g.*, [60,61] for reviews of structures and properties of poly(diacetylenes)).

Diacetylenes and poly(diacetylenes) may form polar crystals if two conditions are fulfilled (*cf.* section 1):

(i) the side groups R_1 and R_2 are different or, at least, have different spatial arrangements of constituent atoms so that molecules are non–centrosymmetric;

(ii) the arrangement of the molecules (in monomer crystals) or of polymer segments in the unit cell is non-centrosymmetric.

Syntheses of asymmetric polymerizable diacetylenes have been reported [63-65] and, to date, the structures of two have been published [64,65]. The compound described in [65] crystallizes in the centrosymmetric space group P$\overline{1}$, thus one cannot expect it to exhibit piezoelectric or pyroelectric properties. The other diacetylene, whose structure was described in [63], seems to be more promising: the compound, 6-(p-toluenesulfonyloxy)-2,4-hexadiynyl-p-fluorobenzenesulfonate (R_1 = $-CH_2-OSO_2-\Phi CH_3$, R_2 = $-CH_2-OSO_2-\Phi F$) was reported to crystallise in the space group Pc. It should also be mentioned that there is indirect evidence for a polar arrangement of molecules in the crystal lattice of the monomer of another diacetylene, R_1 = R_2 = $-CH_2-O-\Phi(NO_3)_2$, abbreviated DNP [62].

4. PIEZOELECTRICITY OF MOLECULAR CRYSTALS

As was discussed earlier, the piezoelectric properties of molecular crystals have been studied by few authors. Mason [5] reported on the piezoelectricity in some organic materials, for example, urea, m-dichlorobenzene, m-dihydroxybenzene, and squaric acid $\{C_4O_2(OH)_2\}$. Piezoelectricity has also been reported in other molecular crystals, e.g., sym-trioxane, benzil, thiourea, m-nitroaniline (mNA), N-isopropylcarbazole (NIPC) (cf. [50,66-70] and references therein). Temperature dependences of the piezoelectric coefficients were measured in resorcinol [71], hexamethylenetetramine (HMTA) [72,73], acetone-camphoric acid [74] and carbazole-trinitrobenzene [75] complexes, m-dinitrobenzene (mDNB) [66,76] and molecular ferroelectric tanane [77]. The number of crystals studied has been surprisingly small taking into account the fact that the piezoelectricity should be observed in all optically nonlinear crystals, at present intensively studied [31-33].

The limited interest in a more systematic study of molecular crystals has been probably due to difficulties usually encountered in growing molecular crystals of a size suitable for measurements. Moreover, one may guess that low piezoelectric and electromechanical coupling coefficients, expected of these materials [5], also contributed to this phenomenon. However, the latter is obviously not correct: it may be seen from Table 4 that both the piezoelectric and electromechanical coupling coefficients of molecular materials are comparable with those of many "conventional" ones.

Qualitatively, the magnitude of the piezoelectric coefficient depends on the ability of molecules in a crystal to undergo minor deformations and/or displacements upon the action of external stresses, and this in turn depends on *inter* and *intra*molecular interactions within the crystal. A simple estimation shows that an uncompensated dipole moment of the order of 10^{-2} to 10^{-1} D appearing upon deformation results in a piezoelectric coefficient of the order of that found in most piezoelectric materials.

Table 4. PIEZOELECTRIC AND ELECTROMECHANICAL COUPLING COEFFICIENTS OF MOLECULAR AND INORGANIC MATERIALS

Material	i, λ	$\|d_{i\lambda}\|$ $(10^{-12}$ C/N$)$	$k_{i\lambda}$	Reference[+]
HMTA	1,4	8.7	0.14	[73]
m-DNB	3,2	11.9	0.16	[66,76]
m-NA	3,1	3.3		[69]
NIPC	3,2	17.3	0.14	[66]
Tanane	3,6	50	0.14	[77]
PVDF	3,3	27.1	0.13	[78]
Quartz	1,1	2.3	0.10	[79]
LiNbO$_3$	1,5	69.2	0.44	[80]
PZT (PbZr$_x$Ti$_{1-x}$O$_3$)	1,5	584	0.68	[81]
CdS	3,3	0.46	0.15	[82]

[+] Values of $k_{i\lambda}$ either taken directly from the paper cited or calculated from parameters given therein. All values are at room temperature.

Far from phase transitions, the piezoelectric coefficients seldom exhibit dramatic temperature dependences. Examples of such a behaviour are shown below: we shall present the piezoelectric coefficients of the crystals, whose structures are discussed in the preceding section.

The highly symmetrical crystal HMTA has been known not to undergo any phase transition from 4.2 K to its melting point. For the point group $\bar{4}3m$ the piezoelectric coefficient has only one independent component, and the d matrix can be written as

$$d = \begin{vmatrix} 0 & 0 & 0 & d_{14} & 0 & 0 \\ 0 & 0 & 0 & 0 & d_{14} & 0 \\ 0 & 0 & 0 & 0 & 0 & d_{14} \end{vmatrix}$$

The piezoelectric coefficient, shown in Fig. 13, is independent of temperature over the entire range covered by the experiment. It is worth noting that the effective elastic coefficient reported in the same paper [73] is also weakly temperature dependent.

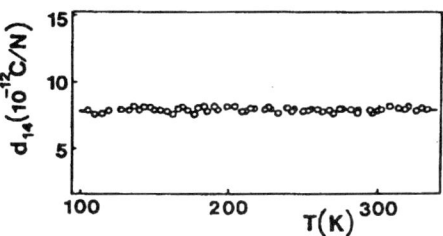

Fig. 13. Temperature dependence of the piezoelectric coefficient of HMTA [73].

The temperature dependences of the piezoelectric and elastic coefficients of mDNB were published in [76]. As mentioned in the preceding section, the crystal belongs to the mm2 point group, being pyroelectric and hence piezoelectric. The d matrix may be written as

$$d = \begin{vmatrix} 0 & 0 & 0 & 0 & d_{15} & 0 \\ 0 & 0 & 0 & d_{24} & 0 & 0 \\ d_{31} & d_{32} & d_{33} & 0 & 0 & 0 \end{vmatrix}$$

thus having 5 independent components. Three of these components (d_{31}, d_{32} and d_{33}) were determined in [76]; all were found to be weak functions of temperature

294

over the entire range covered by the experiment. Their temperature dependences are shown in the next section (Fig. 17).

One may expect the elastic and piezoelectric coefficients to exhibit anomalies around phase transitions occurring in materials. This indeed is the case of several crystals, among them the ferroelectrics (*e.g.*, TGS, Rochelle salt, $BaTiO_3$, KH_2PO_4 *etc.* [9,11]). In molecular crystals, the effect of phase transitions on the elastic properties was studied by many; however, a similar effect on the piezoelectric properties has been reported for few crystals only.

A Curie-Weiss type behaviour of both the elastic and piezoelectric coefficients was found by Legrand [77] in the ferroelectric tanane (see also section 5). A similar temperature dependence was also found in crystals of the piezoelectric complex, carbazole-trinitrobenzene [75]. The crystal belongs to the 222 point group, thus there being 3 independent components of the piezoelectric coefficient

$$d = \begin{vmatrix} 0 & 0 & 0 & d_{14} & 0 & 0 \\ 0 & 0 & 0 & 0 & d_{25} & 0 \\ 0 & 0 & 0 & 0 & 0 & d_{36} \end{vmatrix}$$

The method of measurements employed in [75] gives only relative values of $d_{i\lambda}$ but it is clearly seen from Fig. 14 that two components follow the Curie-Weiss dependence (equation 36) with $T_o = 95 \pm 2$ K. This behaviour may indicate the possibility of a transition to a polar phase at around 100 K. However, no measurements confirming the existence of such a transition have been reported to date.

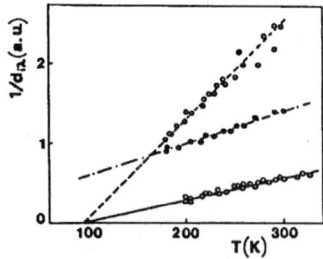

Fig. 14. Temperature dependence of the inverse piezoelectric coefficient in carbazole-trinitrobenzene CT complex [75]. Solid line, d_{14}; dot-and-dashed line, d_{25} ; dashed line, d_{36}.

5.1 PYRO AND FERROELECTRIC MOLECULAR CRYSTALS

5.1. PYROELECTRIC EFFECT

The existence of pyroelectricity has been reported in several polar molecular crystals (e.g., [3,15]). Pyroelectric coefficients of the order of 10^{-6} to 10^{-5} C/m^{2}K have been found in 5-chlorosalicylideneaniline and salicylidene-4-bromoaniline by Lang et al. [83]. Low temperature pyroelectricity of saccharose was studied by Mangin and Hadni [84]. However, only since 1983 has a systematic study of pyroelectricity of molecular crystals been undertaken. Gavrilova et al. [85], and, independently, Asaji and Weiss [86-88] and Giermanska et al. [70,76] reported the temperature dependences of the pyroelectric coefficients in several substituted benzenes: mDNB [70,76], mNA [70,86,88], m-aminophenol [70,86], resorcinol [85], m-chloronitrobenzene [87,88], m-bromonitrobenzene [88], 2,3-dichlorophenol [88], p-cyanonitrobenzene [87] and p-cyanobromobenzene [87]. Crystal structures of all these crystals are known (some of them were discussed in the section 3): except for the latter compound, the direction of the spontaneous polarization is determined by the symmetry, and any rotation of the p vector is not allowed. In most of these crystals, the pyroelectric coefficients were found to increase with temperature, reaching at room temperature values of the order of several μC/m^{2}K, which may be considered typical of polar molecular crystals far from phase transitions (see Fig. 16). Higher values, inversions of sign and/or discontinuities of the pyroelectric coefficients are usually associated with phase transitions occurring in these materials. The well known example is the temperature dependence of the spontaneous polarization in thiourea [50], shown in Fig. 15. The pyroelectric coefficient (estimated as a temperature derivative of P_s - cf. equation 12) varies from ca. -10^{-4} C/m^{2}K at 100 K to ca. -10^{-2} C/m^{2}K at the ferroelectric phase transition temperature (see the next subsection). Rapid change of the pyroelectric coefficient, associated with the phase transition, was also observed in N-isopropylcarbazole at 140 K [66,89]; a similar effect, although less pronounced, was reported in p-cyanobromobenzene [87]. However, sign inversion of the pyroelectric coefficients has also been observed in *m*DNB, *m*NA and 2,3-dichlorophenol [69,86,88] in which no phase transitions were reported. The temperature dependence of the pyroelectric coefficient in *m*NA is shown in Fig. 16. Inversion of the sign is discussed later in this section.

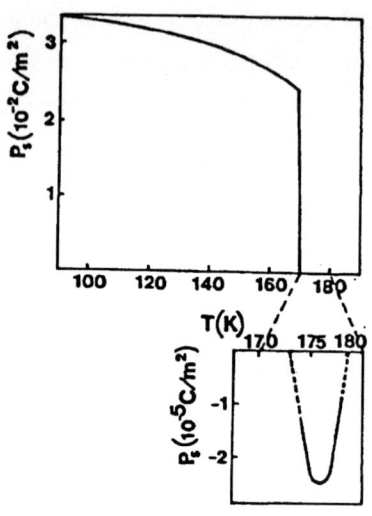

Fig. 15. Temperature dependence of the spontaneous polarization in phases I and III of thiourea. Note different scales of P_s (after [50]).

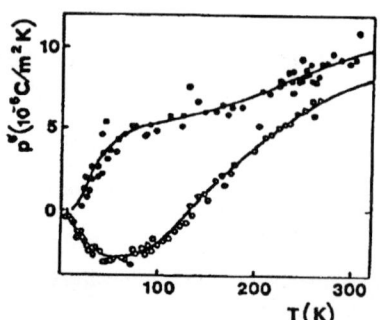

Fig. 16. Temperature dependences of the pyroelectric coefficients of *m*-bromonitrobenzene (full circles) and *m*-nitroaniline (open circles). The full lines are fits to the experimental results with two Einstein functions. See text for discussion (after [88]).

The temperature dependence of the pyroelectric effect may be controlled by either the primary or the secondary effect (*cf.* equation 16). From the point of view of the lattice dynamics, the former effect is associated with the temperature dependence of the average population of anharmonic *polar* phonons of a given symmetry, whereas the latter one with *all* anharmonic phonons, contributing to the thermal expansion and elastic coefficients. Microscopic approaches to the temperature dependence of the pyroelectricity are usually limited to the *primary* coefficient, whereas the direct measurements yield in most cases the total coefficient.

The earliest attempts to explain pyroelectricity are due to Boguslawski [90], who employed Einstein's theory of the specific heat of solids. Within his model, the polar crystal is described as a set of independent anharmonic oscillators characterized by their Einstein temperatures $\Theta_i = h\nu_i/k$, and the spontaneous polarization may be written as

$$P_s = P_o + \sum_i A'_i \bar{n}_i \tag{40}$$

where P_o is a constant (the spontaneous polarization at 0 K), A'_i are coefficients of the linear combination, n is the average occupancy of an i-th vibration

$$\bar{n}_i = \{exp(\Theta_i/T) - 1\}^{-1} \tag{41}$$

and the summation runs over all polar lattice vibrations. The primary pyroelectric coefficient may thus be explicitly calculated as

$$p^\varepsilon = \sum_i A_i (\Theta_i/T)^2 \frac{exp(\Theta_i/T)}{\{exp(\Theta_i/T) - 1\}^2} \tag{42}$$

This simple model was later extended by Born [91] and Szigeti [92] who considered the influence of non-linearity of the electrical response of samples and of the anharmonicity of lattice vibrations. According to Born [91], at low temperatures the primary coefficient should be proportional to temperature (see also [93]), whereas Szigeti [92] predicted that the primary pyroelectric coefficient at low temperatures should follow the specific heat, being proportional to T^3.

All microscopic models, when applied to molecular materials, must be modified and take into account features of the "molecular architecture". In particular, changes of relevant *molecular* parameters should enter a modified model, *i.e.*, appropriate polar internal vibrations should be taken into account.

The procedure of fitting experimentally determined temperature dependences of the pyroelectric coefficients was employed to several polar molecular crystals. Asaji *et al.* [88] fitted the experimental curves determined for *m*NA, *m*−chloronitrobenzene and *m*-bromonitrobenzene with equation (42), using a sum of two Einstein functions (*i.e.*, taking four adjustable parameters). The fits, shown in Fig. 16, although reasonably accurate, should be accepted with some caution for two reasons: (*i*) the Einstein temperatures (and, consequently, the vibration frequencies) obtained from the fit are some *effective* values without any direct correspondence to the data available from spectroscopic measurements, (*ii*) the equation describing the *primary* coefficient was used to fit the experimentally determined total coefficient. The secondary coefficient should therefore be either negligibly small or at least proportional to the primary one over the temperature range covered by the experiments, which may (but does not necessarily have to) be true (*cf.* [93,94]).

Giermanska *et al.* [76] employed a similar procedure to explain the inversion of sign of the total pyroelectric coefficient in *m*DNB. The fit to equation (42) was made after the separation of the primary and secondary effect, and the Einstein temperatures, instead of being parameters of the fit, were taken from available spectroscopic results. The secondary coefficient was calculated from independently determined temperature dependences of the elastic, piezoelectric and thermal expansion coefficients, as is shown in Fig. 17. Both primary and secondary effects exhibit inversions of signs, the secondary effect contributing to *ca.* 30 % of the total effect. Unfortunately, the experiments reported in [76] do not cover the low-temperature region.

An inversion of sign of the pyroelectric coefficient was attributed in all cases to competition between vibrations (*i.e.*, the components of the sum in equation 42). According to [76], the inversion occurs as a result of competition between the lattice modes (modifying the average orientations of molecular dipoles with respect to the polar axis) and internal vibrations (changing the value of the molecular dipole moment).

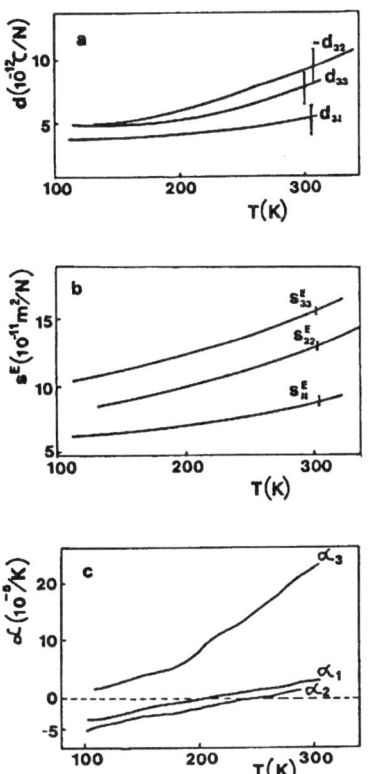

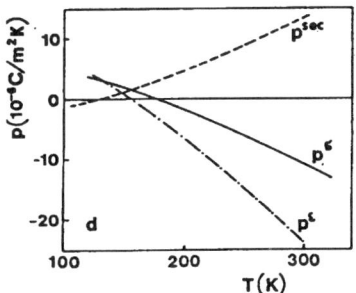

Fig. 17. Temperature dependences of the piezoelectric coefficient (a), elastic compliance (b) and thermal expansion coefficient (c), and the separation of the experimentally measured total pyroelectric coefficient (p^σ) into the primary (p^ε) and secondary (p^{sec}) coefficients in mDNB (d) [76].

A different approach was put forward by Fleck and Weiss [95] extending the model of Mopsik and Broadhurst [96] for piezoelectricity and pyroelectricity in polymers (see section 6). The authors considered a solid built of molecules possessing a non-zero dipole moment and non-zero polarizability. The spontaneous polarization may be written as

$$P_s = P_o + \frac{N_A}{3v}\left(\frac{\epsilon_p - \epsilon_o}{\epsilon_o}\right)\mu\cos\xi + (\epsilon_p - \epsilon_o)E_{N,p} \tag{43}$$

where N_A is Avogadro's number; $\bar{v}$ is the molar volume; ϵ_p and $E_{N,p}$ stand for the electric permittivity and electric field on a given lattice site due to the presence of neighbouring molecules, measured in the direction of the polar axis; ξ is the angle between the directions of the vectors of dipole moment and polarization. The polarization is a function of temperature, orientation of molecules with respect to the polar axis and intermolecular distances (entering *via* the volume); the total pyroelectric coefficient p^σ may be thus calculated as a derivative

$$\left(\frac{\partial P}{\partial T}\right) = \left(\frac{\partial P}{\partial T}\right)_{\delta,B,\epsilon} + \left(\frac{\partial P}{\partial T}\right)_{v,B,\epsilon} + \left(\frac{\partial P}{\partial T}\right)_{v,\delta,\epsilon} + \left(\frac{\partial P}{\partial T}\right)_{v,\delta,B} \tag{44}$$

where B is the angle between the vectors of E_N and P. These equations have been successfully employed to fit the results obtained for *m*-chloronitrobenzene, *m*-bromonitrobenzene and *p*-cyanobromobenzene [87].

5.2. FERROELECTRICITY IN MOLECULAR CRYSTALS

It was mentioned in section 1 that ferroelectricity has been found in few molecular materials. Among the low-molecular weight crystals, only in two, thiourea and tanane, has the phenomenon been documented by independent measurements.

Ferroelectricity in thiourea was discovered by Solomon [49]. Systematic studies undertaken by Goldsmith and White [50] showed that the transition from the low-temperature ferroelectric phase I to the high-temperature paraelectric phase V (*cf.* Section 3) occurs between 169 and 202 K, *via* a sequence of phases:

    phase I $\xrightarrow{\text{169 K}}$ phase II $\xrightarrow{\text{173 K}}$ phase III $\xrightarrow{\text{176 K}}$
    ferroelectric   *non-polar*   *ferroelectric*

    phase IV $\xrightarrow{\text{202 K}}$ phase V
    non-polar   *non-polar*

The sequence is best observed on the temperature dependence of the electric permittivity (Fig. 18). The temperature dependences of the spontaneous polarization and pyroelectric coefficient, given in [50] (see Fig. 15), show that phases I and III are ferroelectric with the spontaneous polarization directed along the a axis, the transition from phase I to phase II being of the first order. Moreover, it has been established (*e.g.*, [97]) that phases II, III and IV exhibit a structural modulation along the b direction. The spontaneous polarization and its temperature dependence may be understood as arising essentially from reorientations of the thiourea molecules in the crystal lattice without any major distortion of their conformations [98].

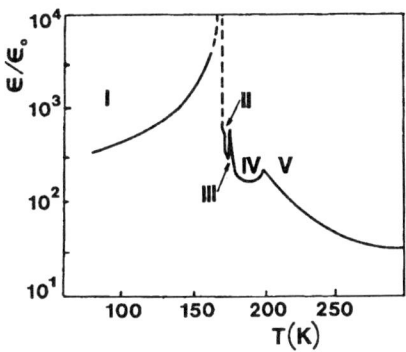

Fig. 18. Temperature dependence of the static electric permittivity measured in the thiourea crystal. The regions of stability of phases I to V are indicated with Roman numbers [50].

A similar sequence of the phase transitions was found in deuterated thiourea $SC(ND_2)_2$ [50]: the deuteration results primarily in shifting the transition temperatures by *ca.* 15 K (the transitions taking at 185, 192, 196 and 214 K respectively). Shifts of the transition temperatures have been observed in hydrogen-bonded ferroelectrics, the best known example being KH_2PO_4 (*cf.*, *e.g.*, [9] and references therein), thus the shift observed in thiourea seems to confirm the importance of the hydrogen bond in the stabilization of a ferroelectric arrangement of molecules in its low-temperature phase [52]. A simple qualitative model explaining the appearance of the spontaneous polarization and its temperature dependence is outlined in [52]. The spontaneous polarization may be estimated as a vectorial sum of the molecular dipole moments (taking the moments measured in a solution, one obtains $P_s = 2.3 \times 10^{-2}$ C/m^2 in a reasonable agreement with the experimental value - *cf.* Fig. 15). The temperature dependence may be fully accounted for if one assumes that, on lowering temperature, the molecules of thiourea are rotated at 169 K by *ca.* $7°$ from positions characteristic of the high-temperature phase (phase V), and then the angle changes continuously reaching *ca.* $10°$ as the temperature approaches 0 K.

The kinetics of the ferroelectric switching were also studied by Goldsmith and White [50]. The switching time was found to range from 10^{-2} s at 5×10^3 V/m to 10^{-6} s at 4×10^5 V/m, following a $\tau_{s,o}^{(const/E)}$ dependence below 2×10^4 V/m and an $E^{-3/2}$ dependence at higher fields.

Recent research on hydrogenated and deuterated thiourea aims mainly at establishing the nature of the incommensurate phases II-IV and the transitions between them (*e.g.*, [99-104]).

The ferroelectric behaviour of tanane (2,2,6,6-tetramethylpiperidine-1-oxyl, $C_9H_{18}NO$), a free radical, was reported in 1973 [105]. Tanane exhibits an order-disorder (second-order) phase transition at 286.7 K, *ca.* 25 K below its melting point. Perdeuterated tanane undergoes an analogous transition at almost the same temperature (287.5 K), thus indicating that hydrogen bonding is not involved in the stabilization of the crystal structures. The high temperature phase is piezoelectric (tetragonal point group $\bar{4}2m$) whereas the low-temperature phase is orthorhombic (point group mm2) [106]. A dielectric anomaly has been observed at the transition temperature [107]. The behaviour of tanane may be phenomenologically described assuming the polarization P being linearly coupled to the shear strain ε_6 and to external electric fields (piezoelectric coupling) [108,77]. The model predicts a Curie-Weiss behaviour of the inverse electric susceptibility, and the piezoelectric and elastic coefficients reasonably fit the experiments (*cf.* section 4). In general, however, the quantity of experimental data describing the properties of tanane is rather limited, most probably due to difficulties associated with the relatively low thermal, mechanical and chemical stability of the material.

6. PIEZO, PYRO AND FERROELECTRICITY IN POLYMERS

It has been recognized for many years that many biopolymers exhibit piezoelectric activity (*e.g.,* [19,109] and references therein). Piezoelectric and pyroelectric activity has also been reported for several synthetic polymers (see, *e.g.*, [22-26] for reviews) raising questions as to the nature of the observed effects. Hayakawa and Wada [110,22] distinguish three mechanisms of piezoelectricity and pyroelectricity in polymers:

(i) intrinsic piezoelectricity due to internal (microscopic) strain;

(ii) intrinsic piezo and pyroelectricity due to strain and temperature dependence of the spontaneous polarization;

(iii) piezo and pyroelectricity due to the presence of charges trapped in a spatially inhomogeneous dielectric.

Extending the above classification, one should also mention a fourth mechanism of field-induced orientation of dipoles at elevated temperatures followed by a subsequent "freezing" of the built-in polarization. The latter two mechanisms are not related directly to the microscopic properties of unit cells of polymer crystallites and will not be discussed here. Thus only crystalline (or partially crystalline) polymers are considered. However, it should be realized that synthetic "crystalline" polymers, when solidified from the melt or cast from solutions, usually consist of randomly oriented crystallites (*cf.* Fig. 10), and must be oriented by, for example, mechanical deformation (stretching) and/or application of high electric fields (poling).

In principle, the arrangement of polymer segments in unit cells of intrinsically pyroelectric and/or piezoelectric polymers, groups *(i)* and *(ii)* in the above classification, should meet the symmetry requirements discussed in earlier sections of this chapter, *i.e.*, the unit cell should allow for a non-zero spontaneous polarization or, at least, should lack an inversion centre. Naturally, polymers with large dipole moments (vinyl and vinylidene polymers, polyamides *etc.*) are the obvious candidates.

6.1. POLY(VINYLIDENE FLUORIDE) AND ITS COPOLYMERS

The piezoelectric activity of stretched and poled PVDF was first reported by Kawai in 1969 [20], and since then extensive studies have been carried out by several groups, the number of papers published so far on the properties of this polymer well exceeding a thousand. Initially, there was controversy about the origin of the piezoelectric and pyroelectric activity of PVDF but at present the dominant role of the polar structure of PVDF crystallites has been recognized beyond any doubt (see, *e.g.*, [23,111] and references therein). The structural origin of the piezo and pyroelectricity was additionally confirmed by the discovery of ferroelectricity and second harmonic generation [21].

As was already mentioned in section 3, four crystalline forms have been identified. The non-polar form II is obtained directly from the melt or by casting from PVDF solutions in common solvents (*e.g.* acetone) [112]. When poled with fields in excess of 10^8 V/m, form II transforms into its polar modification (form IV) [113]. The transformation was postulated to occur by the propagation of a kink along the polymer chain [114]. Mechanical orientation of form II at temperatures slightly above room temperature (*ca.* 320-330 K) yields the polar phase I, the degree of transformation dependent upon the stretching conditions. Form I may also be obtained under high pressures at elevated temperatures (*ca.* 5×10^8 Pa and 565 K) [115]. Unoriented phase I has also been obtained directly from the melt by the so called pressure quenching technique [116], or by solvent casting at room temperature from, for example, dimethylformamide (DMF), dimethylacetamide (DMA) or hexamehylphosphoramide [117]. Phase III may be obtained from the melt by slow cooling, by the annealing of phases I or II or by casting from DMA at elevated temperatures [115,117].

Conditions and mechanisms of interconversions of various forms of PVDF are still subject to extensive studies. However, most important for practical purposes are the properties of polar phase I and non-polar phase II. Usually, electroactive samples containing phase I are obtained by the procedure shown schematically in Fig. 19: samples obtained by solidification from the melt and containing mostly phase II, are stretched several (typically 3 to 5) times their original length, and then poled with ac fields of the order of 10^8 to 10^9 V/m or by corona charging. As a result, one obtains samples containing crystallites of the phase I with chains extending preferentially along the direction of drawing and with the dipoles oriented preferentially in the direction of poling. It should be noticed that the commonly employed convention of indexing the directions in oriented PVDF samples shown in Fig. 19 is not fully consistent with the structure shown in Fig. 11. The spontaneous polarization of phase I is directed along the "b" axis of the unit cell but in the "3" direction in macroscopic samples.

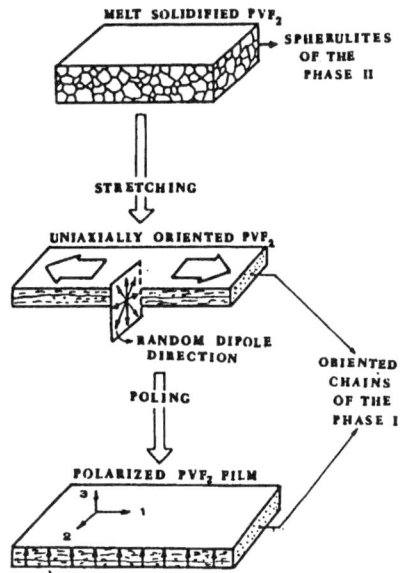

Fig. 19. Schematic representation of the processes commonly employed to obtain piezoelectrically and pyroelectrically active PVDF films. The numbering of arrows at the bottommost part of the figure corresponds to the convention of indexing directions in oriented PVDF films [103] (copyright 1973 by AAAS; reprinted with permission).

As mentioned in Section 3, the point group of phase I of PVDF is mm2, thus in a hypothetical single-crystalline sample the pyroelectric coefficient has only one component

$$p = (0 \quad 0 \quad p_3)$$

and the piezoelectric coefficient has 5 independent components

$$d = \begin{vmatrix} 0 & 0 & 0 & d_{24} & 0 & 0 \\ 0 & 0 & 0 & d_{24} & 0 & 0 \\ d_{31} & d_{32} & d_{33} & 0 & 0 & 0 \end{vmatrix}$$

The spontaneous polarization at room temperature can be estimated as a vectorial sum of molecular dipole moments to amount to 0.132 C/m². Obviously, this "single crystal" value is never reached because of the partial crystallinity and imperfect orientations of crystallites in macroscopic samples.

The ferroelectric behaviour of PVDF was postulated in [21] and confirmed in several experiments showing a hysteresis of the piezoelectric coefficients, electric permittivity, intensity of IR bands and electric displacement (*e.g.*, [113,118-122]). Fig. 20 shows the temperature dependence of the hysteresis loops measured on a pre-poled PVDF sample at various temperatures [121]. The coercive field was found to decrease markedly as temperature increased, whereas the remnant polarization ($D_{p=0}$) was almost temperature-independent.

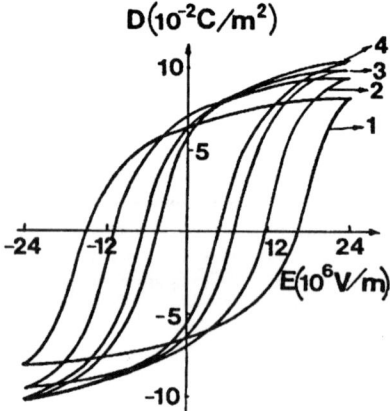

Fig. 20. Hysteresis loops in PVDF, pre-poled at 293 K under 2.4×10^8 V/m and measured at 173 K, 213 K, 253 K and 293 K (curves 1 to 4, respectively) [121].

In spite of several efforts, the temperature of the ferroelectric/paraelectric phase transition in PVDF has not been established. Herchenroder *et al.* [123] have located it at *ca.* 413 K, but the measurements of Hicks *et al.* [122] clearly demonstrate that at this temperature PVDF is still ferroelectric. At present, it is commonly accepted that the phase transition temperature in PVDF is located above its melting point. It should also be noted that neither the temperature dependence of the coercive field nor that of the remnant polarization measured in [122] follow the trends determined by Furukawa *et al.* [121]. Most probably, these discrepancies are associated with differences in sample preparation.

The kinetics of polarization reversal was reported by Furukawa and Johnson [124]. The switching times (τ_s) varied between several seconds and several microseconds, following the exponential law

$$\tau_s = \tau_{s,o} \, exp(E_a/E) \tag{45}$$

where E_a and $\tau_{s,o}$ are constants, E_a ranging between 1×10^9 and 2×10^9 V/m, dependent upon temperature.

The well established existence of ferroelectricity in PVDF raises the question of the nature of the states involved in polarization reversal. The unit cell of phase I is orthorhombic, hence a natural assumption would be a switching of the PVDF chains by 180° ("2-site model"). However, Kepler and Anderson [125] have pointed to the fact that the orthorhombic symmetry of the unit cell results from a small (*ca.* 1 %) distortion of a hexagonal primitive lattice, thus switching by a 60° jump would also be conceivable ("6-site model"). Broadhurst and Davis [24,126] employed the Bragg-Williams approach to both models to explain features of the ferroelectric behaviour of PVDF, arriving at a qualitative agreement with the experimental data.

The piezoelectric and pyroelectric activity of PVDF has been studied by several groups; the experimentally determined (macroscopic) coefficients differ substantially, apparently dependent upon the conditions of preparation and the history of the samples. In general, the piezo and pyroelectric coefficients depend on the time of poling, but after a sufficiently long time saturation values are reached, dependent upon poling field and poling temperature. It was demonstrated that the piezoelectric coefficients of unoriented samples are linear functions of the amount of phase I.

308

Both the piezoelectric coefficients and pyroelectric coefficient are weakly temperature dependent below *ca.* 220 K (*i.e.* below the glass transition temperature of the amorphous phase of PVDF; above, to at least 350 K, both increase with temperature [78,128,129]. As an example, the temperature dependences of the piezoelectric, pyroelectric and electromechanical coefficients of oriented samples containing phase I are shown in Fig. 21, and typical room temperature values are listed in Table 5 [78].

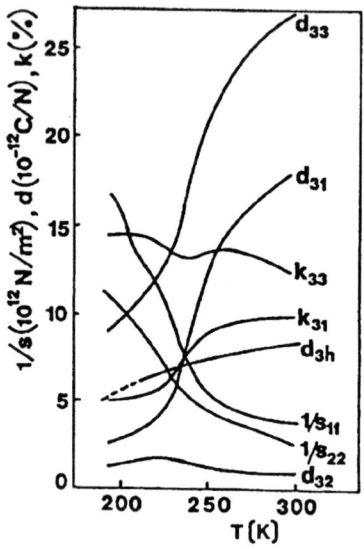

Fig. 21. Temperature dependences of the piezoelectric, elastic and electromechanic coupling coefficients in oriented phase I of PVDF. The film was drawn 510 % at 423 K and poled at 383 K with 10^8 V/m for 2 hours [78]. Index "h" denotes hydrostatic pressure.

The temperature dependences measured by Klaase and van Turnhout [78] reasonably agree with results obtained by Chung *et al.* [130] for samples containing oriented and unoriented phase I (see Fig. 22). It may be noticed from Fig. 22 that the pyroelectric and piezoelectric coefficients are mutually proportional over almost the entire temperature range. This feature has also been reported by other authors: at a given temperature, both coefficients are proportional to the remnant polarization of the samples [131-133].

Table 5. PIEZOELECTRIC, ELECTROMECHANICAL COUPLING AND
PYROELECTRIC COEFFICIENTS OF THREE FORMS OF PVDF
AT ROOM TEMPERATURE [78]

Phase	Conditions of drawing K - %	poling§ K - 10⁷ V/m	d_{31}	d_{32}	d_{33}	k_{31}	k_{33}	p_3 10⁻⁶
				10⁻¹² C/N			%	C/m²K
I^u		373 - 5	4.2	4.2	-11.1	1.8	3.8	
I^d	373-400	393 - 10	17.9	0.9	-27.1	10.3	12.6	30
	348-400	438 - 10	37			14.7		
II^u		383 - 10	5.3	5.3	-13.6	3	4.4	13
	440-350	383 - 10	9.8	2.2	-14.8	4.5		14
II^d	440-350	438 - 10	16.8			7.8		
III^u		423 - 9	4.9	4.9	-13.3	5.3		16
III^d	373-400	483 - 5	19			9.6		

§ samples poled for 2 hours; ᵘ undrawn; ᵈ drawn.

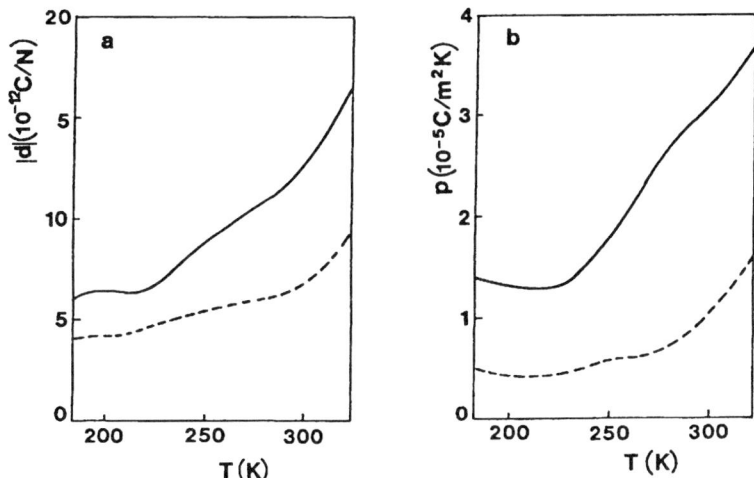

Fig. 22. Comparison of the temperature dependences of the hydrostatic piezoelectric coefficient and pyroelectric coefficient for unoriented (dashed lines) and uniaxially oriented (full lines) phase I PVDF [130].

The origin of pyroelectric activity was examined by Glass and Negran [134]. Their results demonstrate that the pyroelectricity is associated primarily with the thermal expansion of the samples, *i.e.* it is mostly a secondary effect. This conclusion agrees with that of Kepler and Anderson [135] who found the primary coefficient to be of negligible importance in PVDF; however, the secondary effect can account only for about 50-60 % of the total observed effect. To explain the discrepancy, reversible field-induced crystallization of the amorphous phase was postulated, and then confirmed by X-ray experiments.

Models explaining the piezoelectric and pyroelectric activity of PVDF, taking into account its dipolar nature, were proposed by Hayakawa and Wada [110,22], and by Broadhurst *et al.* [96]. The phenomenological model put forward by Hayakawa and Wada starts from a free energy polynomial of the form

$$F = F_o + A\varepsilon^2 + f(P) - Q\varepsilon P^2 \tag{46}$$

where A is an expansion coefficient, Q is the electrostriction coefficient (*cf.* equation 4), and $f(P)$ is a polynomial expanding the free energy with respect to polarization

$$f(P) = P^2 \sum_0^\infty a_j P^j \tag{47}$$

Resulting equations predict a linear dependence between the piezoelectric coefficient and the remnant polarization, as was indeed observed in experiments.

Broadhurst and Davis [96] considered a system of lamellar crystals of various sizes and orientations, dispersed in amorphous matrix, the spontaneous polarization of the crystallites being compensated by a space charge of carriers localized on the crystal surfaces. The polarization due to the presence of a set of dipoles having the dipole moment μ is given by the equation

$$P_o = \upsilon \frac{(\epsilon_c + 2\epsilon_o)N}{3\epsilon_o v_c} \mu J_o(\phi_o) <\cos\theta_o> \tag{48}$$

where N is a number of dipoles in the volume v, υ is the mole fraction of the crystalline phase, ϕ_o is the average amplitude of librations of molecular dipoles, Θ_o is a mean orientation angle between the direction of the spontaneous polarization of crystallites and the direction "3" of the sample, J_o is the Bessel function of the zeroth order and first kind and the index "c" labels the values characteristic of crystalline phase. Broadhurst et al. [96], taking into account thermal fluctuations of dipoles, inhomogeneity of macroscopic samples and temperature variation of parameters, arrived at the equations for the hydrostatic piezoelectric coefficient and pyroelectric coefficient

$$d_p = P_o \beta_c \left(\frac{\epsilon_c - \epsilon_o}{3\epsilon_o} + \frac{\phi_o^2 \gamma}{2} + \frac{\partial \ln l_x}{\partial \ln v_c} \right) \tag{49}$$

$$p = -P_o \alpha_c \left(\frac{\epsilon_c - \epsilon_o}{3\epsilon_o} + \frac{\phi_o^2 (\gamma + \frac{1}{2} T^{-1} \alpha_c^{-1})}{2} + \frac{\partial \ln l_x}{\partial \ln v_c} \right) \tag{50}$$

where α and β are the volume thermal expansion coefficient and volume compressibility, γ is the Gruneisen coefficient and l represents a linear parameter denoting either the sample thickness ($l_x = l_s$) or a separation of counter charges on the crystal surfaces ($l_x = l_c$), dependent upon the physical situation. The model provides a good explanation of the basic features of the piezoelectric and pyroelectric activity of PVDF, yielding numerical values in agreement with the experimental results. In particular, the effect due to dipole fluctuations (which may be considered as equivalent to the primary effect) was estimated to contribute to ca. 20 % of the total value of the pyroelectric effect.

The highly polar phase I of PVDF may be obtained at normal pressure only after a sequence of stretching and poling has been applied to the as-solidified samples. Apparently, the conformation trans-gauche$^+$-trans-gauche$^-$ or (at elevated temperatures) trans^3-gauche$^+$-trans^3-gauche$^-$ is more favourable than the all-trans one. Lando and Doll [136] pointed out the possibility of obtaining an all-trans conformation of the chain (and, in consequence, phase I) in copolymers of vinylidene fluoride with trifluoroethylene (TrFE) or tetrafluoroethylene (TFE). The all-trans conformation is stabilized by the presence of comparatively

312

bulky fluorine atoms.

There have been some reports confirming the ferroelectric behaviour of the (PVDF-TFE) copolymer (*e.g.*, [137]); however, most studies have been carried out on the copolymer with TrFE (PVDF-TrFE), with the mole fraction of VDF ranging in most cases between 0.5 and 0.8. PVDF-TrFE copolymers exhibit a well-defined ferroelectric-paraelectric phase transition, the presence of which has not been convincingly demonstrated in PVDF. Existence of the phase transition was confirmed by calorimetric, dielectric and structural measurements (*e.g.*, [138-142]). The transition is of the first order [143], its temperature being dependent upon the composition of the polymer.

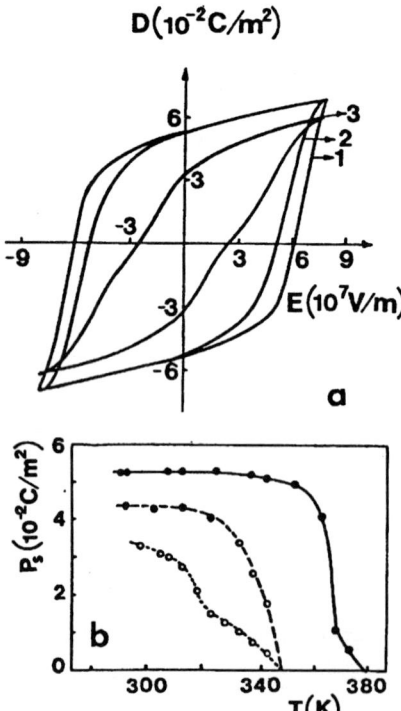

Fig. 23. (a) Ferroelectric hysteresis loops in the 55/45 PVDF-TrFE copolymer measured at 293, 313 and 333 K (curves 1 to 3, respectively). (b) Temperature dependences of the spontaneous polarization measured in the 51/49, 55/45 and 66/34 PVDF-TrFE copolymers (dotted, dashed and full lines, respectively). Full circles: values obtained from single hysteresis loops, open circles: values obtained from loops showing features of double loops (*e.g.*, loop 3 in (a)) [140].

Yamada and Kitayama [140] carried out a systematic study of the dielectric and ferroelectric properties of PVDF-TrFE as a function of the polymer composition. The ferroelectric hysteresis loops measured at several temperatures on the 55/45 copolymer (*i.e.*, copolymer containing 55 mole % of VDF), shown in Fig. 23a, resemble those measured on pure PVDF (Fig. 20). The remnant polarization and coercive field were also found to depend on the polymer composition: at room temperature, both the remnant polarization and the saturation polarization of the 13/87 copolymer were about an order of magnitude smaller than those of the 82/18 copolymer. Temperature dependences of the saturation polarization in three copolymers are shown in Fig. 23b. As is seen from the figure, the transition temperature depends on the composition. The temperature, obtained from the measurements of electric permittivity, amounted to *ca.* 333 K for 37 to 49 % of VDF, increasing gradually to 398 K for the 82/18 copolymer, in good agreement with values determined calorimetrically.

The electric permittivity shown in Fig. 24 clearly demonstrates the first-order character of the transition. The permittivity of the paraelectric phase was found to fulfil the Curie-Weiss law (equation 31), with the Curie constant amounting to 2300 K in the 55/45 copolymer.

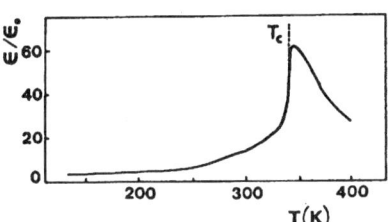

Fig. 24. Temperature dependence of the electric permittivity in the 55/45 PVDF-TrFE copolymer [140].

Furukawa [143] examined the phase transition and switching in the PVDF-TrFE copolymer within the Landau-Devonshire formalism. Having written the expression for the free energy of the samples in the form identical to that given by equations (27) and (29), he calculated the expansion coefficients: $\alpha = 3.5 \times 10^7$ Vm/KC, $B = -6.0 \times 10^{12}$ Vm^5/C^3, $C = 1.1 \times 10^{13}$ Vm^{10}/C^5. With these coefficients, one can reproduce experimentally measured temperature

314

dependences of the remnant polarization and the electric permittivity (cf. Table 3). However, the coercive field predicted by the equations resulting from the Landau-Devonshire theory was found to be much higher than the experimentally determined one. This discrepancy was explained assuming that the polarization reversal may occur below the "thermodynamic" coercive field via a nucleation of reversed domains and their subsequent growth.

The ferroelectric-paraelectric phase transition in PVDF-TrFE copolymers is associated with changes of the polar *all-trans* conformation to a disordered conformation consisting of a random sequence of the *trans-gauche*, *trans-trans* and *trans³-gauche* segments [141,144]. The high temperature phase was also found not to exhibit any piezo or pyroelectricity [140,145] (cf. Fig. 25). It is interesting to note that, according to Furukawa et al. [146], the primary pyroelectricity in copolymers determines the magnitude of the total effect: the secondary effect is reported to contribute to 25 % of the total effect at most.

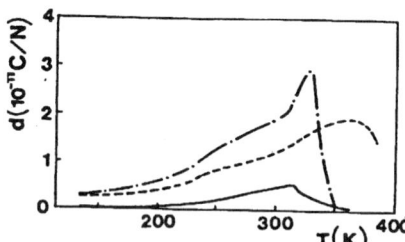

Fig. 25. Temperature dependences of the hydrostatic piezoelectric coefficients in 13/87, 55/45 and 72/18 PVDF-TrFE copolymers (solid, dot-dashed and dashed lines, respectively) [140].

Kinetics of ferroelectric switching in the 65/35 copolymer were studied by Furukawa et al. [147]. The results are similar to those obtained earlier on PVDF: the switching transients (D_s) can be fitted by an equation of the form

$$D_s(t) = D_o + 2P_r\{1 - exp[-(t/\tau_s)^n]\} \tag{51}$$

where D_o is an instantaneous electric displacement, P_r is the remnant polarization, and τ_s is the switching time dependent upon the applied voltage according to equation (45), with $\tau_{s,o}$ of the order of 1 ns, and E_a ranging from 8 x 10^8 V/m at 350 K to 1.8 x 10^9 V/m at 200 K. The switching characteristics were found to depend on the composition, crystallinity and degree of orientation of the crystalline phase, the fastest switching times determined experimentally amounting to *ca.* 100 ns at high fields [147].

6.2. OTHER POLYMERS

Several polymers built of optically active (asymmetric) molecular segments are piezoelectric after uniaxial orientation. Many of the natural and modified biopolymers, as well as the synthetic polypeptides belong to this group. Fukada and his collaborators (*e.g.*, [148]) investigated several of these materials, and among them oriented films of poly(γ-methyl-L-glutamate) and poly(γ-benzyl-L-glutamate), pMG and pBG respectively. The backbones of these polymers have α-helical structures which, in oriented samples, extend preferentially along the "3" direction. The symmetry predicts only one independent component of the piezoelectric matrix: labelling the direction perpendicular to the film surface with the index "1", this will be d_{14}. The experimentally determined values of d_{14} in pBG were found to depend linearly on the degree of orientation, reaching *ca.* 4 x 10^{-12} C/N [148].

Another group of polymers in which substantial piezo and pyroactivity was discovered, are polyamides. As mentioned in Section 3, in some of them (nylon 5, nylon 11 *etc.*) one should expect a large uncompensated dipole moment, thus making these polymers interesting candidates for further study. Films of nylon 11, subject to appropriate poling, were found to exhibit healthy piezoelectric properties, inferior to those of PVDF, but better than most other polymers [149,150]. This property of nylon 11 is probably associated with the phase transition occurring at 368 K, in which the low-temperature triclinic phase α transforms into a pseudohexagonal phase γ [59], which is piezoelectrically active.

The pyroelectric effect has also been observed in crystals of some diacetylenes and poly(diacetylenes). It was reported in both the monomer and polymer of the most extensively investigated diacetylene, bis(p-toluenesulphonate) of 2,4-

hexadiyne-1,6-diol, commonly referred to as pTS ($R_1 = R_2 = -CH_2-OSO_2-\Phi-CH_3$) [152]. However, pTS crystallizes in the monoclinic system (centrosymmetric space group $P2_1/c$ [153]) and, thus, at least in poly(pTS), the effect observed cannot be a true pyroelectric effect, it being most probably associated with the presence of defects created during polymerization [154].

Pyroelectricity has also been observed in another diacetylene: 1,6-bis(2,4-dinitrophenoxy)-2,4-hexadiyne (DNP; $R_1 = R_2 = -CH_2-O-\Phi-(NO_2)_2$). Lipscomb et al. [155] and Schultes et al. [156] reported the existence of a polar low-temperature phase which, in the monomer, appears below 46 K [156], the spontaneous polarization amounting to ca. 10^{-3} C/m^2. Upon polymerization, the transition shifts towards lower temperatures, becoming unobservable in fully polymerized crystals.

A possibility of the existence of a polar phase was also postulated in an asymmetrically substituted diacetylene p(TS-FBS), $R_1 = -CH_2-OSO_2-\Phi-CH_3$, $R_2 = -CH_2-OSO_2-\Phi-F$ [63,64]. However, no direct measurement of spontaneous polarization in asymmetric diacetylenes has been reported to date.

To give the reader an idea of typical values of piezoelectric and pyroelectric coefficients of polymers, some have been listed in Table 6.

Table 6. PIEZOELECTRIC AND PYROELECTRIC COEFFICIENTS OF SOME POLYMERS

Polymer	$\lvert d_{i,\lambda}\rvert$		$\lvert p_i{}^\varepsilon \rvert$		Ref.
	i,λ	10^{-12} C/N	i	10^{-6} C/m^{2}K	
pBG	1,4	4		–	[148]
pMG	1,4	3		–	[148]
PVF[*]	3,1	1	3	10	[151]
nylon 11	3,1	3	3	3	[149]
PVDF[§]	3,1	18	3	30	[80]

[*] PVF - poly(vinyl fluoride); [§] phase I.

7. PYROELECTRIC EFFECT IN LANGMUIR-BLODGETT FILMS

Langmuir-Blodgett (LB) film deposition is discussed in Chapter 3 and elsewhere in this book, thus only a brief reference to work on the pyroelectric effect in LB films is given here. Multilayer LB film structures may be polar if contributions of successive layers to the spontaneous polarization do not cancel each other. Blinov *et al.* [157] demonstrated that polar LB films may be obtained by the deposition of molecules of the type $C_{18}H_{37}$–O–Φ–N=N–Φ–R or $C_{18}H_{37}$–O–Φ–R, where Φ is an aromatic ring, and R stands for OH, NH , CN, COOH or a more complex hydrophilic substituent. Estimated values of the spontaneous polarization were of the order of 10^{-2} C/m^2, the pyroelectric coefficients, however, amounting to only 8 x 10^{-8} to 3 x 10^{-6} C/m^{2}K. The low pyroelectric activity is attributed to small temperature variations of the molecular dipole moment, neglecting other factors. In a polar X-type film,[1] the magnitude of the pyroelectric signal was found to be a linear function of the thickness, whereas in a Y-type film (centrosymmetric) the pyroelectric voltage was found to be a nearly-zero oscillatory function of the number of deposited layers [158] (see Fig. 26).

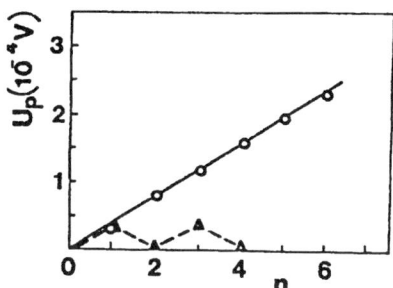

Fig. 26. The pyroelectric voltage as a function of the number of layers for X-type (circles) and Y-type (triangles) LB films [158].

More promising results were obtained on alternating layer LB films of fatty acids and fatty amines [159-163]. Roberts *et al.* [160-163] studied alternate layer films of ω-tricosenoic acid/docosylamine (TCA/DAm) and ω-tricosenoic

---

[1]Langmuir-Blodgett film types are discussed in Chapter 3.

318

acid/p–octadecylaniline (TCA/OAn). For a "thick" sample (99 layers) of the former film they reported the pyroelectric coefficient to be of the order of 10^{-5} C/m^2 K, slightly smaller than that found in PVDF but larger than the values in other pyroelectric polymers (see Table 6); the value of the pyroelectric coefficient was found to be even larger in a TCA/OAn LB film. Measurements were carried out employing two techniques: *static*, consisting of the measurement of the current arising as a result of temperature changes; *dynamic*, measuring a response of the sample after a heat pulse (*cf.* [3]). The static pyroelectric current obtained for the 99-layer sample is shown in Fig. 27. It is interesting to note that the coefficient obtained by the dynamic method is substantially larger than that obtained by the static method, the ratio being dependent upon the nature of substrate. Jones *et al.* [163] studied the temperature dependences of the pyroelectric coefficients in both TCA-DAm and TCA-OAn films, finding them to be dominated by the primary effect below 240 K, but substantially reduced due to a competition from the secondary effect above this temperature. These findings contradict the results of Blinov who postulated that the primary effect dominates in LB films [157].

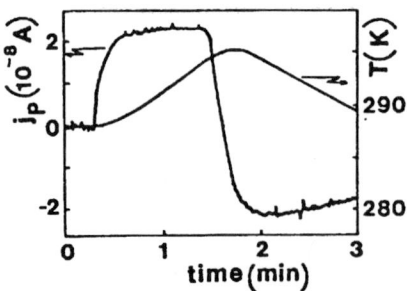

Fig. 27. The time evolution of the temperature and the pyroelectric current in the ω-tricosenoic acid/docosylamine LB film (99 layers) [162] (copyright 1986 by IEEE; reprinted with permission).

From the LB technique has emerged a new class of molecular materials, where new and interesting results may be expected (see, *e.g.*, [164]); among them, multilayers having pyroelectric response comparable with or better than PVDF seem to be a realistic goal. The high pyroelectric activity should be sought among highly polarizable molecules with strong temperature dependences of their dipole moments (*e.g.*, undergoing mild conformational changes in their side groups). One may expect that future work will proceed in this direction.

8. APPLICATIONS OF POLAR MOLECULAR MATERIALS

Potential applications of polar molecular materials may be sought in various fields, only some of them being related directly to the properties discussed in this chapter. It seems that current and potential applications of polar materials built of low-molecular weight molecules will make use of their non-linear optical properties rather than of those related to piezo and pyroelectricity. It should be stressed, however, that this statement is based on presently available information and, as all forecasts, it may be made useless by progress in studies of molecular crystals. For example, a discovery of a molecular fast-switching ferroelectric crystal (which is not inconceivable), would undoubtedly enhance research in this domain and might eventually lead to new applications. Among the three groups of molecular materials reviewed in this chapter, polar polymers (mostly PVDF and its copolymers) have found large-scale applications. The applicability of these relatively inexpensive, easy to manufacture materials has been realized quite early. With respect to conventional piezo and pyroelectric crystals, the polymers offer the following advantages [26]:

   (i)   they are flexible and tough;

   (ii)   they can be made very thin ($<10 \ \mu$m);

  (iii)   they may be obtained in large areas;

  (iv)   they can easily be shaped into a desired form;

   (v)   they have a low mechanical impedance, and thus exhibit good acoustic coupling to water, human body *etc.*;

  (vi)   they can be made locally or heterogeneously active by appropriate poling;

 (vii)   they can be manufactured in the form of composites whose properties can be controlled over a wide composition range.

Table 7. APPLICATIONS OF POLAR POLYMERS [25][*]

Audio-frequency transducers	*Electromechanical transducers and devices*
Microphones	Phonograph cartridges
Noise-cancelling microphones	Contactless switches
Telephone transmitter	Telephone dial
Bimorph transducers	Typewriter keyboard
Headphones	Bimorphs
Loudspeakers (tweeters)	Display device
Accelerometer	Light shutter
	Optical fibre switches
	Air flow fan
Ultrasonic and underwater	Variable focus mirrors
transducers	Mechanical transformer
	Coin sensor
Ultrasonic transmitters/receivers	Impact detector
Bulk-wave transducers	
Nondestructive testing transducers	
Stack transmitters	*Pyroelectric and optical devices*
Rayleigh and Lamb-wave devices	
Delay lines	Infrared detectors
PVDF-MOSFET	Intrusion detectors
Imaging arrays	Vidicons
Hydrophones	Laser-beam profile system
Light modulator	IR-to-visible convertor
Variable-focus transducer	Photocopying machine
Anti-fouling transducer	Optical SHG device
	Reflectivity detector

[*] See the original table for references.

The main disadvantages of polar polymers are associated with their relatively high electrical and mechanical losses: the electrical loss factor exceeds 1 % already at low frequencies, rising sharply in the MHz range.

Any application of PVDF and its copolymers, making use of their ferroelectric properties (*e.g.*, as elements of memory devices) does not seem feasible for several reasons, the main one being insufficient switching rate; the experiments on switching, briefly mentioned in Section 6.1, show that the switching times are at best of the order of fractions of microseconds in the available voltage and temperature ranges. The polymer films, however, have been widely applied as piezo and pyroelectrically active elements. These applications have been competently reviewed in several publications (*e.g.*, [25,27,165,166]), thus only a brief list has been given here (see Table 7), demonstrating the variety of devices that have been (or, at least, can be) constructed basing on PVDF.

Finally, it appears from the brief discussion put forward in the preceding section that among the several potential applications of LB films, those making use of the pyroelectric response may be expected in the nearest future.

9. FINAL REMARKS

This chapter gives a brief summary of work on piezo, pyro and ferroelectric behaviour of molecular materials, aiming at presenting a simple and possibly unified approach to the phenomena observed. The limited extent of this review made it impossible to discuss several important topics related to the research on polar materials, and in particular a more profound approach to interactions and phase transitions in molecular crystals. The measuring techniques, important in getting information about properties of materials, have also been left unmentioned. The reader is referred to more exhaustive monographs dealing with these topics; some of them have been cited in this chapter.

ACKNOWLEDGEMENTS

The author is indebted to Dr A.Miniewicz for permission to use results of his work (Fig. 9) prior to publication, and for helpful comments. This work was partly supported by the Polish Academy of Sciences within the Programmes CPBP 01.12 and CPBP 01.14.

REFERENCES

[1]   J. Curie and P. Curie, *Compt. Rend. Acad. Sci. Paris*, **91**, 294 (1880); **91**, 383 (1880).

[2]   J. Valasek, *Phys. Rev.* **17**, 475 (1921).

322

[3] S.B. Lang, *'Sourcebook of Pyroelectricity'*, Gordon and Breach, New York (1974).

[4] W.G. Cady, *'Piezoelectricity'*, McGraw-Hill, New York (1946).

[5] W.P. Mason, *'Piezoelectric Crystals and Their Application to Ultrasonics'*, van Nostrand, Princeton (1950).

[6] D.A. Berlincourt, D.R. Curran and H. Jaffe, in: W.P. Mason (Ed.), *'Physical Acoustics, Principles and Methods'*, Academic Press, New York (1964).

[7] E. Dieulesaint and D. Royer, *'Ondes Elastiques dans les Solides'*, Masson, Paris (1974).

[8] J. Zelenka, *'Piezoelectric Resonators and their Application'*, Academia, Prague (1986).

[9] F. Jona and G. Shirane, *'Ferroelectric Crystals'*, Pergamon Press, London (1962).

[10] E. Fatuzzo and W.J. Merz, *'Ferroelectrics'*, North-Holland, Amsterdam (1967).

[11] G.A. Smolenskii, V.A. Bokov, V.A. Isupov, N.N. Krainik, R.R. Pasynkov and M.S. Shur, *'Ferroelectrics and Antiferroelectrics'* (in Russian), Nauka, Moscow (1971).

[12] R. Blinc and B. Zeks, *'Soft Modes in Ferroelectrics and Antiferroelectrics'*, North-Holland, Amsterdam (1974).

[13] M.E. Lines and A.M. Glass, *'Principles and Application of Ferroelectrics and Related Materials'*, Clarendon Press, Oxford (1977).

[14] B.A. Strukov and A.P. Levanyuk, *'Physical Background of Ferroelectric Phenomena'* (in Russian), Nauka, Moscow (1983).

[15] S.B. Lang, *'Literature Guide to Pyroelectricity'*, published semiannually in *"Ferroelectrics"*.

[16] S.B. Lang, *'Bibliography of Piezoelectricity and Pyroelectricity of Polymers'*, published semiannually in *"Ferroelectrics"*.

[17] K.Toyoda, *'Bibliography of Ferroelectrics'*, published semiannually in *"Ferroelectrics"*.

[18] A.I. Kitaigorodsky, *'Molecular Crystals and Molecules'*, Academic Press, New

323

York (1973).

[19]   E. Fukada, *Ultrasonics* **6**, 229 (1968); *Adv. Biophys.* **6**, 121 (1974).

[20]   H. Kawai, *Japan. J. Appl. Phys.* **8**, 975 (1969).

[21]   J.G. Bergman, J.H. McFee and G.R. Crane, *Appl. Phys. Lett.* **18**, 203 (1971); A.M. Glass, J.H. McFee and J.G. Bergman, jr., *J. Appl. Phys.* **42**, 5219 (1971); J.H. McFee, J.G. Bergman and G.R. Crane, *Ferroelectrics* **3**, 305 (1972).

[22]   Y. Wada and R. Hayakawa, *Japan. J. Appl. Phys.* **15**, 2041 (1976).

[23]   R.G. Kepler, *Ann. Rev. Phys. Chem.* **29**, 497 (1978); R.G. Kepler and R.A. Anderson, *CRC Critical Rev. Solid State Mater. Sci.* **9**, 399 (1980).

[24]   G.T.Davis and M.G. Broadhurst, in: M.Campos (Ed.), *'Proc. Int. Symposium on Electrets and Dielectrics'*, Academia Brasileira de Ciencias, Rio de Janeiro (1977); M.G. Broadhurst and G.T. Davis, in: G.M. Sessler (Ed.), *'Electrets'*, Springer, Berlin (1980); (2nd edition 1987).

[25]   G.M. Sessler, *J. Acoust. Soc. Am.* **70**, 1596 (1981).

[26]   Y. Wada, in: J. Mort and G. Pfister (Eds.), *'Electronic Properties of Polymers'*, Wiley, New York (1982).

[27]   T. Wang, J.M. Herbert and A.M. Glass (Eds.), *'The Applications of Ferroelectric Polymers'*, Blackie, Glasgow (1987).

[28]   B.G. Broadhurst, F. Micheron and Y. Wada (Eds.), *'Special Issue on PVDF and Associated Polymers'*, *Ferroelectrics* **32** (1981); K.D. Pae and Y. Wada (Eds.), *'Proceedings of the US-Japan Seminar on Piezoelectric Polymers'*, *Ferroelectrics* **57** (1984).

[29]   R. Gerhard-Multhaupt, *IEEE Trans. Electr. Insul.* **EI-22**, 531 (1987).

[30]   G.M. Sessler and R. Gerhard-Multhaupt (Eds.), *'Proceedings of the 5th International Symposium on Electrets'*, IEEE Service Center, Piscataway, USA (1985); D.K. Das-Gupta and A.W. Patullo (Eds.), *'Proceedings of the 6th International Symposium on Electrets'*, IEEE Service Center, Piscataway, USA (1988).

[31]   L.G. Korenyeva, V.F. Zolin and B.L. Davydov, *'Molecular Crystals in Nonlinear Optics'* (in Russian), Nauka, Moscow (1975).

324

[32] D.J. Williams (Ed.), *'Nonlinear Optical Properties of Organic and Polymeric Materials'*, Am. Chem. Soc., Washington (1983).

[33] D.S. Chemla and J. Zyss (Eds.), *'Nonlinear Optical Properties of Organic Molecules and Crystals'*, Academic Press, Orlando (1987).

[34] J.F. Nye, *'Physical Properties of Crystals: their Representation by Tensors and Matrices'*, Clarendon Press, Oxford (1964); Y.Sirotine and M. Chaskolskaia, *'Fondements de la Physique des Cristaux'* (in French), Mir, Moscow (1984).

[35] *'IRE Standards on Piezoelectric Crystals'*, Proc. IRE **37**, 1378 (1949); **45**, 354 (1957); **46**, 765 (1958); **49**, 1162 (1961).

[36] D.F. Nelson, *Phys. Rev. Lett.* **60**, 608 (1988).

[37] L.D. Landau, *Phys. Z. Sovietun.* **11**, 26 (1937); **11**, 545 (1937); A.F. Devonshire, *Adv. Phys.* **3**, 85 (1954).

[38] J.C. Toledano and P. Toledano, *'The Landau Theory of Phase Transitions'*, World Scientific, Singapore (1987).

[39] K. Imai, *J. Phys. Soc. Japan* **49**, 2263 (1980).

[40] S. Triebwasser, *IBM J. Res. Develop.* **2**, 212 (1958).

[41] B.A. Strukov, *Phys. Tv. Tela* **6**, 2862 (1964).

[42] J. Trotter and S.L. Williston, *Acta Cryst.* **21**, 285 (1966).

[43] T.L. Charlton and J. Trotter, *Acta Cryst.* **16**, 313 (1963).

[44] E.M. Gopalakrishna, *Z. Kristall.* **121**, 378 (1965).

[45] J.L. Stephenson and A.C. Skapski, *J. Phys. C: Solid State Phys.* **5**, L233 (1972).

[46] T. Bjorvatten and O. Hassel, *Acta. Chem. Scand.* **13**, 1261 (1959); **16**, 249 (1962); T. Bjorvatten, *Acta Chem. Scand.* **16**, 749 (1962); **17**, 689 (1963).

[47] A.I. Kitaigorodskii, T.L. Khotsyanova and Y.T. Struchkov, *Dokl. Akad. Nauk. SSSR* **78**, 1161 (1951).

[48] Y. Iwata and T. Watanabe, *Ann. Rep. Res. React. Inst. Kyoto Univ.* **7**, 87 (1974).

[49] A.L. Solomon, *Phys. Rev.* **104**, 1191 (1956).

[50] G.J. Goldsmith and J.G. White, *J. Chem. Phys.* **31**, 1175 (1959).

[51] L. Demeny and I. Nitta, *Bull. Chem. Soc. Japan* **3**, 128 (1928).

[52] M. Elcombe and J.C. Taylor, *Acta Cryst.* **A24**, 410 (1968).

[53] F. Bechtel, D. Chasseau and J. Gaultier, *Crystal Struct. Commun.* **5**, 297 (1976).

[54] L. Toupet and A. Miniewicz, in preparation.

[55] L.N. Becka and W.D.J. Cruickshank, *Proc. Roy. Soc. London* **A273**, 435 (1963).

[56] Ye.L. Galperin, Yu.V. Strogalin and M.P. Mlenik, *Vysokomol. Soed. (USSR)* **7**, 933 (1965); Ye.L. Galperin and B.P.Kosmynin, *Vysokomol. Soed. (USSR)* **A11**, 1432 (1969).

[57] J.B. Lando, H.G. Olf and A. Peterlin, *J. Polym. Sci. A-1*, **4**, 941 (1966); J.B. Lando and W.W. Doll, *J. Macromol. Sci.-Phys.* **B2**, 219 (1968); **B4**, 309 (1970).

[58] R. Hasegawa, M. Kobayashi and H. Tadokoro, *Polym. J.* **3**, 591 (1972); R. Hasegawa, Y. Takahashi, Y. Chatani and H.Tadokoro, *Polym. J.* **3**, 600 (1972).

[59] B.A. Newman, T.P. Sham and K.D. Pae, *J. Appl. Phys.* **48**, 4092 (1977).

[60] H.J. Cantow (Ed.), *'Polydiacetylenes'*, Springer, Berlin (1984).

[61] D. Bloor and R.R. Chance (Eds.), *"Polydiacetylenes: Synthesis, Structure and Electrical Properties'*, M. Nijhoff, Dordrecht, Netherlands (1985).

[62] P. Strohriegl, *Macromol. Chem., Rapid Commun.*, **8**, 437 (1987).

[63] M. Bertault, L. Toupet, J. Canceill and A. Collet, *Macromol. Chem., Rapid Commun.* **8**, 443 (1987).

[64] P. Strohriegl, H. Schultes, D. Heindl, P. Gruner-Bauer, *Ber. Bunsenges. Phys. Chem.* **91**, 918 (1987).

[65] M. Bertault, J. Canceill, A. Collet and L. Toupet, *J. Chem. Soc., Chem. Commun.* 163 (1988).

[66] R. Nowak, *'A Study of Elastic and Piezoelectric Properties of Molecular Crystals'* (in Polish), Thesis, Tech. Univ. Wroclaw (1984).

[67] M. Kobayashi, *J. Chem. Phys.* **76**, 1187 (1982).

[68] A. Yoshihara, W.D. Wilber, E.R. Bernstein and J.C. Raich, *J. Chem. Phys.* **76**, 2064 (1982).

326

[69] H. Terauchi, T. Kojima, K. Sakaue and F. Tajiri, *J. Chem. Phys.* **76**, 612 (1982).

[70] J. Giermanska, *'Pyroelectric and Electret Properties of Polar Organic Crystals'* (in Polish), Thesis, Tech. Univ. Wroclaw (1984).

[71] V.A. Koptsik, *Kristallografiya (USSR)* **4**, 219 (1959).

[72] V.E. Bottom and E. Rodrigues, *J. Chem. Phys.* **44**, 2201 (1966).

[73] R. Nowak, R.J. Wycisk and M. Samoc, *Mater. Sci. (Poland)* **9**, 277 (1983).

[74] J. Minhua and H.J. Weber, *Z. Kristall.* **138**, 19 (1973).

[75] A. Miniewicz, M. Samoc, J. Sworakowski, B. Jakubowski, Z. Zboinski and A. Cehak, *Chem. Phys. Lett.* **76**, 442 (1980).

[76] J. Giermanska, R. Nowak and J. Sworakowski, *Mater. Sci. (Poland)* **10**, 77 (1984); *Ferroelectrics* **65**, 165 (1985).

[77] J.F. Legrand, *J. Physique (Paris)* **43**, 1099 (1982).

[78] P.T.A. Klaase and J. van Turnhout, *Proc. 3rd Int. Conf. 'Dielectric Materials, Measurements and Applications'*, IEE Conf. Publ. No. 177, p.411 (1979).

[79] R. Bechman, *Phys. Rev.* **110**, 1060 (1958).

[80] R.T. Smith and F.S. Welsh, *J. Appl. Phys.* **42**, 2219 (1971).

[81] E. Mattiat, *'Ultrasonic Transducer Materials'*, Plenum Press, New York (1971).

[82] D.A. Berlincourt, H. Jaffe and L.R. Shiozawa, *Phys. Rev.* **129**, 1009 (1963).

[83] S.B. Lang, M.D. Cohen and F. Steckel, *J. Appl. Phys.* **36**, 3171 (1965).

[84] J. Mangin and A. Hadni, *Phys. Rev.* **B18**, 7139 (1978).

[85] N.D. Gavrilova, S.N. Drozhdin, V.K. Novik and E.G. Maksimov, *Solid State Commun.* **48**, 129 (1983).

[86] T. Asaji and A. Weiss, *Z. Naturforsch.* **40a**, 567 (1985).

[87] S. Fleck and A. Weiss, *Z. Naturforsch.* **42a**, 645 (1987).

[88] T. Asaji, M. Taya and D. Nakamura, *Phys. Stat. Sol. (a)* **102**, 815 (1987).

[89] R. Nowak, J. Sworakowski, R. Kowal, J. Dziedzic and R. Poprawski, *Ferroelectrics* **65**, 79 (1985).

[90] J. Boguslawski, *Phys. Z.* **15**, 569 (1914).

[91] M. Born, *Rev. Mod. Phys.* **17**, 245 (1945).

[92] B. Szigeti, *Phys. Rev. Lett.* **35**, 1532 (1975).

[93] S.B. Lang, *Phys. Rev. B* **4**, 3603 (1971).

[94] A.S. Bhalla and R.E. Newnham, *Phys. Stat. Sol. (a)* **58**, K19 (1980).

[95] S. Fleck and A. Weiss, *Z. Naturforsch.* **41a**, 1289 (1986); S. Fleck, M.C. Bohm and A. Weiss, *Z. Naturforsch.* **42a**, 57 (1986).

[96] F.J. Mopsik and M.G. Broadhurst, *J. Appl. Phys.* **46**, 4202 (1975); M.G. Broadhurst, G.T. Davis, J.E.McKinney and R.E. Collins, *J. Appl. Phys.* **49**, 4992 (1978).

[97] Y. Shiozaki, *Ferroelectrics* **2**, 245 (1971).

[98] C. Calvo, *J. Chem. Phys.* **33**, 1721 (1960).

[99] P. Figuiere, M. Ghelfenstein and M. Szwarc, *Chem. Phys. Lett.* **33**, 99 (1975).

[100] V. Winterfeldt and G. Schaack, *Z. Physik B* **36**, 303 (1980); **36**, 311 (1980).

[101] A.H. Moudden, F. Denoyer, M. Lambert and N. Fitzgerald, *Solid State Commun.* **32**, 933 (1979).

[102] K. Gesi, *J. Phys. Soc. Japan* **51**, 710 (1982).

[103] A.H. Moudden, E.C. Stevenson and G. Shirane, *Phys. Rev. Lett.* **49**, 557 (1982).

[104] W. Rehwald and A. Vonlanten, *J. Phys. C: Solid State Phys.* **15**, 5361 (1982).

[105] D. Bordeaux, J. Bornarel, A. Capiomont, J. Lajzerowicz-Bonneteau, J. Lajzerowicz and J.F. Legrand, *Phys. Rev. Lett.* **31**, 314 (1973).

[106] D. Bordeaux, A. Capiomont, J. Lajzerowicz-Bonneteau, M. Jouve and M. Thomas, *Acta Cryst.* **B30**, 2156 (1974); A. Capiomont and J. Lajzerowicz-Bonneteau, *Acta Cryst.* **B30**, 2160 (1974).

[107] J.P. Bachheimer, J. Bornarel, J. Lajzerowicz and J.F. Legrand, *Ferroelectrics* **21**, 365 (1978).

[108] J. Lajzerowicz and J.F. Legrand, *J. Physique (Paris)* **41**, 1375 (1980); J.F. Legrand, J. Lajzerowicz, J. Lajzerowicz-Bonneteau and A. Capiomont, *J.Physique (Paris)* **43**, 1117 (1982).

[109] R.L. Zimmerman, *J. Bioelectricity* **1**, 265 (1982).

328

[110] R. Hayakawa and Y. Wada, *Adv. Polym. Sci.* **11**, 1 (1973).

[111] D.K. Das-Gupta, K. Doughty and S.B. Shier, *J. Electrostatics* **7**, 267 (1979); D.K. Das-Gupta, *Ferroelectrics* **33**, 75 (1981).

[112] G.T. Davis, J.E. McKinney, M.G. Broadhurst and S.C. Roth, *J. Appl. Phys.* **49**, 499 (1978); K. Tashiro, H. Todakoro and M. Kobayashi, *Ferroelectrics* **32**, 167 (1981).

[113] J.I. Scheinbeim, C.H. Yoon, K.D. Pae and D.H. Newman, *J. Appl. Phys.* **51**, 5156 (1980); A.J. Lovinger, *Macromolecules* **14**, 225 (1981).

[114] H. Dvey-Aharon, P.L. Taylor and A.J. Hopfinger, *J. Appl. Phys.* **51**, 5184 (1980); J.D. Clark, P.L. Taylor and A.J. Hopfinger, *J. Appl. Phys.* **52**, 5903 (1981).

[115] B. Servet, D. Broussoux and F. Micheron, *J. Appl. Phys.* **52**, 5926 (1981); Y. Takahashi, Y. Matsubara and T. Tadokoro, *Macromolecules* **15**, 334 (1982).

[116] J.I. Scheinbeim, C. Nakafuku, B.A. Newman and K.D. Pae, *J. Appl. Phys.* **50**, 4399 (1979).

[117] R.P. Tenlings, J.H. Bumbleton and R.L. Miller, *Polym. Lett.* **6**, 441 (1968); M. Kobayashi, K. Tashiu and T. Todakoro, *Macromolecules* **8**, 158 (1975).

[118] A.J. Lovinger, *Science* **220**, 1115 (1973).

[119] M. Tamura, K. Ogasawara, N. Ono and S. Hagiwara, *J. Appl. Phys.* **45**, 3768 (1974).

[120] D. Rezvani and J.G. Linvill, *Appl. Phys. Lett.* **34**, 828 (1979).

[121] T. Furukawa, M. Date and E. Fukada, *J. Appl. Phys.* **51**, 1135 (1980); M. Date, T. Furukawa and E. Fukada, *J. Appl. Phys.* **51**, 3830 (1980); M. Takahashi, M. Date and E. Fukada, *Ferroelectrics* **32**, 73 (1981).

[122] J.C. Hicks, T.E. Jones, M.L. Burgener and R.B. Olsen, *Ferroelectrics Lett.* **44**, 89 (1982).

[123] P. Herchenroder, Y. Segui, D. Horne and D.Y. Yoon, *Phys. Rev. Lett.* **45**, 2135 (1980).

[124] T. Furukawa and G.E. Johnson, *Appl. Phys. Lett.* **38**, 1027 (1981).

[125] R.G. Kepler and R.A. Anderson, *J. Appl. Phys.* **49**, 1232 (1978).

[126] M.G. Broadhurst and G.T. Davis, *Ferroelectrics* **32**, 177 (1981).

[127] J.I. Scheinbeim, K.T. Chung, K.D. Pae and B.A. Newman, *J. Appl. Phys.* **50**, 6101 (1979).

[128] M. Ohigashi, *J. Appl. Phys.* **47**, 949 (1976); S.B. Lang, *J. Appl. Phys.* **50**, 5554 (1979).

[129] G.R. Davies, *Inst. Phys. Conf. Ser. No. 58*, The Institute of Physics, London (1981).

[130] K.T. Chung, B.A. Newman, J.I. Scheinbeim and K.D. Pae, *J. Appl. Phys.* **53**, 6557 (1982).

[131] G. Pfister, M. Abkowitz and R.G. Crystal, *J. Appl. Phys.* **44**, 2064 (1973).

[132] N. Murayama, T. Oikawa, T. Katto and K. Nakamura, *J. Polym. Sci., Polym. Phys. Ed.* **13**, 1033 (1975); N. Murayama and H. Hashizume, *J. Polym. Sci., Polym. Phys. Ed.* **14**, 989 (1976).

[133] D.K. Das-Gupta and K. Doughty, *J. Phys. C: Solid State Phys.* **11**, 2415 (1978).

[134] A.M. Glass and T.J. Negran, *J. Appl. Phys.* **50**, 5557 (1979).

[135] R.G. Kepler and R.A. Anderson, *J. Appl. Phys.* **49**, 4490 (1978); *J. Appl. Phys.* **49**, 4918 (1978); R.G. Kepler, R.A. Anderson and R.R. Lagasse, *Ferroelectrics* **33**, 91 (1981); *Phys. Rev. Lett.* **48**, 1274 (1982); R.G. Kepler and R.A. Anderson, *Mol. Cryst. Liq. Cryst.* **106**, 345 (1984).

[136] J.B. Lando and W.W. Doll, *J Macromol. Sci. Phys.* **B2**, 205 (1968).

[137] J.C. Hicks, T.E. Jones and J.C. Logan, *J. Appl. Phys.* **49**, 6092 (1978).

[138] T. Yagi and M. Tatemoto, *Polym. J.* **11**, 429 (1979); T. Yagi, M. Tatemoto and J. Sako, *Polym. J.* **12**, 209 (1980).

[139] T. Furukawa, G.E. Johnson, H.E. Bair, Y. Tajitsu, A. Chiba and E. Fukada, *Ferroelectrics* **32**, 61 (1981); T. Yamada, T. Ueda and T. Kitayama, *J. Appl. Phys.* **52**, 948 (1981).

[140] T. Yamada and T. Kitayama, *J. Appl. Phys.* **52**, 6857 (1981).

[141] K. Tashiro, K. Takano, M. Kobayashi, Y. Chatani and H. Tadokoro, *Polymer* **25**, 195 (1984); *Ferroelectrics* **57**, 297 (1984).

330

[142] M. Latour, R.M. Faria and R.L. Moreira, in: D.K. Das-Gupta and A.W. Patullo (Eds.), *'Proceedings of the 6th International Symposium on Electrets'*, IEEE Service Center, Piscataway, USA (1988), p.467.

[143] T. Furukawa, *Ferroelectrics* **57**, 63 (1984).

[144] A.J. Lovinger, G.T. Davis, T. Furukawa and M.G. Broadhurst, *Macromolecules* **15**, 323 (1982); **15**, 329 (1982); A.J. Lovinger, T. Furukawa, G.T. Davis and M.G. Broadhurst, *Polymer* **24**, 1225 (1983); **24**, 1233 (1983).

[145] H. Yamazaki, J. Ohwaki, T. Yamada and T. Kitayama, *Appl. Phys. Lett.* **39**, 772 (1981).

[146] T. Furukawa, J.X. Wen, K. Suzuki, Y. Takashina and M. Date, *J. Appl. Phys.* **56**, 829 (1984).

[147] T. Furukawa, M. Date, M. Ohuchi and A. Chiba, *J. Appl. Phys.* **56**, 1481 (1984); T. Furukawa, H. Matsuzaki, M. Shina and Y. Tajitsu, *Japan. J. Appl. Phys.* **24**, L661 (1985); Y. Tajitsu, H. Ogura, A. Chiba and T. Furukawa, *Japan. J. Appl. Phys.* **26**, 554 (1987); Y. Tajitsu, T. Masuda and T. Furukawa, *Japan. J. Appl. Phys.* **26**, 1749 (1987).

[148] E. Fukada and S. Takashita, *Japan. J. Appl. Phys.* **10**, 722 (1971); M. Date, S. Takashita and E. Fukada, *J. Polym. Sci. A-2* **8**, 61 (1970); T. Konaga and E. Fukada, *J. Polym. Sci. A-2* **9**, 2023 (1971); T. Furukawa and E. Fukada, *J. Polym. Sci., Polym. Phys. Ed.* **14**, 1979 (1976); T. Furukawa, K. Ogiwara and E. Fukada, *J. Polym. Sci., Polym. Phys. Ed.* **18**, 1697 (1980).

[149] B.A. Newman, P. Chen, K.D. Pae and J.I. Scheinbeim, *J. Appl. Phys.* **51**, 5161 (1980).

[150] J.I. Scheinbeim, *J. Appl. Phys.* **52**, 5939 (1981).

[151] R.J. Phelan, jr, R.L. Peterson, C.A. Hamilton and G.W. Day, *Ferroelectrics* **7**, 375 (1974).

[152] H. Kiess and R. Clarke, *Phys. Stat. Sol. (a)* **49**, 133 (1978); D.Q. Xiao, D.J. Ando and D. Bloor, *Chem. Phys. Lett.* **90**, 247 (1982).

[153] J.P. Aime, J. Lefebvre, M. Bertault, M. Schott and J.O. Williams, *J. Physique (Paris)* **43**, 307 (1982).

[154] M. Bertault, M. Schott and J. Sworakowski, *Chem. Phys. Lett.* **104**, 605 (1984).

[155] G.F. Lipscomb, A.F. Garito and T.S. Wei, *Ferroelectrics* **23**, 161 (1980).

[156] H. Schultes, P. Strohriegl and E. Dormann, *Ferroelectrics* **70**, 161 (1986).

[157] L.M. Blinov, N.N. Davydova, V.V. Lazarev and S.G. Yudin, *Phys. Tv. Tela* **24**, 2686 (1982); *Thin Solid Films* **120**, 161 (1984).

[158] L.M. Blinov, L.V. Mikhnev, C.P. Palto and S.G. Yudin, in: M. Borissov (Ed.), *'Molecular Electronics. Proc. of 4th Int. School on Cond. Matter Phys. Varna'*, World Scientific, Singapore (1987).

[159] G.W. Smith, M.F. Daniel, J.W. Berton and N. Ratcliffe, *Thin Solid Films* **132**, 125 (1985).

[160] P. Christie, G.G. Roberts and M.C. Petty, *Appl. Phys. Lett.* **48**, 1101 (1986).

[161] P. Christie, C.A. Jones, M.C. Petty and G.G. Roberts, *J. Phys. D: Appl. Phys.* **19**, L167 (1986).

[162] C.A. Jones, M.C. Petty and G.G. Roberts, *'Proc. 6th IEEE Int. Symp. Applications of Ferroelectrics'*, IEEE 1986, p. 195.

[163] C.A. Jones, M.C. Petty and G.G. Roberts, *Thin Solid Films* **160**, 117 (1988); C.A. Jones, M.C. Petty, G.G. Roberts, G. Davies, J. Yarwood, N.M. Ratcliffe and J.W. Parton, *Thin Solid Films* **155**, 187 (1987).

[164] G.G. Roberts, *Adv. Phys.* **34**, 475 (1985); S.T. Kowel, R. Selfridge, C. Elderling, N. Matloff, P. Stroeve, B.G. Higgins, M.P. Srinivasan and L.B. Coleman, *Thin Solid Films* **152**, 377 (1987).

[165] M.A. Marcus, *Ferroelectrics* **40**, 29 (1982).

[166] W.N. Lawless (Ed.), *'Proc. IEEE Int. Symp. on Applications of Ferroelectrics'*, *Ferroelectrics* **50** (1983).

CHAPTER 6
Holography
K. Firth

1. HISTORY AND BASIC CONCEPTS

The invention of holography by Gabor in 1948 [1] came about as a means of improving the resolution of the electron microscope, one of its limitations being a large spherical aberration associated with electron lenses. This can be corrected relatively easily in ordinary optical systems whereas, in electron optical systems, complete correction is not possible; even today, electron lenses used at apertures of more than $f/100$ suffer from unacceptable aberration. Gabor realised that if it were possible to record an electron wavefront and then optically reconstruct it (suitably scaled) it would be possible to correct for spherical aberration of the electron lens. In the event, other factors were equally important in limiting the resolution and, even now, although electron holography is used for other purposes, improved resolution has not been achieved because of the limited coherence of available electron sources. Nevertheless, the concept of recording a wavefront, rather than a focused image, has found considerable use since the invention of the laser in 1960.

Recording media respond only to variations in the intensity of light whereas a wavefront is characterised by values of both amplitude and phase at each point on a surface. Thus, phase information is lost in an ordinary photograph but it can be recorded by converting its variations into variations in intensity. This can be achieved by introducing a reference beam which interferes with wavefronts generated by light scattered from the object but, since the recorded intensity corresponds to the square of the amplitude, the phase recorded is not unique as, for example, $\cos^2\Phi = \cos^2(\pi + \Phi)$. This ambiguity often results in two waves being reconstructed when the reference beam is incident on the hologram; one represents a virtual image of the object and the other a real image.

1.1 IN-LINE HOLOGRAPHY

The concept used by Gabor is illustrated in Figure 1. Its success depends on the area occupied by an object obscuring a relatively small part of the incident wavefront so that unscattered light constitutes the reference beam. In the general case, the incident wavefront may be converging, diverging or parallel. If the object is removed and the hologram occupies the same position as in recording, then light diffracted by the hologram will reconstruct the wavefronts scattered by the object; these will be seen as a virtual image, as if viewed through a window (the holographic plate). In addition, wavefronts with conjugate phase will also be produced creating an out of focus real image. Its effect can be minimised by using a parallel illuminating beam with point scattering objects, such as aerosol particles, at a sufficient distance from the plate that they produce a Fraunhofer rather than a Fresnel diffraction pattern.

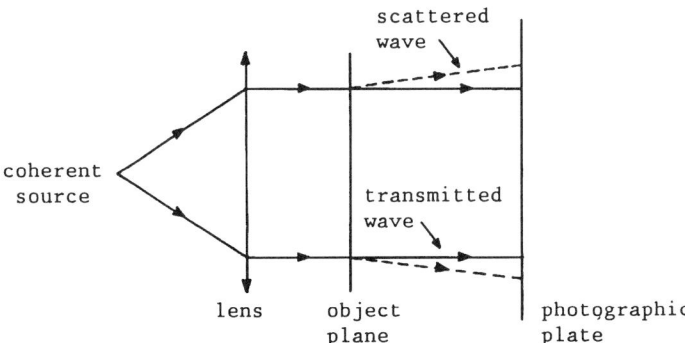

Fig. 1. Optical system for constructing an in-line (Gabor) hologram.

334

1.2 OFF-AXIS HOLOGRAPHY

Optical holography found little application until the early 1960's. This was in part due to the limitations of in-line holography as well as to the absence of suitable light sources. The advent of the laser then made possible the concept of off-axis holography [2], its development signifying the beginning of the use of holography for display purposes. Its basis, shown in Figure 2, illustrates the recording and reconstruction of a transmission off-axis hologram.

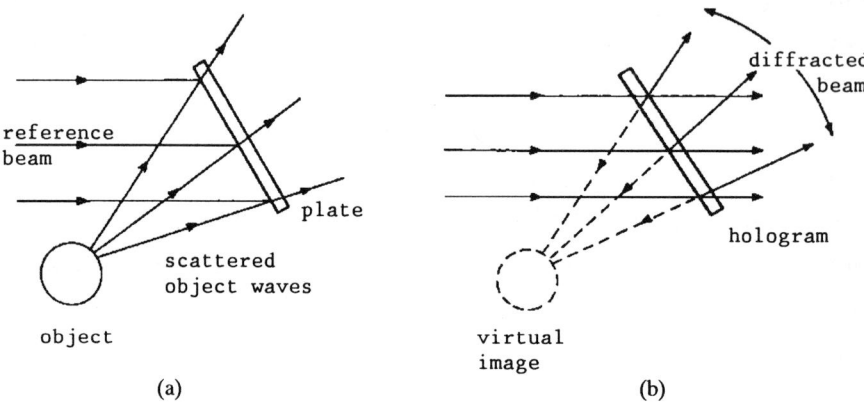

Fig. 2. (a) Hologram recording with off-axis reference beam and (b) the reconstruction of a virtual image from an off-axis hologram.

Light scattered from the object interferes with the reference beam to form an interference pattern which is recorded on a photographic plate. In the case of a diffusely scattering object, some light from any point on the object reaches every part of the plate; thus, if the hologram is broken, any part of it will reconstruct the object which is usually viewed as a virtual image. However, the smaller the piece, the less the detail and the less noticeable the three-dimensional effect which is characteristic of such holograms. It should be pointed out that this 'redundancy' effect is not characteristic of all holograms but only of diffuse holograms.

It can easily be seen that the larger the angle between the reference beam and the centre of the rays scattered by the object, the smaller the fringe period and the higher the resolution. It is clearly necessary that any relative movement between the fringe pattern and the recording medium during exposure must be small compared to the fringe spacing, otherwise the fringes will not be recorded. Also, the laser used for exposure must have adequate coherence, both longitudinal and transverse, and for objects with a depth of more than a few centimetres, specially equipped lasers must be used. This reduces the available power and increases the cost.

Leith and Upatnieks [2] showed that it is possible to record several holograms on the same plate and to read them separately, provided that the directions of the reference and object beams are incident at sufficiently different angles of incidence with respect to the plate and to each other. For display purposes, if the reference beam is incident on the plate at 180° with respect to the direction during recording, as shown in Fig. 3, then a real image, suspended in space, is observed; the ordinary laws of perspective, whereby parts of the object furthest away appear to move less than close to, do not apply when the angle of view is varied. In the real image, parts nearest the eye correspond to parts of the object furthest away and, thus, the perspective view is anomalous and is said to be pseudoscopic (normal perspective is said to be orthoscopic). If a reconstructed pseudoscopic image is used as the object for making a further hologram, then the real image resulting from replaying this second hologram is orthoscopic.

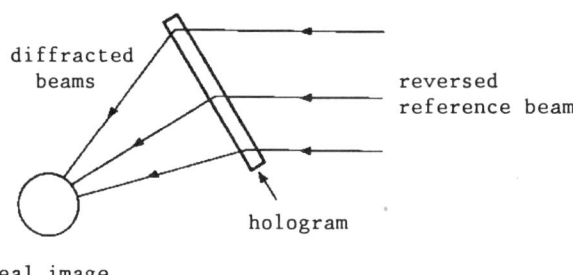

Fig. 3.   Reconstruction of a real image from an on-axis hologram.

1.3 LATER DEVELOPMENTS

Since the invention of off-axis holography, the principal developments in the methods of recording holograms have been the invention of the reflection hologram by Denisyuk [3], the invention of holographic interferometry by Stetson & Powell [4] and the invention of the *Rainbow hologram* by Benton [5]. In the development of recording materials for holograms, the use of dichromated gelatin for recording volume holograms by Shankoff [6] and others has probably been the most important milestone leading to the possibility of producing high quality 'holographic' optical elements, although the development of photothermoplastic media which can give near real-time holograms, developed in situ, has been equally important in the field of non-destructive testing.

2. HOLOGRAPHIC RECORDING

When a holographic material is being exposed, it is important that there should be no relative movement between the object, the recording plate and all optical components after the beam splitter which is used to generate the object and reference beams. These components usually comprise microscope objectives to generate diverging beams large enough to expose objects of reasonable size and small pinholes, 5 to 20 μm in diameter, accurately positioned at the focus of each objective. These act as 'beam cleaners', removing unwanted light generated by reflections in the laser output mirror *etc*, which would otherwise interfere with the main beam and cause unwanted fringes to appear on the hologram reconstruction. If a polarizing beam splitter is used, then it is necessary to use a half wave plate to ensure that both beams have the same polarization, otherwise there will be no interference. A half wave plate may also be used before the beam splitter to rotate the input polarization so as to vary the object to reference beam ratio, this ratio being important if it is necessary to have faithful recording of relative intensities in the reconstructed image. It is then necessary for the intensity of the object beam to be much less than that of the reference beam. This is easy to arrange since the object is usually a diffuse scatterer.

There are two types of holographic recording, producing holograms which work by either transmitting or reflecting the reconstruction beam.

2.1 TRANSMISSION HOLOGRAMS

A typical optical arrangement for recording transmission holograms is shown Fig. 4. These are characterised by the mean angle between the reference and object beams being less than 90° (typically 20 to 30°) so that both impinge on the recording plate from the same side. The average fringe period is typically *ca.* 1 μm so that resolution requirements for the recording medium are not extreme but it is possible to make transmission holograms with even smaller angles and correspondingly lower resolution media. A disadvantage of this type of hologram, sometimes called a Fresnel transmission hologram, is that a laser is required to reconstruct it. In this case the object appears as a virtual image as if seen through a window. If illuminated as shown in Fig. 3, a real image is produced exhibiting pseudoscopic parallax as mentioned in section 1.3 (see Fig. 5).

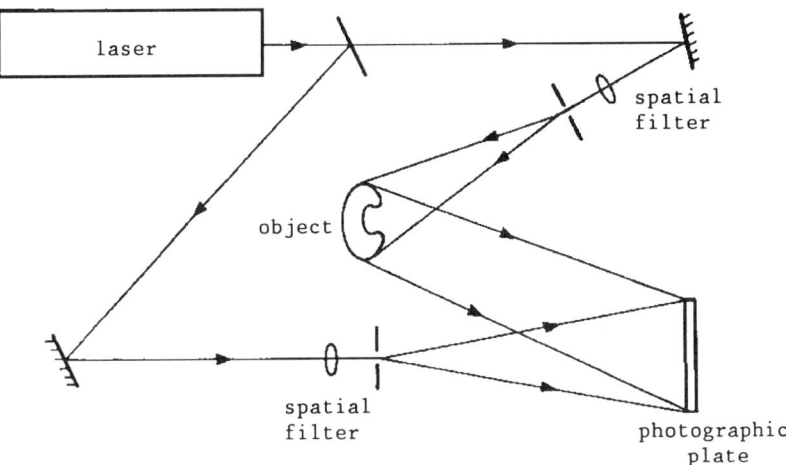

Fig. 4. Optical system for recording a transmission hologram.

(a)

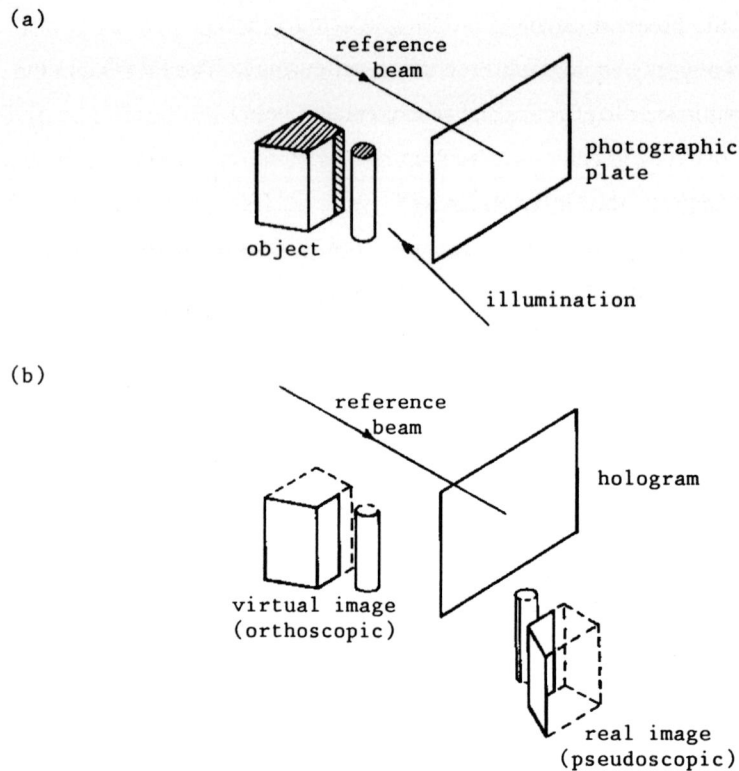

Fig. 5. Holographic formation of orthoscopic and pseudoscopic images: (a) hologram recording and (b) image reconstruction.

2.2 IMAGE PLANE HOLOGRAMS

A typical arrangement for recording image plane holograms is shown in Fig. 6. The lens used to image the object needs to be of about the same size as the object and the image, which straddles the recording plane, should have a magnification of unity if the perspective is to appear correct. The only way in which the hologram can represent a magnified view of the original object and retain a proper spatial relationship is by recording at one wavelength, scaling appropriately and then by replaying the hologram at another. The reason for this is that while the magnification is M in the image plane, perpendicular to this, in the longitudinal direction, it is M^2. Thus, the longitudinal and transverse magnifications are only equal when both are equal to unity. The motivation for using an image plane hologram in display applications is that it is possible to use a light source of reduced coherence, such as a filtered mercury discharge lamp, to replay the object; this gives an image which is free from the speckle phenomenon observed with highly coherent sources, at a substantially lower cost.

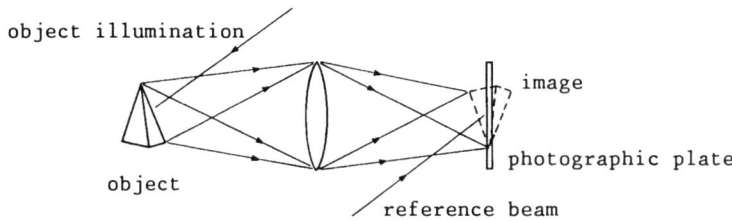

Fig. 6. Formation of an image hologram.

2.3 RAINBOW HOLOGRAMS

The invention of the *rainbow hologram* by Benton [5] was a major advance in display holography, since it enabled bright reconstruction of transmission holograms to be obtained using white light sources; the reconstruction is coloured and the colour varies with the angle of view, hence the name rainbow. This advance was obtained at the expense of retaining only horizontal parallax so that it is, in fact, a variety of holographic stereogram. The rainbow hologram is usually produced by a two step process (Fig. 7): (i) the use of the real image from a portion of conventional transmission hologram selected by a horizontal slit; (ii) the use of the reconstructed image, which has lost vertical parallax information, to record a secondary hologram. The reference beam for this hologram is a convergent beam inclined in the vertical plane. When the second hologram is illuminated with the conjugate of the reference beam used to make it, it forms an orthoscopic image of the object near the hologram, together with a real image of the slit used to limit the illuminated area of the first hologram. All light

340

diffracted by the hologram passes through this slit pupil; the object can only be seen if viewed close to this pupil and at the optimal distance a very bright image may be observed. If a white light source is used the different colours are dispersed vertically, as mentioned above, with different coloured images occurring at different heights.

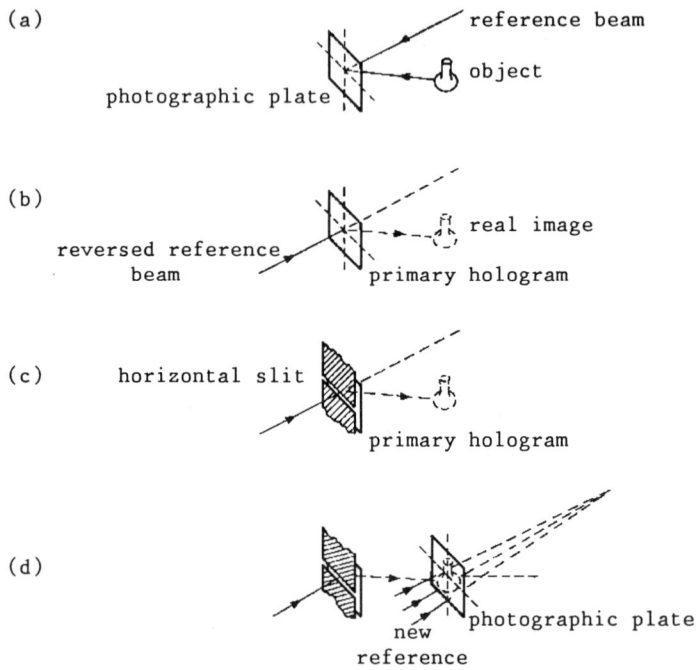

Fig. 7. Steps involved in the production of a rainbow hologram: (a) recording the primary hologram; (b) projecting the real image; (c) real image with no vertical parallax; (d) recording the final hologram [8].

2.4 REFLECTION HOLOGRAMS

In reflection holograms, the angle between the object and reference beams is greater than 90° (usually approaching 180°), the beams being incident from opposite ends of the recording plate. In the simplest set-up for recording reflection holograms, for example, when recording from a coin, the recording plate rests on the object and stability problems due to the very closely spaced fringes (typically < 0.2 μm) are minimized. The more complex arrangement, shown in Fig. 8, is used for objects with considerable depth or when the object beam is the real (pseudoscopic) image reproduced from a transmission hologram, arranged to straddle the recording plate, as mentioned earlier. This arrangement produces startling effects in display holograms when, for example, the object is a sword or dagger!

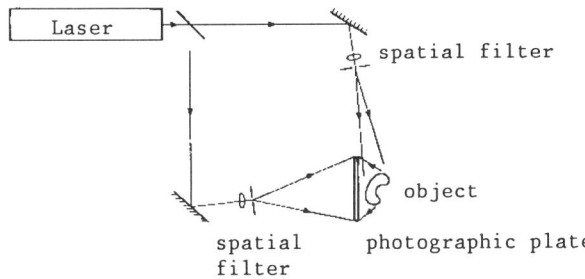

Fig. 8. Typical optical arrangement for recording a reflection hologram.

3. RECORDING METHODS

Methods of recording holograms can be classified in a number of ways, though not all are mutually exclusive. The two most basic classifications depend on whether the fringes are recorded by modulating the intensity of the transmitted light during recording (often referred to inaccurately as amplitude recording materials) or whether the phase of the transmitted light is modified by the recording material (phase recording). The phase recording materials may be classified as to whether they modulate the phase of the reconstructed beam by means of variations in the refractive index at constant thickness (*volume phase recording*) or by means of variations in thickness (*surface relief phase recording*). The ideal volume phase recording materials would have a large, preferably linear, change of refractive index with input intensity and, also, have low scatter and high resolution and sensitivity. In most cases, phase recordings give higher intensities during reconstruction; amplitude recording gives efficiencies of only a few per cent, at best. However, amplitude recording was the earliest method to be used and, using readily available materials, it is still the simplest. These should have a square law characteristic (photographic gamma = 2) as well as being free from scatter and having high resolution (>1000 line pairs/mm) and sensitivity.

3.1 THICK AND THIN VOLUME PHASE RECORDING

When recording transmission holograms (but not reflection holograms) it is possible to work in a regime in which more than one reconstructed wave is produced, just as a diffraction grating can have many significant diffracted orders. This regime occurs when the angle between the recording beams (Θ) and the thickness of the recording medium (d) are both low. For obvious reasons, it is a condition that is best avoided in most instances. A criterion for determining whether one is likely to be in the *thin* regime is by calculating the so called P parameter given by:

$$P = \lambda^2/L^2\eta_0\eta_1 \tag{1}$$

where λ is the free space wavelength, η_0 is the average refractive index of the medium, η_1 is the peak to average change in refractive index, and L is the mean grating period given by $L = \frac{1}{2}\lambda \sin\Theta/2$. (The expression for L applies when the two beams are symmetrically disposed at an angle Θ with respect to the normal to the recording medium). If $P > 10$, the holograms are classified as *thick* whereas if $P < 1$, they are classified as *thin*. The efficiencies of thick and thin gratings, produced by interfering two plane waves, can be described by relatively simple analytic functions. For thin gratings, the efficiency η_n for light diffracted into the n-th order is given by

$$\eta_n = J_{2n}(2\pi\eta_1 d/\lambda\cos\Theta) \tag{2}$$

where J_n is the n-th order Bessel function. For thick gratings only one diffraction order occurs and the efficiency, η, is given by

$$\eta = \sin^2[\pi\eta^1 d/\lambda\cos\Theta] \tag{3}$$

In both cases the symbols have been defined previously. It can be seen that 100 % efficiency may be obtained using thick gratings since, as mentioned above, all the light is diffracted into one order. This regime is often known as the Bragg regime, as the theory is formally similar to the reflection of X-rays from the Bragg planes of atoms in crystalline materials.

3.2 SURFACE RELIEF RECORDING

Surface relief structures have been used to produce the majority of holograms made to date, because of the ease with which they can be replicated by embossing. An ideal surface relief medium would have a path length change proportional to the incident light intensity, low scattering and high resolution and sensitivity. Although the distinction is not often made in the literature, surface relief holograms can be made with properties similar to those of thin and thick volume holograms, dependent upon the ratio of the fringe depth to fringe period. When the depth is small compared to the period, it behaves like a thin volume hologram whereas when the depth is greater than the period, it is more like a thick volume hologram. Surface relief media only record transmission holograms but reflective surface relief holograms can be produced by metallising with an improvement in efficiency. However, unlike true reflection holograms, they reflect over a wide range of angles and wavelengths and can exhibit multiple orders.

4. PROPERTIES OF IMPORTANT RECORDING MEDIA

4.1 SILVER HALIDE MEDIA

Photographic media have a sensitive layer made from an emulsion of silver halide particles embedded in gelatin together with sensitizing dyes *etc*. The ideal silver halide recording medium for holographic work is a true 'Lippman' emulsion, *i.e.* one in which the silver halide grains are small (on average 5 nm or less in diameter) and where the size distribution is narrow with few grains as large as 10 nm. This ensures high resolution and low scattering. Commercially available emulsions did not approach this performance until recently when specially developed emulsions from Ilford became available. Holographic emulsions developed by Agfa Gevaert and Kodak tend to have a greater fraction of the larger grains and even at low concentrations these can cause significant scattering. Typical sensitivities range from 1 to 100 $\mu J/cm^2$.

When making 'amplitude' holograms normal development procedures are used. For transmission holograms an average optical density of *ca.* 0.7 is optimum, whereas for reflection holograms an average density of *ca.* 2.0 is required. The bleach techniques required to make phase holograms are of two basic kinds: (i) those in which the emulsion is initially developed and fixed before bleaching (*conventional bleaching*) and (ii) those in which fixing is omitted (*reversal bleaching*). In the former, the developed silver is converted to a transparent salt with a higher refractive index than gelatin. Typical bleaching baths contain strong oxidizing agents such as mercuric chloride, cupric bromide, ferric bromide, potassium dichromate or potassium ferricyanide.

Volume holograms, made using conventional bleaching, tend to exhibit high levels of scattering due to grain growth. For this reason, reversal bleaching, in which the fixing bath is omitted and the developed silver is dissolved leaving a phase hologram made up of undeveloped silver halide crystals, is now the preferred process. The silver halide crystals are much smaller than the developed silver grains and, therefore, exhibit less scattering.

A problem with bleached holograms of this type is their tendency to darken when exposed to ambient light because of silver formation but this effect can be minimized by using a reversal bleach containing potassium iodide. Suitable formulae for reversal bleaching agents are usually provided by the manufacturers of the photographic plates. Another effect which occurs in halide media is that of shrinkage during processing. This shows itself most dramatically in reflection holograms which 'replay' at shorter wavelengths than the construction wavelength, *e.g.* holograms made using a red HeNe laser at 632.8 nm replay in the green. The most favourable aspects of halide media are that they are commercially available, have considerably higher sensitivities than most other media and are easily made panchromatic by the use of appropriate sensitizing dyes.

4.2 DICHROMATED GELATIN

The medium consists of a gelatin layer incorporating dichromate ions, usually in the form of either an ammonium or potassium salt. The dichromate may be incorporated either during or after formation of the gelatin layer, in the latter case by bathing the gelatin layer in a solution containing typically 1 to 5 % by weight of the dichromate salt. The material is unavailable commercially because of its limited shelf life (a few days at most), due in part to a dark reaction which causes hardening and, in the case of layers sensitized by ammonium dichromate, to the evolution of nitrogen. Although layers sensitized using the potassium salt do not suffer this problem, they are somewhat less sensitive.

Dichromated gelatin (DCG) has been used in printing for more than 100 years as a material for recording images by a surface relief process. A positive image is projected (or contact printed) on to the sensitized layer using a light source with a high blue content, the DCG being only sensitive for wavelengths up to about 530 nm. Light hardens the gelatin and, thus, in running water the unexposed area (*corresponding to the dark areas of the picture*) may be washed away, the exposed areas remaining unchanged. A similar process may be used to record surface relief holograms though better media, *e.g.* photoresists, are available and, thus, it would be unusual.

In 1968, Shankoff [6] reported that DCG could be used for recording volume holograms; the gelatin layer is pre-hardened prior to exposure and then, after, it is processed by immersing in water (or in aqueous solution of an inorganic salt) followed by dehydration by immersion in baths containing increasing concentrations of propan-2-ol. High quality transmission and reflection holograms may be produced using DCG. Its performance has not yet been surpassed by any other material but to obtain consistent results, a high degree of environmental control (temperature & humidity) is necessary at all stages of fabrication. For this reason, the material has only been used seriously in the production of high value items such as holographic head-up displays (HUD's) for military aircraft. Other disadvantages of DCG are its low sensitivity (0.1 to 1 J/cm^2, dependent upon wavelength), limited spectral sensitivity and, in particular, the necessity to encapsulate the holograms to prevent ingress of moisture, which causes the holograms to deteriorate.

The mode of action of DCG is only partly understood; the hardening of gelatin upon exposure is due to (i) photoreduction of Cr^{6+} to Cr^{3+} and (ii) the consequent cross-linking process which involves the Cr^{3+}. Upon exposure the change in refractive index is very small (<0.001) but this is amplified to -0.1 during the dehydration process due to the formation of sub-microscopic voids in the gelatin. If the pre-hardening is insufficient, then the gelatin exhibits a milky appearance in the unexposed areas, and the holograms become spectrally wide and exhibit a 'blotchy' appearance. By means of variations in the storage and processing employed it is possible to produce thermally stable reflection holograms which reflect at wavelengths below or above the wavelength of construction.

4.3 PHOTOPOLYMERS

Holographic photopolymers are not available commercially, although the Polaroid Corporation has an experimental formulation of undisclosed composition (styled DMP128 and requiring wet processing) which can be used for reflection holography. It has a higher index change capability than DCG (to ±0.1) and is not very sensitive to moisture. However, it has the drawback that it exhibits extreme 'frequency chirp', a phenomenon also observed in DCG but to a much lesser extent; in the reflection hologram, the spacing between layers of equal refractive index alters through the layer thickness, thus broadening and lowering the response curve. Because of the magnitude of this effect it does not appear to be possible to make narrow holograms (<60 nm halfwidth) and, thus, the range of possible applications is limited although DMP128 is ideal for display-type applications. It has moderate sensitivity (5 to 10 mJ/cm^2) and can be made panchromatic and achieve low scattering.

Another holographic photopolymer has been described by DuPont [7] but again its composition is not disclosed. Optimum exposure is typically 30 mJ/cm^2 in the blue-green region of the spectrum but the resolution limit (1000 lines/mm) is insufficient to record reflection holograms. The medium comprises at least two components, one active (*i.e.* photopolymerizable) and the other not; the properties may be varied significantly by altering the proportions of the two. Such a medium would be of interest if the resolution and index modulation could both be increased, in particular, because it can be used without wet processing, the potential applications being in holographic interferometry, since post exposure processing, which consists of exposure to incoherent ultraviolet light, can occur in situ.

4.4 PHOTORESISTS

The materials used to record surface relief holograms are a small selection from those used for microlithography in the semiconductor industry, all of which have proprietary formulations. The Shipley series 1350 and 1450 resists are often mentioned in the literature and layers of appropriate thickness, usually 1 μm or less, can be produced by spin coating on, for example, glass or fused silica substrates. The exposure characteristic, *i.e.* the amount of material removed on development after a certain level of exposure from an incoherent light source such as a mercury arc lamp, would be advantageous if a linear recording characteristic is required for a particular application. They require large exposures (*ca.* 1 J/cm^2 at $\lambda < 480$ nm) but have good resolution (several thousand lines/mm) and can be used to make very low scatter gratings for use in monochromators. They also find use in making masters for embossed holograms for security printing and low-end display applications.

348

4.5 PHOTOTHERMOPLASTICS

This method of recording holograms from photothermoplastics utilises some of the techniques developed in photocopying, *i.e.* photoconduction in a corona charged medium to produce a charge image of the fringe pattern, but in this case a thin layer (*ca.* 0.5 to 1 μm) of a thermoplastic medium is coated onto the organic photoconductor layer, typically a few microns thick. The medium is uniformly charged, selectively discharged during exposure and then recharged (see Fig. 9); this results in a charge image across the thermoplastic layer. The medium is developed by heating above the flow point of the thermoplastic and then allowing it to cool rapidly. A surface relief image is produced by the interaction of electrostatic and surface tension forces in the liquid. It is necessary to ensure that the electric forces are not greater than those of surface tension otherwise a noisy (frosty) recording is obtained. Under optimal conditions, the spatial frequency response shows a bandpass characteristic, peaking at 1000 to 2000 lines/mm with efficiencies up to *ca.* 30 % and sensitivities down to *ca.* 0.1 mJ/cm^2 or less. Attractions of this type of material are that it can be processed in situ in less than 1 second and is reusable; recordings can be erased by heating to a higher temperature than for normal development with no electric field present. Its main use is in holographic interferometry.

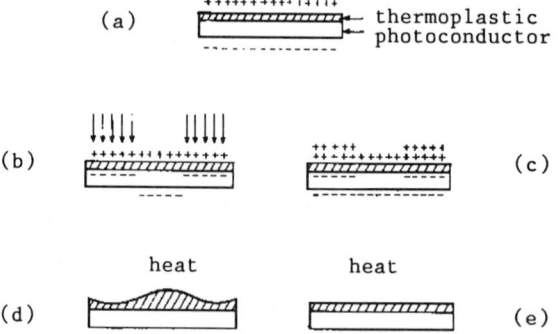

Fig. 9. Steps in the record-erase cycle of a photothermoplastic recording material: (a) first charging; (b) exposure; (c) second charging; (d) development; (e) erasure.

5. APPLICATIONS

5.1 DISPLAY HOLOGRAMS

Display holograms range in size from a few cm to approaching 1 m^2 and in cost from a few pence to several hundred pounds or more. Security printing of embossed surface relief holograms on credit cards is the largest single application of holography in terms of the number produced, with those on book covers being second. There is a sizeable market for holographic 'jewellery' and novelty items, the prices currently ranging from about £15 to £20 for a pendant to £100 for a 20 cm square display hologram. Such holograms are usually made using bleached silver halide media, though there is a small quality market for copies made from DCG. There is also a limited market for larger holograms for advertising in lieu of three dimensional models for sales demonstration purposes. The largest size holograms, made in silver halide media, are rarely seen outside exhibitions. Holographic portraiture requires masters to be exposed using pulsed ruby lasers, with diffused illumination of the subject to safe levels. Optimum exposure and processing of such large plates is a specialist 'art' and the plates themselves are no longer produced.

5.2 HOLOGRAPHIC INTERFEROMETRY

This technique finds most of its applications in non-destructive testing and is the only significant industrial application of holography. It takes two main forms, *live fringe* and *double exposure* interferometry.

In live fringe interferometry, the hologram (preferably processed in situ which, of course, is easier with dry processing media such as photopolymers and photothermoplastic media) is replaced in exactly the same position as it occupied during recording together with the original object. When viewed through the hologram, the fringe pattern shows displacement, for example, when the object is stressed. The second technique, that of double exposure interferometry, uses two exposures of the same object made at different times to indicate defects or high stress areas. As with live fringe interferometry, a fringe pattern is produced by interference between the reconstructions from the two exposures. This technique is more suited to industrial environments than the live fringe method which requires a high degree of vibration isolation, (somewhat better than that required to record a single hologram). The 'frozen fringe' technique can be used in an industrial environment if a dual pulse laser is used for the two exposures.

The various specialized applications of these techniques are beyond the scope of the present book, but may be found in *"Optical Holography"* by Hariharan [8].

5.3 HOLOGRAPHIC OPTICAL ELEMENTS

Holographic techniques offer the possibility of making optical components such as scanners, beam splitters, lenses and mirrors which can be lightweight, low cost, and have properties which cannot be realized using conventional optics, *e.g.* as aspheric reflectors on a spherical or even planar surface. Such components are ideally suited for use with monochromatic light sources such as lasers.

Scanners are the earliest and still the most widespread application of holographic optical elements, since they are used in large numbers in laser printers *etc*. The simplest type of scanner, which comprises an array of holographically produced diffraction gratings, mounted on the periphery of a rotating disk, is shown in Fig. 10. The term 'hologon' has been coined to describe such a scanner since it replaces a polygonal mirror scanner used in conventional systems of this type [9]. The best type of hologram for this application is the thick surface relief type, since it has high efficiency and can be replicated easily at low cost on lightweight plastic discs. It also maintains a more constant efficiency across large scan angles, far better than gratings made using volume hologram materials. Another sizeable application is for bar code scanners used, for example, in supermarket checkouts, as described by Dickson *et al.* [10]. This was the earliest commercial use of holographic optical elements on any scale.

An advantage of the simple transmission grating scanner (shown in Fig. 10) is that it confers a degree of immunity from the effects of wobble of the scanner axis, which means that a lower precision mechanism can be used to achieve the desired accuracy. Another type of scanner [11] combines the functions of a scanner and focusing lens in a single holographic optical element, though the immunity from wobble is lost. A recent review article summarizes the present position on holographic scanners [12].

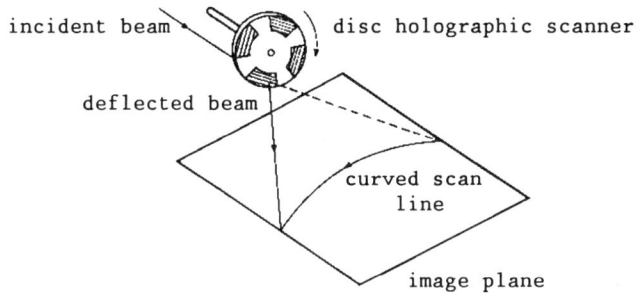

Fig. 10. Disc holographic scanner.

Holographic lenses do not appear to have found much application as yet, but holographically produced beam splitters have been proposed for use in the read-write heads of optical disc information storage systems, analogous to the audio Compact Disc system. The reduced head weight, possible using such techniques, allows a more rapid access to data.

The most important application of holographic reflecting elements to date is in *head up displays* (HUD's) used in military aircraft, in the pilot's line of sight, to present essential information such as height, heading, airspeed and targeting information. A conventional HUD consists of a collimator lens (to project an image of the information on the cathode ray tube to infinity) and a neutral beam splitter to project the image at infinity to the pilot, while still allowing him to see the outside world. The beam splitter typically has a transmission of 70% and a reflectance of 20%.

The constraints on the conventional HUD limit the field of view attainable to about 12° vertical by 15° horizontal. This is due to limitations on the position,

352

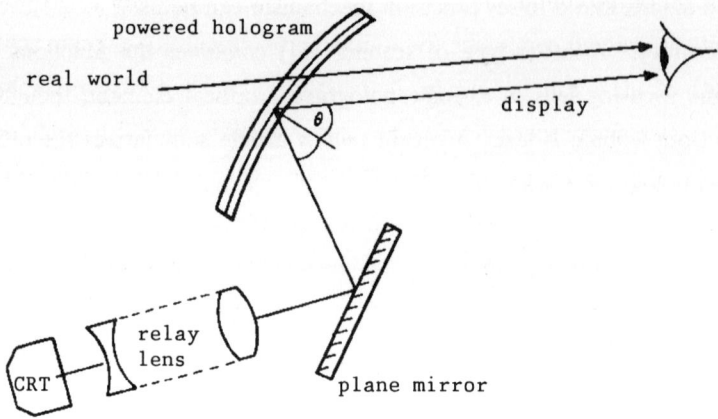

Fig. 11. Off-axis combiner diffraction HUD.

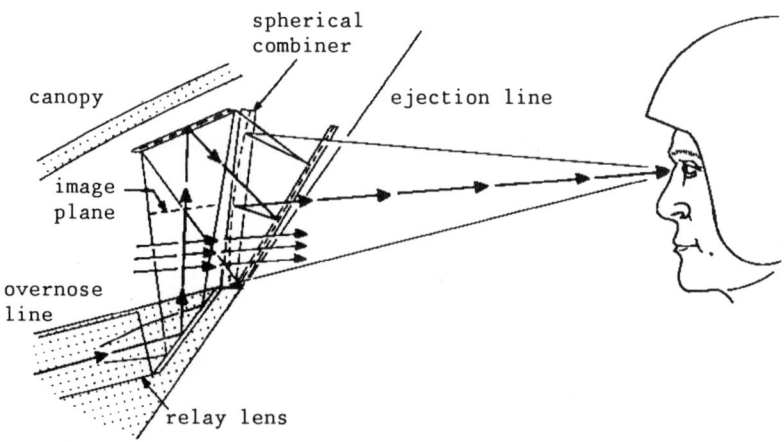

Fig. 12. Quasi axial system.

size and weight of the collimating optics. By arranging a relay lens to provide an intermediate image and combining the function of collimator and beam splitter, larger fields of view can be obtained. However, this requires an aspheric, asymmetric reflector in order to correct the severe aberrations which would occur if a spherical reflector was used at a large off axis angle, as would necessarily be the case. Such a reflector would be much too expensive to produce. Thus one solution would be to use a hologram, made on a spherical surface, but which behaved in the same manner as the complex reflector described above. While this is possible (Fig. 11), another approach using holographic mirrors (two planar and one spherical − shown in Fig. 12), is so far the only type for which large production orders have been placed. This arrangement uses the angular selectivity of holograms working over a small range of wavelengths (obtained using the narrow band phosphor type P43) to fold the ray paths sufficiently so that the spherical mirror is only working at 11° off axis, compared to more than 30° for most designs using the configuration shown in Fig. 11. This means that the holographic mirror on the spherical surface in Fig. 12 need not have any power, since all aberrations can be cancelled by the relay lens. The mirror has no chromatic aberration, easing the overall optical design greatly. Other possible applications for holographic reflectors are in Helmet Mounted Displays (similar in concept to HUD's) and in flight simulation systems.

REFERENCES

[1]   D.Gabor, *Nature*, **161**, 777 (1948).

[2]   E.N.Leith and J.Upatnieks, *J. Opt. Soc. Am.*, **52**, 1123 (1962).

[3]   Y.N.Denisyuk, *Sov. Phys. Doklady*, **7**, 543 (1962).

[4]   K.A.Stetson and R.L.Powell, *J. Opt. Soc. Am.*, **55**, 1694 (1965).

[5]   S.A.Benton, *J. Opt. Soc. Am.*, **59**, 1545 (1969).

[6]   T.A.Shankoff, *Applied Optics*, **7**, 2101 (1968).

[7]   B.L.Booth, *Applied Optics*, **14**, 593 (1975).

[8]   P.Hariharan, *Optical Holography*, Cambridge University Press, Cambridge

354

(1984).

[9]   C.J.Kramer, *Laser Focus*, **17**, 70 (1981).

[10]   L.Dickson, G.Sincerbox and A.Wolfheimer, *IBM J. Res. & Develop.*, **26**, 228 (1982).

[11]   Y.Ono and N.Nishida, *Applied Optics*, **21**, 4542 (1982).

[12]   H.Ikeda, F.Yamagishi and T.Inagaki, *Fujitsu Sci. Tech. J.*, **23**, 125 (1987).

INDEX

Peter Anderson

19, Ashbrook Drive

Belfast BT4 2FG

Ph. (0232) 652202

£44·50